Emerging Nanotechnology Applications in Electrical Engineering

Ahmed Thabet Mohamed
Aswan University, Egypt

IGI Global
PUBLISHER of TIMELY KNOWLEDGE

A volume in the Advances in Computer and Electrical Engineering (ACEE) Book Series

Published in the United States of America by
IGI Global
Engineering Science Reference (an imprint of IGI Global)
701 E. Chocolate Avenue
Hershey PA, USA 17033
Tel: 717-533-8845
Fax: 717-533-8661
E-mail: cust@igi-global.com
Web site: http://www.igi-global.com

Library of Congress Cataloging-in-Publication Data

Names: Mohamed, Ahmed Thabet, 1974- author.
Title: Emerging nanotechnology applications in electrical engineering / by Ahmed Thabet Mohamed.
Description: Hershey, PA : Engineering Science Reference, 2021. | Includes bibliographical references and index. | Summary: "This book is a collection of innovative research on the methods and application of nanoparticles in electrical engineering that discusses the wide array of uses nanoparticles have within electrical engineering and the diverse electric and magnetic properties that nanomaterials help make prevalent"-- Provided by publisher.
Identifiers: LCCN 2021009265 (print) | LCCN 2021009266 (ebook) | ISBN 9781799885368 (h/c) | ISBN 9781799885375 (s/c) | ISBN 9781799885382 (eISBN)
Subjects: LCSH: Nanoelectronics. | Electronic apparatus and appliances--Materials. | Electric apparatus and appliances--Materials. | Nanostructured materials.
Classification: LCC TK7874.84 .M63 2021 (print) | LCC TK7874.84 (ebook) | DDC 621.31028/4--dc23
LC record available at https://lccn.loc.gov/2021009265
LC ebook record available at https://lccn.loc.gov/2021009266

This book is published in the IGI Global book series Advances in Computer and Electrical Engineering (ACEE) (ISSN: 2327-039X; eISSN: 2327-0403)

British Cataloguing in Publication Data
A Cataloguing in Publication record for this book is available from the British Library.

Advances in Computer and Electrical Engineering (ACEE) Book Series

Srikanta Patnaik
SOA University, India

ISSN:2327-039X
EISSN:2327-0403

MISSION

The fields of computer engineering and electrical engineering encompass a broad range of interdisciplinary topics allowing for expansive research developments across multiple fields. Research in these areas continues to develop and become increasingly important as computer and electrical systems have become an integral part of everyday life.

The **Advances in Computer and Electrical Engineering (ACEE) Book Series** aims to publish research on diverse topics pertaining to computer engineering and electrical engineering. **ACEE** encourages scholarly discourse on the latest applications, tools, and methodologies being implemented in the field for the design and development of computer and electrical systems.

COVERAGE

- Applied Electromagnetics
- Chip Design
- Circuit Analysis
- Sensor Technologies
- Algorithms
- Digital Electronics
- VLSI Design
- Computer Science
- Optical Electronics
- Analog Electronics

IGI Global is currently accepting manuscripts for publication within this series. To submit a proposal for a volume in this series, please contact our Acquisition Editors at acquisitions@igi-global.com or visit: https://www.igi-global.com/publish/.

The Advances in Computer and Electrical Engineering (ACEE) Book Series (ISSN 2327-039X) is published by IGI Global, 701 E. Chocolate Avenue, Hershey, PA 17033-1240, USA, www.igi-global.com. This series is composed of titles available for purchase individually; each title is edited to be contextually exclusive from any other title within the series. For pricing and ordering information please visit https://www.igi-global.com/book-series/advances-computer-electrical-engineering/73675. Postmaster: Send all address changes to above address.

Titles in this Series

For a list of additional titles in this series, please visit:
https://www.igi-global.com/book-series/advances-computer-electrical-engineering/73675

Modeling and Applications of Solar Photovoltaic Thermal Technology
Chandra Shekhar Malvi (Madhav Institute of Technology & Science, India) and Siddhartha Kosti (Rajkiya Engineering College, India)
Engineering Science Reference • © 2021 • 315pp • H/C (ISBN: 9781799879169) • US $225.00

Hydrogen Fuel Cell Technology for Stationary Applications
Gheorghe Badea (Technical University of Cluj-Napoca, Romania) Raluca-Andreea Felseghi (Ştefan cel Mare University of Suceava, Romania) and Ioan Aşchilean (Technical University of Cluj-Napoca, Romania)
Engineering Science Reference • © 2021 • 332pp • H/C (ISBN: 9781799849452) • US $215.00

Computational Methodologies for Electrical and Electronics Engineers
Rajiv Singh (G.B. Pant University of Agriculture and Technology, Pantnagar, India) Ashutosh Kumar Singh (Indian Institute of Information Technology, Allahabad, India) Ajay Kumar Dwivedi (SIET, India & Dr. A.P.J. Abdul Kalam Technical University, Lucknow, India) and P. Nagabhushan (Indian Institute of Information Technology, Allahabad, India)
Engineering Science Reference • © 2021 • 281pp • H/C (ISBN: 9781799833277) • US $225.00

Innovations in the Industrial Internet of Things (IIoT) and Smart Factory
Sam Goundar (British University Vietnam, Vietnam) J. Avanija (Sree Vidyanikethan Engineering College, India) Gurram Sunitha (Sree Vidyanikethan Engineering College, India) K. Reddy Madhavi (Sree Vidyanikethan Engineering College, India) and S. Bharath Bhushan (Sree Vidyanikethan Engineering College, Tirupati, India)
Engineering Science Reference • © 2021 • 311pp • H/C (ISBN: 9781799833758) • US $225.00

Innovative Applications of Nanowires for Circuit Design
Balwinder Raj (National Institute of Technical Teachers Training and Research, Chandigarh, India)
Engineering Science Reference • © 2021 • 263pp • H/C (ISBN: 9781799864677) • US $225.00

Optimizing and Measuring Smart Grid Operation and Control
Abdelmadjid Recioui (Université M'hamed Bougara de Boumerdes, Algeria) and Hamid Bentarzi (Université M'hamed Bougara de Boumerdes, Algeria)
Engineering Science Reference • © 2021 • 400pp • H/C (ISBN: 9781799840275) • US $225.00

IGI Global
PUBLISHER of TIMELY KNOWLEDGE

701 East Chocolate Avenue, Hershey, PA 17033, USA
Tel: 717-533-8845 x100 • Fax: 717-533-8661
E-Mail: cust@igi-global.com • www.igi-global.com

Table of Contents

Preface

Nanotechnology has the ability to increase the rate of construction and make it a faster, cheaper and more versatile process, as it allows the process of automating nanoscale structures at a much lower cost and with environmentally friendly systems. Nanotechnology is no longer included in the realm of young imagination but has become a reality that enjoys great interest around the world, especially the developed countries of the world, as it is now one of the most prominent trends and priorities of scientific research. The actual developments and serious discoveries in this new and promising field have helped to strengthen its image and the increase investments for many laboratories in universities and research and industrial institutions. The world is now witnessing a frantic race between research centers and industry in order to employ nanotechnology in new industries and products, and many scientists confirm that nanotechnology will form the future of this world within the framework of a new industrial revolution. The Nanotechnology plan is the nucleus of the next industrial revolution. Countries of the world now see that the future will be for nanotechnology and its products, and the importance of this modern technology is that it is inexpensive, compared to other traditional technologies and its economic returns are very high. It represents a mixture between science and technology, directed towards scientific applications. It begins from the basic components of matter, namely atoms and molecules. Hence, it has an impact on all fields of science and technology, as well as other applied disciplines. Hence, physical, chemical, biological, mechanical and electronic sciences, as well as materials science and information technology are all intertwined. Therefore, scientists from different disciplines are active in the field of nanotechnology; including physicists, chemists, engineers, materials experts and scientists. This new technology holds huge promises for various and rapidly evolving nanoscale applications in various fields of science, research and development because scientists see that nanotechnology will paint the image of the world over the next 30 years, in addition to its being a new technological revolution that will radically change aspects of life. All sectors of the economy will be deeply affected by nanotechnology. The tremendous scientific progress and the great technological development have contributed to highlighting materials with advanced specifications that help the industrial fields in enhancing their products and their competitive position in terms of quality, efficiency and accuracy, since costs will turn into indirect industrial costs. Furthermore, the dependence of industrial units on nanomaterials addresses the economic problem represented by its call for natural resources because nanotechnology enables access and misses the required properties as long as the properties of the material and its specifications are determined within the economic unit. Thus, the industries in which nanotechnology will be introduced will wipe out from the market all the industries that are produced with the current traditional technology and methods because they will consume less energy and materials and will be better, cheaper and cleaner, since they do not pollute the environment. Additionally, they will be hundreds of times more robust and durable than the latter.

THE CHALLENGES

Welcome to the amazing nano world! God willing, you will find in this book scientific and laboratory applications of nanoscience in the field of industrial applications and in the scope of high-voltage engineering. This book will encourage researchers and investors to follow the applications received for scientific and laboratory use in all environmental fields, and modern education must cope with nanoscience and nanotechnologies. It is also necessary to train new generations of researchers, engineers and technologists in the fields of nanoscience. As a contribution from us in the education of nano scopes, we present this book which represents a framework that combines the basic ideas necessary to understand modern developments, based on nanoscience and nanotechnology in the fields of electrical engineering. Nanotechnology is the science of modifying molecules or atoms to make new products. The new nanotechnology revolution aims to develop a new type of industrial electronic applications used in various environmental fields, where scientists turn to engineers designing wires and equipment at the atomic level. In such system, the fields of physics are intertwined and overlap, and the fields of chemistry, biology, electricity, electronics, and mechanics are forcefully put together. This interconnectedness and overlap can promise many new innovations. The future will be the Nanotech industry which could lead the world to a new industrial revolution. Nano is a billionth of a meter, and the diameter of a human hair is 80,000 nanometers. At this level of technology, all normal physical and chemical laws do not exist. For example, nano-carbon tubes are 100 times stronger than steel and six times lighter than it. Therefore, this technology is distinguished in the areas of its human use, since the thing made of it is smaller, cheaper, lighter and better able to perform the functions assigned to it. Nanotechnology is a very new, active and very fast research field, and many scientists working in this field confirm that nanotechnology will bring about a new industrial breakthrough in various areas of life, leading to fundamental transformations in economics and technology. Nanotechnology applications have attracted the attention of many scientists, industrialists and funders, and the assistance of governments and developed countries to nanotechnology research has increased in recent years. Nanotechnology can give developing countries the opportunity, as what happened in China, to invest in research and development, support university programs aimed at finding scientists and engineers specializing in nanotechnology and send them to study in nanotechnology institutes.

SEARCHING FOR A SOLUTION

Nowadays, the energy sector has been receiving great attention from both the consumer and the producer because of its impact on all aspects of life. The progress of countries is measured by the amount of energy they use. Electrical energy is one of the types of energy that has become more demanding as a result of the delivery of the service to a large percentage of consumers, in addition to the progress and increase of industrial production. Hence, it is necessary to find advanced systems capable of transferring these huge amounts of electrical energy with superior efforts and safety. Nanotechnology is one of the new and very rapid research fields which is studied by many scientists all over the world in governments and commercial, academic and research laboratories. Many scientists working in this field confirm that nanotechnology will bring about a new industrial revolution in the near future in various areas of life, as well as radical transformations in economy and technology. Nanotechnology applications attract the attention of many scientists, industrialists and funders. The new nanotechnology revolution aims to develop a new type of atomic electronics that adopts quantum mechanics and the movement of individual particles, producing equipment faster and many times smaller than anything around us now. Consequently, scientists turn to engineers designing equipment at the atomic level, and in such a system the fields of engineering, electricity, electronics, mechanics, physics, and chemistry intertwine, overlap and become stronger together. This interconnectedness and overlap can promise a lot of new inventions and innovations.

ORGANIZATION OF THE BOOK

The energy sector has received incredible consideration from both the shopper and the maker, in view of its effect on all parts of life. The advancement of nations is estimated by the measure of energy they use. Electrical energy is one of the sorts of energy, and it turned out to be more requested because of the conveyance of administration to a huge level of buyers regardless of the advancement and increment of mechanical creation. Therefore, it is important to discover propelled frameworks equipped for moving these tremendous measures of electrical energy with prevalent endeavors securely. It is likewise fundamental that they are financial frameworks with a long operational life and that they meet the specialized particulars and general conditions of safety. We are aware in the study of electrical building that the measure of exertion influences legitimately on the measure of the separation transmitted and the sum additionally moved from energy. Remarkable endeavors should be confined to methods for protection that promise us continuity of transport effectively and securely, as the electrical force in the transmission frameworks is at present moved to endeavors that may arrive at 1MV now and then. Moreover, there are some nanoparticles in which endeavors might be utilized to reenact the measure of reluctant and nonstop electric detonators or a heartbeat that may at some point arrive at 120 kV. One of the utilizations of these tests is to create perfect protection from all the conditions during the time of utilization in the frameworks of transmission or dispersion of electrical force, regardless of whether it is a situation because of the section of gigantic measures of energy. Hence, this book clarifies this NEW vision in ten parts which are summed up as follows:

Chapter 1 presents dielectrics and nanoparticles, effects of nanoparticles on life-time characteristics, theory and effective parameters of electrical treeing in nanodielectrics theory and effective parameters of water treeing in nanodielectrics.

Chapter 2 discusses nanodielectrics in ac and dc power cables, multiple nanodielectrics in power cables, effects of nanoparticles and multiple nanoparticles on power cables, and partial discharges in nanodielectrics of power cables, as well as forecast and recommendations.

Chapter 3 reviews traditional power capacitors, types of ac and dc power capacitors, capacitive charge of nanodielectrics, nanoparticles in power capacitors insulation, effect of nanoparticles on power capacitors characteristics, theory and effective parameters of high voltage metallized film capacitors, trends of nanoparticles and multiple nanoparticles in power capacitors insulation, as well as forecast and recommendations.

Chapter 4 presents ceramic insulators, polymeric insulators, theory and effective parameters of Nano-Tech OHTL insulators, effects of nanoparticles on OHTL insulators, high voltage distribution over suspension insulators, multiple nanocomposites of non-ceramic Insulators, effects of multiple nanoparticles on ceramic and non-ceramic OHTL insulators and trends of nanoparticles inside suspension insulators.

Chapter 5 discusses the new technologies for filling nanoparticles inside tradition fluids of transformers, alternative fluids for transformers, modeling and simulation techniques for nanofluids dielectrics, effect of nanoparticles on nanofluids dielectrics, theory and effective parameters in dynamic charging of nanofluids, comparative study for using individual and multiple nanoparticles in nanodielectrics, electrical conductivity of nanofluids, nanofluids in laboratory, as well as forecast and recommendations.

Chapter 6 explains traditional magnetic materials, theory and effective parameters of magnetic composite materials, particles and magnetic materials, theory and effective parameters of soft magnetic nanodielectrics composites, theory and effective parameters of magnetic composite crystalline core, magnetic particles and core materials, preparation of nanocomposites magnetic materials, theory of magnetic nanocomposites using individual and multiple nanoparticles, as well as forecast and recommendations.

Chapter 7 examines the traditional transformer core, transformer core design, theory and effective parameters of nanocomposite transformer cores, nanoparticles structure in transformer core, theory and effective parameters of magnetization curves and losses of nanocomposites transformer core, trends of nanoparticles in cores of single phase and three-phase transformers, as well as forecast and recommendations.

Chapter 8 overviews properties of electronic materials, fabrication of conducting nanodielectrics, effects of nanoparticles in electronic dielectrics materials, optical light nanodielectrics, as well as forecast and recommendations.

Chapter 9 addresses surface plasmon resonance sensor, theory and parameters of organic light Emitting diode "OLED", theory and parameters of thin-film transistors TFTs, theory and design of compact microstrip patch antenna, effects of nanoparticles in microstrip patch antenna, semiconductors applications, integrated nanoscale electronics and optoelectronics, as well as forecast and recommendations.

Ahmed Thabet Mohamed
Aswan University, Egypt

Acknowledgment

In the Name of Allah, Most Gracious, Most Merciful

First and foremost, praises and thanks to Allah, the Almighty, for His showers of blessings throughout my research work to complete the book successfully. All deepest thanks are due to Almighty Allah, the Merciful and the Compassionate for the uncountable gifts granted to me.

I am extremely grateful to my parents for their love, prayers, caring and sacrifices for educating and preparing me for my future. I am very much thankful to my lovely wife for her love, understanding, prayers and continuing support to complete this book. My sincere appreciation goes to my parents and my wife for their generous advice, insightful suggestions and support during the whole scope of this work. Without their help, love and care, this work would never have been accomplished. Furthermore, I express my special thanks to all the research teamwork of Nanotechnology Research Center at Aswan University.

The present work was supported by Nanotechnology Research Center at Aswan University that is established with the aid of Science and Technology Development Fund (STDF), Egypt, Grant No: Project ID 505, 2009-2011.

Ahmed Thabet Mohamed
Aswan University, Egypt

Chapter 1
Nanotechnology in Industry

ABSTRACT

Nanotechnology development has a rapid progress for material science that effect on variant applications in our life. For this reason, we have focus on the importance of nanotechnology in electrical applications. This chapter describes the classification of nanomaterials, dielectrics, nanodielectrics and historical background of nanotechnology materials in our life. This chapter contains the concepts of nanodielectrics, nanofluids development and shed in brief on the electrical and industrial applications of nanotechnology in our life. Also, the purpose of present work is presented in this chapter.

1 HISTORICAL BACKGROUND

Nanotechnology is one of the new and very rapid research fields, which is studied by many scientists all over the world in governmental, commercial, academic and research laboratories, and many scientists working in this field confirm that nanotechnology will bring about a new industrial revolution in the near future in various areas of life, as well as radical transformations in economy and technology. Nanotechnology applications attract the attention of many scientists, industrialists and funders. The nanotechnology revolution aims to develop a new type of atomic electronics that adopts quantum mechanics and the movement of individual particles, which will produce equipment faster and many times smaller than anything around us now. Scientists turn to engineers designing equipment at the atomic level, and in such a system the fields of engineering, electricity, mechanics, physics, and chemistry intertwine and overlap becoming stronger together. This interconnectedness and overlap can promise a lot of new inventions and innovations. Simultaneously, it is perceived that the modern applications may bring about new difficulties in various fields, such as medication advancement, water sterilization, data and correspondence advances, and the creation of more grounded, lighter materials (Royal Society and Royal Academy of Engineering, 2004). Power transformer is a primary hardware in power transmission stations, so upgrading adequacy and unwavering quality is essential to improve dependability of system. Nanofluids are created by adding nanoparticles to transformer dielectric liquid with the purpose of improving the dielectric liquid's securing and warm attributes. Nanofluids have some potential highlights which make them unique for various mechanical applications. Countless research bunches concentrated

DOI: 10.4018/978-1-7998-8536-8.ch001

on the radically improved warm properties of nanofluids, particularly warm conductivity and convective warmth move (Agarwal et al., 2005; Kochetov et al., ; Material Safety Data Sheet, 2010; Nijenhuis & Butselaar-Orthlieb, 2008; Roustaei et al., 2013; Schulte & Salamanca-Buentello, 2007; Shin & Banerjee, 2011; Uhlig & Marsmann, 2008). From among research examining the warm properties of nanofluids, only a little piece of the investigation bundles focused on the electrical properties of nanofluids. In 1998, an estimated extended AC motivation breakdown quality of a magnetite nanofluid considering transformer oil was conducted (Du, 2010).

In light of this outcome, O'Sullivan reproduced the decoration proliferation in mineral oil and in mineral oil-based nanofluids. The reproduction results showed that conductive nanoparticles hinder the decoration spread, since conductive nanoparticles can go about as electron traps in mineral oil (Aksamit et al., 2010; National Research Council, 2002). Along these lines, it is believed that nanofluids dependent on either mineral transformer oil or vegetable oil can be utilized to violently supplant mineral oil in power transformers. Nanoparticles are neither another revelation nor a development of science, their definition is reliant on their molecule size or on specialized planning. Nanoparticles are commonly framed by two unique procedures dependent on coarser particles, the top-down procedure prompts littler particles for the most part by mechanical denunciation. In base up forms, Nano-measured particles are shaped by the gas stage or the fluid stage responses (Charrière & Dunning, 2014; Rao et al., 2004; Walter, 2013).

2 CLASSIFICATION OF NANOMATERIALS

There are various strategies for request of nanomaterials. Nanoparticles are characterized dependent on one, two and three measurements and on regularity of nanoparticle; natural or inorganic (Feynman, 1960). Nanomaterials of zero measurements (0-D) have nano-measurements in every one of the three headings. Metallic nanoparticles, silver and gold nanoparticles and semiconductors are the perfect instance of this kind of nanoparticles. A large portion of these nanoparticles are round, and the distance across these particles will be in the 1-50 nm go. One measurement nanoparticles (1-D); for example, nanowire and nanotubes of metals, oxides and different materials have been utilized for a considerable length of time in gadgets, science and building. One measurement nanoparticles are created with width or thickness of (sizes 1-100nm) (Pal et al., 2011). If there is an occurrence of two measurement nanoparticles (2-D), Carbon nanotubes are hexagonal systems of carbon molecules, 100 nm long and 1 nm in breadth, as a layer of graphite folded up into chamber. There are two sorts of carbon nanotubes: Single walled carbon nanotubes and multi-walled carbon nanotubes (Köhler & Fritzsche, 2008). Furthermore, three measurement nanoparticles (3-D) fuse mass materials made up of the the individual squares which are in the nanosized scale (1-100 nm). Quantum Dots, Dendrimers, (QDs), Fullerenes (Carbon 60) are models for these materials (Tomalia, 2004). Inorganic Nanoparticles present the metals that can be created the most in nanometric measurements like nanowires which have been created dependent on various metals, oxides, sulfides and nitrides. Most metal oxides created for an enormous scope are silica, titanium dioxide and zinc oxide (Griffiths et al., 2011; Hansen, 2009; International Council on Nanotechnology, 2008; Ostiguy et al., 2010). Hence, Organic Nanoparticles present numerous normal natural materials that can be created in nanometric measurements. Some biodegradable natural nanofibres could be utilized in medication for tissue reengineering or bone recovery, or to control the arrival of medications and design creation materials (Kas et al., 2010; Wong & De Leon, 2010). On the other hand, the sources of nanomaterials can be classified as follows:

First: the atmosphere, the natural sources of nanoparticles in the atmosphere include volcanic eruptions, forest fires, ventilation of hydrothermal systems, physical and chemical erosion of rocks and others. Inorganic nanoparticles can be diffused into the atmosphere by nuclear radiation.

Second: the aquatic environment, nanoparticles are naturally dispersed in the aquatic environment in the form of several colloidal fluids (such as clusters of metal sulfide from hydrothermal systems, aqueous iron and manganese oxides). The small size of the nanoparticles and their wide surface area render them important in linking the organic and inorganic phases of pollutants. The association and stability of nanoparticles (both natural and unnatural) with organic matter, including pollutants, is closely related to their toxic effect on aquatic ecosystems.

Third: Soil, the nanoparticles present in the soil are mostly metallic nanomaterials (such as iron, manganese and their oxides and hydroxides), in addition to natural organic materials and microbial assemblies. A typical example of nanomaterials found in soil is hydrated iron (Ferrihydrite), which is found everywhere in soil.

3 DIELECTRICS AND NANODIELECTRICS

This book surveys recent advances in the field of creating mechanical polymers by nanocomposites materials. The examined materials have been taken into both scholastic and mechanical consideration, since they show sensational improvement in properties at chosen low filler substances. In this regard, the sorts and types of mechanical polymer, structure, and properties of polymer–nanocomposites are addressed. Generally, polymeric materials have been loaded up with engineered or regular inorganic mixes so as to improve their properties, or basically to lessen cost. Additionally, this book researches different components influencing the electrical properties of the polymers protecting materials and the ideal properties of the novel nanoscience materials. Finally, this book discusses the correlation between chosen nanoparticles on modern applications, and creation techniques organizing polymer nanocomposites. Polymeric materials have received an elective decision for electrical applications and gadgets. These applications require multi-usefulness in a solitary material which isn't normally found in an average polymer. A prudent method to manufacture multi-practical materials is to blend polymer with different materials. In the nanocomposite frameworks where the fillers regularly have nano-meter-scale measurements, the property improvement in such frameworks typically accompanies tradeoffs as follows:

The crack inception and proliferation of polyethylene (PE)/clay nanocomposite is described as extension of fibrils (fibrillation) bury scattered with smaller scale voids. The low durability of the earth fortified polyethylene corresponding to flawless polyethylene is identified with the gem structure and interfacial cooperation between the filler and the polymer network. Anyway, the nearness of the alumina particles expands the thermo-oxidative solidness of the composites for perfect Low-density polyethylene (LDPE). The little molecule size of the ZnO nanoparticles does not influence the permittivity of the composite. Along these lines, the interfaces among ZnO and LDPE don't seem to add to the dielectric properties. Acrylonitrile butadiene styrene (ABS)/clay nanocomposites are obtained by utilizing OMLS with various types of natural surfactants in an answer procedure. TGA information demonstrates that handling temperature impacts the conduct of the muds adjusted with quaternary ammonium salts that may debase and catalyze the decay of ABS, while in imidazolium-based nanocomposites, no synergist disintegration is seen. An epic half breed nanocomposites joining ABS with silica nanoparticle has been effectively arranged in watery media utilizing a solid corrosive base collaboration between hydroxyl gatherings of silica surfaces and cyanogroups of ABS. Polyimide (PI)/mud nanocomposite can be delegated a peeled nanocomposite. In view of this uncommon structure, the PI-earth nanocomposites indicate a stamped improvement in mechanical and warm properties. Hence, the mechanical properties of PI/SiO_2 nanocomposite slim movies at low temperature are introduced. PI/SiO_2 nanocomposite films with various silica substance are integrated with another sol–gel. Epoxy nanocomposite frameworks with inorganic oxide fillers show some worthwhile dielectric practices at low nanofiller loadings. The permittivity and tan delta esteems in the nanocomposites are seen as lower than that of microcomposites, just as unfilled frameworks (for scarcely any filler loadings). The expansion of earth to an epoxy sap goes about as a physical boundary, which keeps a portion of the potential responses from happening. The diminishing in the quantity of responses implies a decrease in the cross-connecting. A negligible decrease in the epoxy dc volume resistivity is likewise seen by fusing a limited quantity of nano-filler into the base material. As for the air conditioner dielectric quality, despite the fact that the qualities in nanocomposites are lower than those of unfilled epoxy frameworks, when contrasted with microcomposites, the kind of filler appears to impact the outcomes. The water take-up of nano-silica epoxy composite can be significantly diminished by guaranteeing that the filler particles have surfaces that are functionalized to be hydrophobic. This may likewise improve the composites protection from the maturing handling method. Polyvinyl chloride (PVC) composite warm security, as estimated by a great HCl advancement strategy explicit for Polyvinyl chloride, is fairly brought down for the nanoclay composites when contrasted with fitting control details. Despite the debasement model, the way that nanoclay can smother smoke age is critical for various PVC applications where smoke concealment is of concern. Instances of utilizations incorporate, among others, wire and link, and frothed protection. The warm and mechanical properties of the composites were notably altered by filler expansion, paying little attention to the alumina filler properties. Specifically, a critical increment of the firmness of the polymer within the sight of the alumina particles was discovered, along with an upgraded scraped spot obstruction. Dielectric steady and misfortune digression of the Polyethylenimine (PEI)/$BaTiO_3$ nanocomposites increment with expanding $BaTiO_3$ content. The variety of temperature created little impact on the dielectric properties of PEI/$BaTiO_3$ nanocomposites. The joining of ZnO nanoparticles into Polypropylene (PP) network can grant critical enhancements for the photograph corruption opposition of PP to UV-illumination. The warm strength of Poly (methyl methacrylate) (PMMA) was expanded relying upon dirt stacking. As an afterthought, the PMMA/earth nanocomposites hold great optical properties. This book takes a gander at the improvement of exceptionally leading polymers with great strength and satisfactory handling qualities. A conductive cement has been made by adding nano estimated particles to various sorts of polymers. Additionally, this book addresses the impacts of options of nano-sized particles on the electrical directing properties and shows that electrical resistivity diminishes and assists with shaping the conductive way. It is granted new strategies for improving conductivity to a certain extent.

The creation systems for assembling nanocomposites modern directing materials have been shown as follows:

- The electrical resistivity of polymers has been estimated, and it indicated a progress from a protecting to a directing polymer. There has been an improvement in the electrical resistivity by as much as eight sets of size.
- The electrical and warm conductivities accomplished by the peeled graphite filled polymers are shown to be higher than carbon nanofiber and High heat treated (HHT) nanofiber based polymer composites.

- Conducting materials have various degrees to become the ideal filler. Silver has high conductivity and significant expense while copper has great properties for ease, but not great as silver regarding conductivity. Conductive cements are made by adding Nano measured particles to various kinds of polymers. Film resistivity is then estimated as an element of nanoparticles volume or fixation.
- Micro-sized silver drop with Polyvinyl acetic acid derivation (PVAc) is close to the permeation limit, the expansion of a modest quantity of Nano-sized silver colloids would assist in setting up the conductive way and thus bringing down resistivity. In any case, when the smaller scale measured silver drop is adequate to shape the conductive way without anything else, expansion of Nano-sized silver colloids would show a negative impact on conductivity because of commitment to contact protections. Warming the film to a higher temperature can lighten this negative impact, presumably because of the high action related to Nano-sized particles.
- FTIR ghostly investigation shows the arrangement of BGE–LiF6P buildings. This prompts an expansion in conductivity in the PMMA. More expansion of LiF6P expands the thickness of PMMA/LiF6P arrangement and indicates lower estimation of conductivity.
- The scattering of conductive filler (aluminum powder) in epoxy grid and cement properties of such composites coatings can be improved through adjustment by polythiourethane hardener just for explicit definitions and ideal warm conditions.

In brief, this book presents ongoing advances in the field of creating modern polymers by nanocomposites and multiple nanocomposites materials. The discussed materials have been taken into both scholastic and mechanical consideration since they show emotional improvement in properties at extremely low filler substance. Thus, the sorts and types of mechanical polymer, as well as the structure and properties of polymer–nanocomposites are examined. Customarily, polymeric materials have been loaded up with engineered or common inorganic mixes so as to improve their properties, or just to decrease cost. Likewise, this book addresses different components influencing the electrical properties of the polymers protecting materials. It examines the creation methods designing copolymer nanocomposite. The diagnostic properties and utilizations of nano-sized molecularly of engraved polymers are just like those of the creation of little gadgets. The 0–1 dimensional Nano objects have been broadly contemplated, and have been utilized currently to acquire improved diagnostic outcomes. The 2–3D designed structures are explored by concentrating on their sensor and microarray creation. A few designing methods that are at present utilized in the manufacture are examined; nonetheless, the present status of these applications is still in their creating stages to demonstrate the idea. Concerning the improvement of exhibited sensors, there remains numerous unsolved issues, both actually and reasonably. Particularly, low selectivity of the receptors regularly hampers the acknowledgment of the rich idea. In expansion to this basic issue, conveying pre-polymerization blends, each containing an alternate format, on a coordinated transducer surface in a profoundly exact way is one of the fundamental issues of creation procedure that is presently confronting. Finally, an ongoing distribution about some different nanomaterials (metal and semiconductor nanoparticles), which are perfect with MIP receptors, is examined. These particles as optical and fluorescent tests are significantly more vigorous in rehashed and constant use than natural colors, and can be utilized as a change to report investigative take-up. With the expansion of this usefulness, the material turns out to be substantially more intelligent, receiving an implicit criticism unit. There have not been numerous reports accessible in this regard. Hence, nonstop endeavors ought to be made for the improvement of the materials with journalists.

4 NANOFLUIDS DEVELOPMENT

The base thought of scattering strong particles in fluid to redesign the warmth move conductivity isn't new. It dates back to 1873, when Maxwell introduced a hypothetical reason for foreseeing the powerful warm conductivity of fluid/strong suspension (Maxwell, 1873). As of late, solid particles are used for light for the way to move heat much better than liquids. For more than 100 years, analysts have made inconceivable endeavors to redesign the intrinsically poor warm conductivity of traditional warmth move liquids, such as ethylene glycol, transformer oil and water (Ahuja, 1975). A great deal of exploratory and hypothetical investigations of the compelling warm conductivity of suspensions with strong particles have been led (Bonnecaze & Brady, 1991; Hamilton & Crosser, 1962; Mangrulkar & Kriplani, 2013). Additionally, the entirety of the investigations has been applied on micrometer or millimeters size particles. The primary issue with the utilization of millimeters or micrometers size particles is the quick settling of the particles in liquids (Ahuja, 1975). The enormous size of the particles and the trouble underway for little particles are the constraining components of fluid/strong suspension to be researched for down to earth applications. A significant advance in the improvement of nanoscience was the appraisal of the nano-meter size of atoms in the start of the twentieth century (Feynman, 1959). The improvement of checking burrowing magnifying instrument sets off the advancement of nanotechnology (Lee et al., 1999; Sreelakshmy & Nair Aswathy, 2014). Stephen Choi introduced a wonderful idea of expanding the convection heat move coefficients by utilizing high-conductivity nanofluids, as opposed to expanding siphon power (Eastman et al., 2001). Starting from now and into the foreseeable future, it has been endeavored to suspend distinctive metal oxides and metal nanoparticles in different fluids. The outcomes indicated that nanoparticles still suspended longer than millimeters or micrometers size particles and that nanofluids give incredible warm properties and cooling limit (Eastman et al., 1996; O'Sullivan, 2007; Ramakoteswaa et al., 2014; Vafaei et al., 2006; Wang et al., 1999).

Voltage excitations of both positive and negative polarities can prompt the arrangement of streamers in dielectric fluids. Decoration commencement requires very high electric field levels, normally in the 10^2 to 10^3 (MV/m) run. The specific field level required to start a decoration changes relying on the dielectric fluid's substance structure, its immaculateness, the extremity of voltage excitation and a range of different parameters (Beroual et al., 1998). Positive streamers, if there should arise an occurrence of transformer oil, engender at higher speeds and travel further from their places of commencement for a given voltage excitation than negative streamers. As expressed, beforehand, decoration inception and development can prompt the electrical breakdown of transformer oil (Devins et al., 1981; O'Sullivan, 2007). In all cases, negative streamers in transformer oil have a rugged structure with considerably more parallel spreading than the situation with positive streamers. Higher electric field levels can prompt the improvement of auxiliary mode negative streamers. These proliferate at speeds of up to 5 km/s. The generally low engendering speeds of negative streamers in transformer oil implies that they will generally pose, to a lesser extent, a danger of causing electrical breakdown than positive streamers (Devins et al., 1981; Vicki, 2003).

5 NANO-TECH ELECTRICAL APPLICATIONS

Nanotechnology is a modern technology that has many applications that include the fabrication of particles or particles in the nanoscale range. Nanotechnology is defined as an advanced field that involves the manufacture, preparation and application of many installations, devices and systems consisting of micro-units. The idea of nanotechnology originated from the Greek word nano, meaning dwarf, and nanoparticles are also defined as individual particles whose dimensions do not exceed 100 nanometers. The unique properties and features of nanoparticles are due to their small size, chemical composition and surface structure. The distinctive properties and physical changes of the different materials in the nanoscale have given rise to the properties of industrial products, which resulted in a real and influential increase in industrial and medical applications. In this book, the focus will be on the applications of nanotechnology in the field of electrical applications, especially industrial applications. As a result of the unlimited use of nanotechnology products and nanomaterials, nanoparticles have begun to find their way into the industrial environment surrounding us. This is why the sources of nanoparticles, their behavior and their effects on the electrical industries are mentioned. The different techniques for assessing the spread, fate and behavior of nanomaterials in different electrical industries are also mentioned, in addition to the potential risks of particles and nanomaterials. Industry in all its forms is the main engine for growth because it has raised production and the standard of living and improved the health and education of people. This is because it provided them with most of their needs, which increased with their increase, and preserved their environment as much as possible through its continuous interaction with various fields of science, especially information technology, which is mainly based on its achievements. These achievements are obvious in computer science and electronics which is the science that examines electrical circuits taking closed paths through a group of elements, such as valves, resistors, transistors, and capacitors.

When some materials are processed at an atomic level, they acquire unique and important properties. These properties may be visual, electrical, or magnetic, as well as other various properties. That is why the applications of nanotechnology vary in different areas of life, such as energy, medicine, agriculture, electronics, etc., and we will now review some of these uses:

In electronics: Nanotechnology helps to increase the efficiency of electronic devices, but at the same time it reduces their size and the energy consumed to operate these devices.

In solar cells, scientists have been able to manufacture nanoscale solar cells at a much lower cost than conventional solar cells.

In food: Nanotechnology contributes to the development of food science, from growing grains to their packaging, and helps improve the quality of food, increase its nutritional benefit and maintain its safety.

In medicine: Researchers are currently working on developing nanoparticles with the size of particles to help transport the drug directly to disease-affected cells, and scientists from the Russian National University of Technological Research designed hybrid nanoparticles against cancer and bacteria a few months ago.

In the battery industry: By using nanotechnology in the battery industry, some companies have been able to develop batteries that will not spoil if they are not used for a long time.

In space: The use of nanotechnology will make space travel more practical. By using nanomaterials in the manufacture of spacecrafts, we will be able to reduce their weight and can thus greatly reduce the amount of fuel needed to propel them, and these applications will reduce the cost of space travel.

Air purification: Nanotechnology can be used to convert the exhaust that comes out of cars and factories into harmless gases.

In water treatment: Nanotechnology is used to remove the industrial waste from groundwater, where nanoparticles can be used to convert the polluting chemical into harmless substances. Studies have shown that this method can be used successfully to access pollutants scattered in earthen ponds at a much lower cost than methods that require pumping water from the ground for treatment.

In sports: The applications of nanotechnology are used in the sports arena mainly to increase the strength of tennis rackets, and to improve the quality of tennis balls as it reduces the rate of loss of air.

Nanotechnology is still invading different fields, and what we have mentioned is only the tip of the iceberg for the applications of that technology in the areas of our life. The future of this technology still holds more and more, as the field is fertile and suitable for developing this technology and harnessing it to serve humanity more.

5.1 Nano-Tech Power Cables

Nanotechnology has been widely used in many engineering industries, for example, raw materials can be transformed into materials with higher engineering quality specifications and with pinpoint accuracy by effecting a change in the physical properties of the molecular sizes that make up their structure. Here, benefit is made from the obvious increase in space surface to volume ratio. Then its optical properties, including fluorescence, become a function of particle diameter. By incorporating them into their bulk material, nanoparticles strongly affect the mechanical properties of the material, especially the hardness or ductility. The properties of conventional polymers can be changed by using nanoparticles in the new materials that may be used as lightweight alternatives to the mineral. Thanks to these processes, the specifications of important raw materials can be improved. The nano-supported materials reduce the original weight of the material by increasing the stability properties and all other requirements. Practical nanotechnology represents the increased ability to deal with high-quality precision with matter according to standards modernization, which leads to a reduction in material costs and human labor at the same time. Therefore, the electrical power cables are an important part of the components of the electrical network because they are used as mediums for transmitting power from generators and distributing it to different loads, whether domestic or industrial loads, as the cable's installation, shape and cross-sectional area change according to the amount of electrical energy required, transported and connected. The function of the cables is to properly transfer energy from the source to the devices for use in cities and densely populated areas since is difficult to lay overhead lines. The use of electric power cables goes back to 1926, when a conductor was placed between two layers of glass for insulation inside a duct, and the group consisting of conductor and insulation was known as Cable. Cables are generally classified according to the insulating material used or the implicit voltage. In this book, developing electrical cables using nanotechnology or engineering nanotechnology products is covered under the name of nanometer as a unit of measurement equal to one billionth of a meter; that is a part of a thousand million parts of a meter.

5.2 Nano-Tech Energy and Power Capacitors

In light of scientific warnings of exposing natural energy sources to depletion in the future, intensive studies are being conducted to use renewable energy sources based on modern technologies. In the largest scientific gathering of energy experts in Egypt, nuclear energy, in its two parts (nuclear fusion and biofuels), dominated discussions of the latest technologies that lead to raising the efficiency in energy use of all kinds, of how to save them and ways to reduce harmful emissions to become clean and environmentally friendly energy. The importance of nanotechnology lies in the importance of technological development for many industries, for example the manufacture of solar cells, and considering its low efficiency and high cost as the main obstacles to its use and spread. The recent trends of using nanotechnology and new materials in the manufacture of cells, which jump in efficiency from 5% to 6% to about 50-60%. And with the purpose of automatic steering of solar PV cells to increase energy efficiency. Biofuels made from plants - soybean oil in diesel may be used for heavy transport and as a fuel for trains to reduce exhaust emissions and air pollution by controlling and modifying engine designs to take advantage of the natural and chemical characteristics of this new fuel that are different from the fuel currently used for diesel. Also, research is continuing on developing methods of producing diesel as a biofuel to reduce prices by using the latest nanotechnology technologies. On the other hand, one of the main objectives of developing this book is to provide the engineer with a clear and direct way to use capacitors developed using nanotechnology sciences to improve the power factor in industrial systems in general, the content of this book is an appropriate field to clarify the concept of the necessity to raise the value of nanometric grains in the development of industrial capacitors that is often unclear.

5.3 Nano-Tech Insulators

The transmission and distribution lines are installed through solid insulators to prevent the leakage of electrical current to the ground through the towers and columns carrying them, where the solid insulators used are made of several different materials, have high isolation and ensure successful, safe and cost-effective operation in electric power systems. In recent times, the energy sector has been receiving great attention by consumers and producers alike, due to its impact on all aspects of life. The progress of countries has become measured by the extent of its energy consumption. Electric energy is one of the types of energy which has become the most demanded type as a result of delivering services to a large proportion of consumers in addition to providing and increasing industrial production, as 95% of the industrial sector uses electrical energy. This has impelled us to produce electric energy in very large quantities, so it has become necessary to find advanced systems capable of safely transmitting these huge amounts of electrical energy with super effort and for these systems to be economical with a long operational life and to meet the technical specifications and general safety requirements. It is known that in the science of electrical engineering, the amount of voltage directly affects the amount of distance transmitted and the amount transferred. Hence, in addition to energy and super efforts, you need an isolation mechanism and means of isolation to ensure the continuity of transport in an efficient and safe manner because the electrical power is currently transferred in transportation systems to efforts that simulate the amount of reluctant and continuous electric lightning or pulse that may sometimes reach 120 km. Among the applications of these tests is communication to produce and develop ideal insulators and resistance to all conditions that may occur during the period of their use in transmission or distribution systems of electrical power, whether environmental or resulting from passage of huge amounts of energy.

5.4 Nano-Tech Transformers

The electrical transformer is a very important and useful part of the electrical circuit, and without it, the various vital purposes would not have been achieved in the electrical network and electrical transmission systems.

Electrical power is transferred in huge quantities over very long distances from the sites of its generation to the areas of its use, on very high electrical voltages (hundreds of thousands of volts, reaching a million in some cases), in a way that works to reduce electrical losses in transmission lines, after raising the voltage. The material is still limited due to technological difficulties, despite advanced design theories (it does not exceed a few tens of thousands of volts, and barely reaches half a hundred thousand, or sometimes exceeds it). Distribution of electrical power in areas is to be used with appropriate efforts for the purposes of using it in homes, factories, and other places. The alignment between any electrical device, regardless of the voltage it works with, and the electrical supply spring having a voltage can only be changed by using an electrical transformer of certain specifications. Hence, we can imagine that most electrical devices have an electrical transformer in their basic structure to carry out this task with electrical supply springs various efforts. The electrical transformer is simple in its theories and easy in its uses, and it is assumed that it does not cost the electrical circuit included in its installation except few electrical losses, as all that it does is a process of harmonization between the different parts of the circuit mainly in the efforts. This is the process of converting electrical energy from one file to another in the transformer through the magnetic field in the iron core. This is a very high device in efficiency that may exceed 99% of very large transformers, while we find very small electrical transformers whose capacity does not exceed a part of the volt-ampere. In the infinitesimal devices, the capacity of large transformers may reach tens of thousands of megavolts of amperes in large transformers that are used with electric power transmission lines. Such transformers need special means of transportation when they are transported from one place to another and they are often equipped with wheels that are suitable for exchange between one another in the same place. The transformer industry has advanced and developed in recent times, and its shapes and designs have varied to serve the purpose for which they were made. Transformers are considered one of the basic components in any electrical network, and our topic in this book is to develop the design and installation of electrical transformers using nanotechnology. Technology studies and revolves around the electrical performance of the electrical heart as well as the development of transformer oils. By adding different nanoscale granules to raise the isolation strength theoretically and practically, this book covers everything of interest to engineers and technicians working in the field of electricity in order to expand their knowledge about transformers.

5.5 Nano-Tech Electronic Applications

The term nanoelectronics refers to the use of nanotechnology (electronic components). It includes a variety of elements and materials, with a common feature being that they are of such small size that internal atomic interactions and properties of quantum mechanics need to be studied on a large scale. Advanced molecular electronics and finally generations of complementary symmetry silicon technology (metal-oxide-semiconductor, like the 22 nm structure, are already inside this system)."Nanoelectronics" is sometimes considered a revolutionary technology as it differs significantly from traditional electronics and integrated circuit technologies. The scientific applications of nanoscience and nanotechnology are of great importance, and it relies a lot on training people in these fields. Therefore, modern education must quickly address the changing aspects of nanoscience and nanotechnology. New generations of researchers, technicians and engineers must be trained in the emerging nanoscale fields. This book presents our contribution to the education of the nanoparticles field. In this book, we represent a framework that gathers the basic ideas necessary to understand the recent developments based on nano-sciences and nanotechnology applied to the field of nanoelectronics. It is always noticed that the efficiency of electronic applications is related to the temperatures generated during the operation of these devices, and their increase leads to a decrease in the efficiency of performance and an increase in the amount of electrical power consumed. The size of this component currently puts tremendous pressure on small sized and fast-performing applications.

6 AIM OF WORK

This book has the main role of explaining the effectiveness and reliability of electric and electronic industrial applications, thus, the aim of this work is as follows:

1. Studying the effect of adding individual and multi-nanoparticles into industrial applications to enhance their dielectric and electric properties.
2. Specifying the main theory and parameters affecting the physical and electrical properties of fabricated nanocomposites materials.
3. Specifying optimal nanoparticles for enhancing performance of industrial applications.
4. Assessment of the dielectric tests on experimental fabrication samples of individual and multiple nanocomposites materials.
5. Experimental measurement for electric and electronic characterization of new individual and multiple nanocomposites materials.

REFERENCES

Abhilash, M. (2010). Potential applications of Nanoparticles. *International Journal of Pharma and Bio Sciences*, *1*(1).

Agarwal, S., Tatli, E., Clark, N. N., & Gupta, R. (2005). Potential Health Effects of Manufactured Nanomaterials: Nanoparticle Emission Arising from Incineration of Polymer Nanocomposities. *International Symposium on Polymer Nanocomposities Science and Technology*.

Ahuja, A. S. (1975). Augmentation of Heat Transport if Laminar Flow of Polystyrene Suspensions. I. Experiments and Results. *Journal of Applied Physics*, *46*(8), 3408–3416. doi:10.1063/1.322107

Aksamit, P., Zmarzly, D., Boczarand, T., & Szmechta, M. (2010). Aging Properties of Fullerene Doped Transformer Oils. *IEEE International Symposium on Electrical Insulation*, 1-4. 10.1109/ELINSL.2010.5549834

Beroual, A., Zahn, M., Badent, A., Kist, K., Schwabe, A. J., Yamashita, H., & Torshin, Y. (1998). Propagation and structure of streamers in liquid dielectrics. IEEE Electrical Insulation Magazine, 14(2), 6-17.

Bonnecaze, R. R., & Brady, J. F. (1991). The Effect Conductivity of Random Suspensions of Spherical Particles. *Proceedings of the Royal Society of London*, *A432*, 445–465.

Charrière, & Dunning. (2014). *Timeline: Nanotechnology. Institute for Science, Society and Policy*, University of Ottawa.

Devins, J. C., Rzad, S. J., & Schwabe, R. J. (1981). Breakdown and Prebreakdown Phenomena in Liquids. *Journal of Applied Physics*, *52*(7), 1–7. doi:10.1063/1.329327

Du, Y. (2010). Breakdown Properties of Transformer Oil-based TiO2 Nanofluid. *Conference on Electrical Insulation and Dielectric Phenomena*, 1-4.

Eastman, J. A., Choi, S. U. S., Li, S., Yu, W., & Thompson, L. J. (2001). Anomalously Increased Effective Thermal Conductivities of Ethylene Glycol-based Nanofluids Containing Copper Nanoparticles. *Applied Physics Letters*, *78*(6), 718–720. doi:10.1063/1.1341218

Eastman, J. A., Choi, S. U. S., Thompson, L. J., & Lee, S. (1996). Enhanced Thermal Conductivity through the Development of Nanofluids. *Materials Research Society Symposium Proceedings*, 14, 3-11. 10.1557/PROC-457-3

Feynman, R. P. (1959). There's Plenty of Room at the Bottom. *American Physical Society Meeting*.

Feynman, R. P. (1960). There is Plenty of Room at the Bottom, an Invitation to Enter a New Field of Physics. *Engineering Science Magazine*, *23*, 143.

Griffiths, D., Bernt, W., Hole, P., Smith, J., Malloy, A., & Carr, B. (2011). *Zeta Potential Measurement of Nanoparticles by Nanoparticle Tracking Analysis (NTA)* (Vol. 1). NSTI- Nanotech.

Hamilton, R. L., & Crosser, O. K. (1962). *Thermal Conductivity of Heterogeneous Two Component Systems* (Vol. 1). I&EC Fundamentals.

Hansen, S. F. (2009). *Regulation and risk assessment of nanomaterials: too little, too late* (Ph.D. thesis). Department of Environmental Engineering, University of Denmark.

International Council on Nanotechnology. (2008). *Towards Predicting Nano-biointeractions: An International Assessment of Nanotechnology Environment, Health and Safety Research Needs*. Rice University.

Kas, Choi, Yu, & Pradeep. (2010). Nanofluids: Science and Technology. Wiley-Interscience.

Kochetov, R., Morshuis, P., Smit, J., Andritsch, T., & Krivda, A. (2014). Precautionary remarks regarding synthesis of nanocomposites. *IEEE Electrical Insulation Conference (EIC)*, 2, 51-54.

Köhler, M., & Fritzsche, W. (2008). *Nanotechnology: an introduction to nanostructuring techniques*. John Wiley & Sons.

Lee, S., Choi, S. U. S., Li, S., & Eastman, J. A. (1999). Measuring Thermal Conductivity of Fluids Containing Oxide Nanoparticles. *Transactions of ASME*, 121, 280-288. 10.1115/1.2825978

Mangrulkar, C. K., & Kriplani, V. M. (2013). Nanofluid Heat Transfer-A Review. *IACSIT International Journal of Engineering and Technology*, *3*, 136–142.

Material Safety Data Sheet. (2010). *Shell Diala S3 ZX-IG version 1.0*. Effective Data, Regulation 1907/2006/EC.

Maxwell, J. C. (1873). *A Treatise on Electricity and Magnetism*. Clarendon Press.

National Research Council. (2002). *Small wonders, endless frontiers: A review of the National Nanotechnology Initiative*. National Academies Press.

Nijenhuis, J., & Butselaar-Orthlieb, V. C. L. (2008). *Nanosafety Guidelines. Recommendations for research activities with free nanostructured matter*. Faculty of Applied Sciences, Delft University of Technology.

O'Sullivan, F. M. (2007). *Model for the Initiation and Propagation of Electrical Streamers in Transformer Oil and Transformer Oil Based Nanofluids* (Ph. D. dissertation). Massachusetts Institute of Technology.

Ostiguy, Roberge, Woods, & Soucy. (2010). Engineered nanoparticles: current knowledge about OHS risks and prevention measures. *Chemical Substances and Biological Agents: Studies and Research Projects, IRSST in the Research and Studies Series*, report R-656.

Pal, S. L., Jana, U., Manna, P. K., Mohanta, G. P., & Manavalan, R. (2011). Nanoparticle: An overview of preparation and characterization. *Journal of Applied Pharmaceutical Science*, *1*(6), 228–234.

Ramakoteswaa, N. R., Leena, G., & Ranganayakulu, S. V. (2014). Synthesis, Applications and Challenges of Nanofluids – Review. *IOSR International Conference on Advances in Engineering & Technology*, 21-28.

Rao, C. N. R., Muller, A., & Cheetham, A. K. (2004). *The Chemistry of Nanomaterials* (Vol. 1). Willey-VCH. doi:10.1002/352760247X

Roustaei, A., Saffarzadeh, S., & Mohammadi, M. (2013). An Evaluation of Modified Silica Nanoparticles Efficiency in Enhancing Oil Recovery of Light and Intermediate Oil Reservoirs. *Egyptian Journal of Petroleum*, *22*(3), 427–433. doi:10.1016/j.ejpe.2013.06.010

Royal Society and Royal Academy of Engineering. (2004). *Nanoscience and Nanotechnologies: Opportunities and Uncertainties*. Author.

Schulte, P. A., & Salamanca-Buentello, F. (2007). Ethical and scientific issues of nanotechnology in the workplace. *Ciencia & Saude Coletiva*, *12*(5), 1319–1332. doi:10.1590/S1413-81232007000500030 PMID:18813467

Shin, D., & Banerjee, D. (2011). Enhanced Specific Heat of Silica Nanofluid. *Journal of Heat Transfer*, *133*(2), 1–4. doi:10.1115/1.4002600

Sreelakshmy, K. R., & Nair Aswathy, S. (2014). An Overview of Recent Nanofluid Research. *International Research Journal of Pharmacy*, *4*, 239–243.

Tomalia, D. A. (2004). Birth of a new macromolecular architecture: Dendrimers as quantized building blocks for nanoscale synthetic organic chemistry. *Aldrichimica Acta*, *37*(2), 39–57.

Uhlig, F., & Marsmann, H. C. (2008). Gelest Catalog: Silicon Compounds. *Silanes & Silicones*, *22*, 27–33.

Vafaei, S., Borca-Tasciuc, T., Podowski, M. Z., Purkayastha, A., Ramanath, G., & Ajayan, P. M. (2006). Effect of Nanoparticles on Sessile Droplet Contact Angle. *Nanotechnology Journal*, *17*(10), 2523–2527. doi:10.1088/0957-4484/17/10/014 PMID:21727499

Vicki, L. (2003). The potential environmental impact of engineered nanomaterials. *Nature Biotechnology Journal.*, *21*(10), 1166–117. doi:10.1038/nbt875

Walter, D. (2013). *Primary Particles–Agglomerates–Aggregates* (Vol. 9). Commission for the Investigation of Health Hazards of Chemical Compounds in the Work Area Nanomaterials.

Wang, X., Xu, X., & Choi, S. U. S. (1999). Thermal Conductivity of Nanoparticle-fluid Mixture. *Journal of Thermophysics and Heat Transfer*, *13*(4), 474–480. doi:10.2514/2.6486

Wong, K. V., & De Leon, O. (2010). Applications of nanofluids: Current and future. *Advances in Mechanical Engineering*, *2010*, 1–12.

Chapter 2
Treeing in NanoDielectrics

ABSTRACT

In electrical insulation material designing, treeing is an electrical pre-breakdown marvel in strong protection. Treeing is a typical breakdown component and wellspring of electrical deficiencies in insulation of electrical applications protection. Moreover, water treeing is a diffuse part of the way conductive 3D crest-like a shape inside the utilized dielectrics in covered or water-drenched electrical applications. Nanotechnology techniques have been enhancing the dielectric strength performance with respect to traditional dielectrics. Therefore, this chapter discusses the treeing mechanisms in theoretical models for various nanodielectrics insulation materials. Water treeing in nanodielectrics is also addressed. Furthermore, this chapter contains forecast and recommendations to enhance insulation performance of electrical applications.

1 ELECTRICAL TREEING IN NANODIELECTRICS

Electrical treeing is by and large viewed as the main actual instrument for dielectric breakdown in strong protection materials. Different models are proposed to reproduce the worldly proliferation of electrical trees. The development of electrical trees in heterogeneous strong dielectrics was examined through two-dimensional mathematical recreations. As opposed to customary methodologies, the recently created calculation depends on a deterministic model along with a spatial dissemination of the electrical breakdown strength. With standard needle plane calculations, two primary kinds of treeing, shrub like and branch like trees, were reenacted effectively regarding their shape and the overall voltage reliance. The tree engendering in homogeneous dielectrics was mimicked just like the impact of hindrances after treeing. This tree development was contrasted with tree development in epoxy sap with an inserted glass-mica layer. Moreover, the impact of electrically feeble interfaces between the polymer and the obstruction was explored, and the interim to breakdown was analyzed for various arrangements. Electrical treeing happening at voids and polluting influences is viewed as the principal pre-breakdown component of polymeric insulation materials. Its development is portrayed by three principal stages: commencement, spread and breakdown. During the initiation phase, it is accepted that smaller scale voids are shaped in areas of high electric stress without perceivable partial discharge movement. In the proliferation stage,

DOI: 10.4018/978-1-7998-8536-8.ch002

quantifiable halfway releases cause disintegration and advancement of thin channels over the insulation. This mind-boggling process is influenced by space charge infusion, breaking of polymeric bonds, oxidation and arrangement gases. During Propagation, the electric field appropriation in the channel is constantly changing. In this manner, the release movement will shift inside the electric tree. Electric breakdown happens either when a conductive channel is shaped or when the breakdown quality of the tree structure is diminished beneath the applied pressure. The after effect of this is that the state of electrical tree will rely upon the size and recurrence of the applied electric stress, just as the material properties. Electrical trees are a typical reason for debasement in strong insulation.

Electrical trees are disintegration channels which originate from areas with high nearby electric field (for example, voids, cracks, outside particles and so forth). Tree development contains the arrangement of gas-filled channels in which incomplete releases happen. Propagation of electrical tree into strong insulation can connect the anode hole, causing breakdown of the insulation. The breakdown doesn't happen when slender channels show up at the contrary terminal. A specific measure of time is required for slender channels to transform into thick ones, which are then fit to cause breakdown. The partial discharge action inside the channels is answerable for the channel's breadth development. The attributes of electrical tree proliferation have been discovered delicate to cathode geometry, voltage, recurrence, temperature, etc. During the spread, the electrical tree can embrace differently complex shapes that can be generally described as branch or potentially hedge formed structures. Electrical trees comprise associated diverts, a couple of micrometers in breadth, with branches of many micrometers long. The dividers of the channels are not generally carbonized and just feebly directing along these lines, albeit a short out may happen if the tree connects the insulation (Ding & Varlow, 2004a; Ding & Varlow, 2005; Kurnianto et al., 2006; Noskov et al., 2000; Tanaka et al., 2006).

1.1 Theory and effective parameters

Most inquires up to date have been conducted on observation of the electrical trees development utilizing optical perceptions, pulse distribution (PD) investigation and PD beat examination. In any case, little work has been conducted on researching the pattern of electrical trees before breakdown, such as dissecting and meaning the significant qualities of incomplete release designs because of electrical trees directly before the insulation breakdown. This would be helpful to give a sign of the conduct of the electrical trees before breakdown, for instance, its insulation life is arriving at the cutoff. Examining electrical treeing in polymers, standard test strategies alluded to as needle test strategy give measurable evaluations of the electrical tree commencement stress. Both metallic and semi-leading needles can be applied (Al-Arainy et al., 2006; Bamji, 1999; Kurnianto et al., 2005). Incomplete releases can begin at the interfaces and in the long run advance the beginning and development of electrical trees. The potential purposes behind how littler particles ought to improve the electrical treeing obstruction are discussed, and a model for breakdown-safe of nanocomposite dielectrics is created. Nanoparticles have incredible impact on changing the properties of nanocomposite materials as electrical tree development obstruction is better than the unfilled lattice, thus, particularly focuses on the impact of nanoparticle volume division on the breakdown opposition (Crine, 2007; Ding & Varlow, 2004b; El-Zein et al., 2009; Huuva et al., 2009; Pitsa et al., 2010; Raetzke et al., 2009; Theiler, 1990). This part explores the life-time of tree spread to breakdown and exhibits the tree development pace of electrical tree proliferation by utilizing different sorts and measures of nanoparticles. Additionally, it clarifies the dielectric properties which have been upgraded by including nanoparticles of Clay, Silica, Zinc Oxide, and Aluminum Oxide in the different sorts of epoxy resin modern insulation materials.

An arrangement of a spreading electrical tree and the encompassing damage process zone (DPZ) are exposed to mere rotating flow electrical stress, as shown in Fig. 1. During the tree propagation process, the DPZ is imagined to develop in a self-comparative design and advances through the change of polymer from some underlying morphology to a harmed morphology. Along these lines, the idea of harm is inherently identified with the microstructure of the polymer. Miniaturized scale voids start in the DPZ at the tree tip when the force of the nearby field finally develops to some basic level, as shown in Fig. (1.a). The essential tree channel withdraws from its unique straight direction to interface with a neighboring hole retained, and another neighbor hole is consumed by the fundamental tree channel to frame expanding tree channels. Figure (1.b) shows the microphysical qualities of tree propagation. Harm in the DPZ in front of a tree offers ascension to smaller scale voids, which in the long run combine as an augmentation or branch.

Figure 1. Damage in the DPZ microstructure of the local field around the tree tip

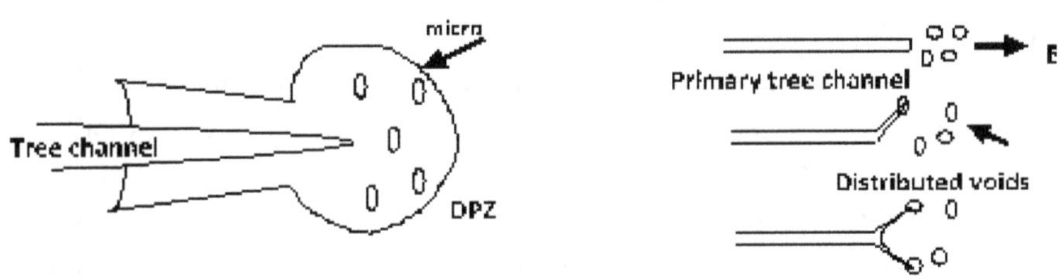

a. Critical damage process zone   b. A propagating tree and its branching process

The trademark small scale void thickness and its dispersion in the DPZ rely on material, example geometry, and stacking conditions. The interim between two continuous additions of tree spread relies likewise upon the arbitrary dispersion of small-scale voids in the DPZ. Hence, the result of this methodology is that tree spread is basically thought to be here as an arbitrary procedure, and the tree structure can be displayed as a fractal group that comprises developing miniaturized scale voids. Fractal bunches are portrayed by the connection between L, the direct size of a group and D, the all-out number of branch components from which the bunch has been shaped. This relationship is a force law with a non-whole number example d_f (Ding & Varlow, 2005).

$$D = \left(L / L_b\right)^{d_f} \qquad (1)$$

Where df is called the fractal dimension, D in the present case is the total number of growing micro-voids which have formed the tree structure, L is the linear length of the tree structure and Lb is the linear length of a growing micro-void.

A fractal, as indicated by Mandelbrot's clarification, is a shape made of parts like the entire here and there. The idea of fractal in the microstructure of materials has been applied with progress to different marvels including fractal conduct of dielectric maturing (Ding & Varlow, 2005), and electrical treeing structures. The physical source of the arrangement fractal structures on account of dielectric maturing and electrical treeing has been investigated and revealed (Kurnianto et al., 2006), and it is contended that the fractal conduct is controlled by the scale-invariance of the corruption development dynamic condition. Fractals may not only give a phenomenological way to deal with portraying structures that have all the earmarks of being like a specific guess, but also uncover significant data for the hypothetical understanding of the designing procedures, and give a valuable apparatus in approving models for the development of the microstructural components. The development of smaller scale voids in the DPZ in front of the tree tip, on a submicroscopic level, can be clarified and depicted as follows: The polymer debasement process begins from chain scission and free extreme arrangement because of charge transporter infusion, and in this way makes a couple of feeble districts. Aggregation of breakages starting from one site during AC voltage application can be considered void. This is the phase of submicroscopic void arrangement. The size of these submicroscopic voids is requested to be 10-1000 A°, and is subject to the particular polymeric structure. At that point, in these submicroscopic voids, the nearby electrical field power increments accordingly cause more progressive bond breakdown, which is the phase of submicroscopic void development. In the long run, some of these submicroscopic voids mix into naturally visible tree limbs.

So as to show the procedure of smaller scale void development in the DPZ, we can think of a miniaturized scale bereft of straight size Lo toward electric field limited by a compelling potential well profundity Uo, as portrayed in detail in (Huuva et al., 2009). The development of such a small-scale void requires the breaking of substance bonds which happens by means of a thermally enacted process with initiation vitality Uo. In a nearby electric field, vitality is put away diminishing the underlying vitality obstruction U_o by the amount $\alpha C_O \pi \varepsilon E^2$. Here, E is the neighborhood electric field strength, ε is the dielectric permittivity, π is a numeric steady. $G^e \equiv \pi \varepsilon E^2$ refers to the measure of electrostatic vitality discharged when the small-scale void develops by a unit length, and is known as the miniaturized scale void augmentation power per unit length. The parameter α is a property of the material, which indicates the initiation territory toward electric field, and αC_0 implies the volume actuated by the nearby electric field. A general reliance of the miniaturized scale void development recurrence K on neighborhood electric field strength E and temperature T has been recently controlled by processing the steady response rate for security breakdown, as follows: (Huuva et al., 2009).

$$K = \left(\frac{kT}{h}\right)\exp\left(\frac{\alpha C_o \pi \varepsilon E^2 - U_o}{kT}\right) \tag{2}$$

Where k is the Boltzmann ($1.38\times10^{-23}JK^{-1}$) and h is the Planck constants ($6.626\times10^{-34}J.s$), respectively.

The active development of an electrical tree structure can now be portrayed quantitatively. By and large, a unit addition of the electrical treeing harm happens in a period interim, which is equivalent to the normal time between two back-to-back augmentations of tree expansion. The pace in general treeing harm (dD/dt) would thus be commonly decided as:

$$\frac{dD}{dt} = \frac{K}{N_b} = \left(\frac{kT}{h}\right).\exp\left(\frac{\alpha C_o \pi \varepsilon E^2 - U_o}{kT}\right) \tag{3}$$

In the simplest case, i.e. Nb=1 in Eq. 3. This implies that a tree proliferates after one developing small scale void in the DPZ is retained for the arrangement of another branch. Clearly, the electrical treeing harm development law depends not just on the particular physical system of the arrangement and development of miniaturized scale voids; an electrical tree development rate condition will be as per the following:

$$\frac{dL}{dt} = \frac{dL}{dD} \cdot \frac{dD}{dt} = \frac{kTL_b^{d_f}}{hN_b d_f} L^{(1-d_f)} \exp\left(\frac{\alpha C_o \pi \varepsilon E^2 - U_o}{kT}\right) \tag{4}$$

By integrating Eq. 4 to establish the electrical tree growth law as:

$$L = L_b \left\{ \frac{kT}{hN_b} \exp\left(\frac{\alpha C_o \pi \varepsilon E^2 - U_o}{kT}\right) \right\}^{1/d_f} t^{1/d_f} \tag{5}$$

The time of tree growth to failure tg, assuming initial tree length Lo=0 to a critical tree length Lc, will be:

$$t_g = \left(\frac{L_c}{L_b}\right)^{d_f} \frac{hN_b}{kT} \exp\left(\frac{U_o - \alpha C_o \pi \varepsilon E^2}{kT}\right) \tag{6}$$

The local electric field strength in the pin-plane sample can be approximately calculated by:

$$E = \frac{2V}{r \ln\left(1 + {4R}/{r}\right)} \tag{7}$$

Where V is the applied voltage across the sample, r is the pin-tip-plane separation, T=300K, and approximately Lc=2mm, i.e., the pin-tip-plane separation is taken as the critical tree length for breakdown r=1μm, *R*=2mm. It is to be noted that the fractal dimension of the nanocomposite will be as follows:

$$d_{fc} = d_f + f \tag{8}$$

Where f is the volume fraction of the nanoparticles.

d_f is the fractal dimension of polymer.

1.2 Dielectrics and Nanoparticles

Clay mineral utilized in the polymer business is kaolinite. Kaolinite is the fundamental constituent of nanoparticle clay; critical amounts of different minerals can be available frequently. The molecule shape is the most significant quality of kaolinite for polymer applications. The platy nature implies that mud nanoparticles greatly affect properties, such as consistency, solidness and quality. Utilizing clay nanoparticles give elevated levels of fire retardancy to the delivered composite, and it has been chosen in this examination. Costing less than Clay, Silica has an impetus to become the best nanoparticle among nanoparticles of mechanical materials. The substance compound silicon dioxide, otherwise called silica nano-silica, with ordinary molecule size of 10~50nm is generally utilized as a nanoparticle to improve the mechanical and warm properties of polymers. Zinc Oxide (ZnO) is a well-known cross-linker for elastic and for different resins, it is additionally utilized as a UV stabilizer, and it has a moderately high refractive file which makes it an effective white shade. Aluminum oxide (Al_2O_3) is utilized for its hardness and quality. It is generally utilized as a coarse or fine rough, including as a substantially less costly substitute for modern precious stones. CT200 Epoxy Resin, Epoxy based insulation, is broadly utilized in electrical building for transformers, turning machines, bushings or cable terminations. It has astounding dielectric and mechanical properties and concoction security, and can be molded into any structure important. Araldite LY5052 and an amine Hardener HY5052 comprises an epoxy resin Araldite LY5052 and an amine Hardener. The LY/HY 5052 epoxy resin is one of the most significant hard cast-sorts of resins utilized in high voltage insulation innovation. Information Data of mechanical materials utilized in this exploration is portrayed in Table (1). Likewise, it is manifested in Table (2) which shows the general permittivity of the flawless CT200 epoxy resin and its nanocomposite materials of slick CT200 epoxy resin with 5wt.% volume fraction.

Table 1. Electric and Dielectric Characterization of Suggested Industrial Materials

Industrial Materials	Input Data
CT200 Epoxy Resin	$c_o=0.56\times10^{-28}m^3$, $U_o=1.382\times10^{-19}$ J, $L_b=10^{-15}$ m, $\varepsilon=5\times8.85\times10^{-12}$ fm^{-1}, $d_f=1.35$
LY/HY 5052 Epoxy Resin	$\alpha c_o=0.71\times10^{-28}m^3$, $U_o=1.387\times10^{-19}$J, $L_b=10^{-15}$ m, $\varepsilon=5\times8.85\times10^{-12}$ Fm^{-1}, $d_f=1.45$

Table 2. Dielectric Constants of CT200/ LY/HY5052 Epoxy Resin and Their Nano-Composites

Industrial Materials	Dielectric Constant
PURE CT200/ LY/HY5052 Epoxy Resin	5
5wt% + Clay / Epoxy Resin	4.24
5wt% + Silica / Epoxy Resin	4.4
5wt% + ZnO/ Epoxy Resin	4.22
5wt% + Al_2O_3/ Epoxy Resin	5.29

1.3 Effects of Nanoparticles on Life-Time Characteristics

Figure (2.a) shows that the fractal measurement is 1.1 up to 1.5 for the mean estimation of 10 unique outcomes for different applied voltages. In any case, Fig. (2.b) shows the life-time qualities of CT200 epoxy resin when including 5wt.% of nanoparticles (clay, silica, ZnO and Al_2O_3) prompts, incrementing

Figure 2. Effect of voltage on the life-time of CT200 Epoxy Resin materials

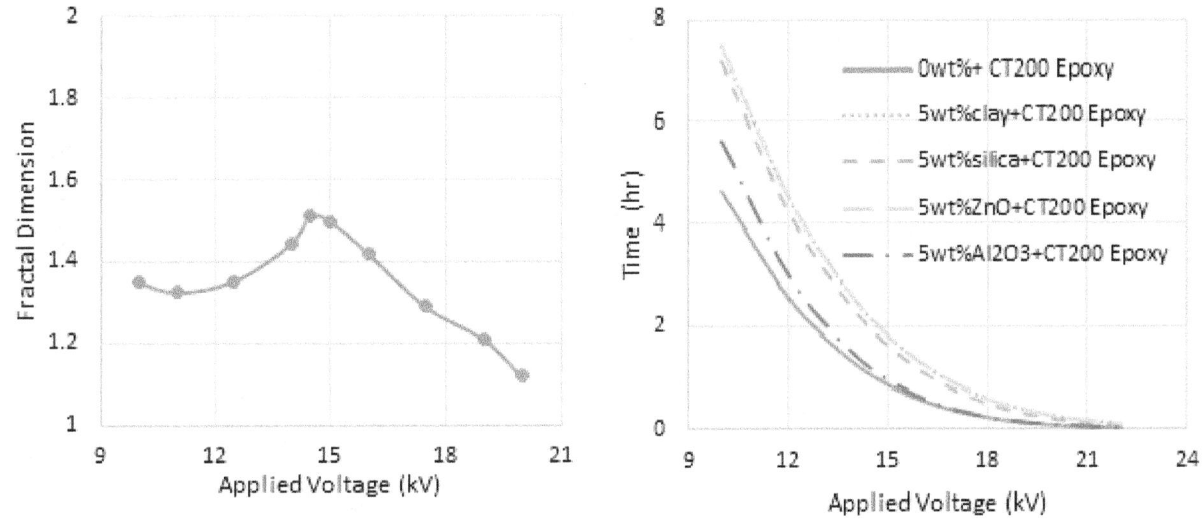

(a) Fractal dimension of electrical tree growth

(b) CT200 epoxy resin nanocomposites

the electrical treeing time to breakdown contrasted with the perfect CT200 epoxy resin insulation materials. The improvement in electrical treeing time to breakdown supports the impacts of nanoparticles (Clay, Silica, Al_2O_3 and ZnO) because of changing the dielectric properties of the slick CT200 Epoxy Resin. According to theoretical calculations, it has been deduced that life-time diminishes exponentially by expanding the applied voltage at different measures of nanoparticles (1 wt.%, 3 wt.% and 5wt.% of Clay, Silica, ZnO and Al_2O_3) to CT200 epoxy resin. It is seen that the improvement has expanded with expanding the rates of nanoparticles (Clay, Silica, ZnO and Al_2O_3); however, the improvement in electrical treeing time to breakdown is as a rule at whatever point by adding Al_2O_3 nanoparticles to the CT200 epoxy resin. This is due to the high dielectric constant estimation of the nanoparticles.

Figure (3.a) shows that the fractal measurement is switched from 1.3 up to 1.6 for 10 distinct estimations of applied voltages. Figure (3.b) shows the life-time of LY/HY5052 epoxy resin nanocomposites which have 5wt% nanoparticles of (Clay, Silica, Al_2O_3 and ZnO). It is seen that the electrical treeing time to breakdown increments with expanding applied voltage contrasted with perfect LY/HY5052 epoxy resin. High improvement occurs by including rate measures of certain nanoparticles (Clay, Silica, and ZnO) to the LY/HY5052 epoxy resin. In any case, low improvement happens at any point, including similar rates of Al_2O_3 nanoparticles to perfect LY/HY5052 epoxy resin. This is due to the high dielectric constant estimation of the nanoparticles. It is to be noted that the types and level of nanoparticles in epoxy resin are the primary explanations behind diminishing or expanding the improvement life-time of the nano-composite. According to theoretical calculations, it is deduced that the life-time to breakdown for different nanocomposites of LY/HY5052 epoxy resin that have different rates of nanoparticles (Clay, Silica, Al_2O_3 and ZnO) diminishes exponentially with expanding voltage at rates 1, 3 and 5wt% individually for all additional nanoparticles (Clay, Silica, Al_2O_3 and ZnO) in the LY/HY 5052 epoxy resin. Life-time to breakdown (hours) is improved at high qualities by expanding the rates of earth, silica, and zinc oxide nanoparticles individually, while low changes in life-time to breakdown (hours) occurs by adding aluminum oxide nanoparticles to the LY/HY5052 epoxy resin.

Figure 3. Effect of voltage on the life-time of insulation, with LY/HY 5052 Epoxy Resin

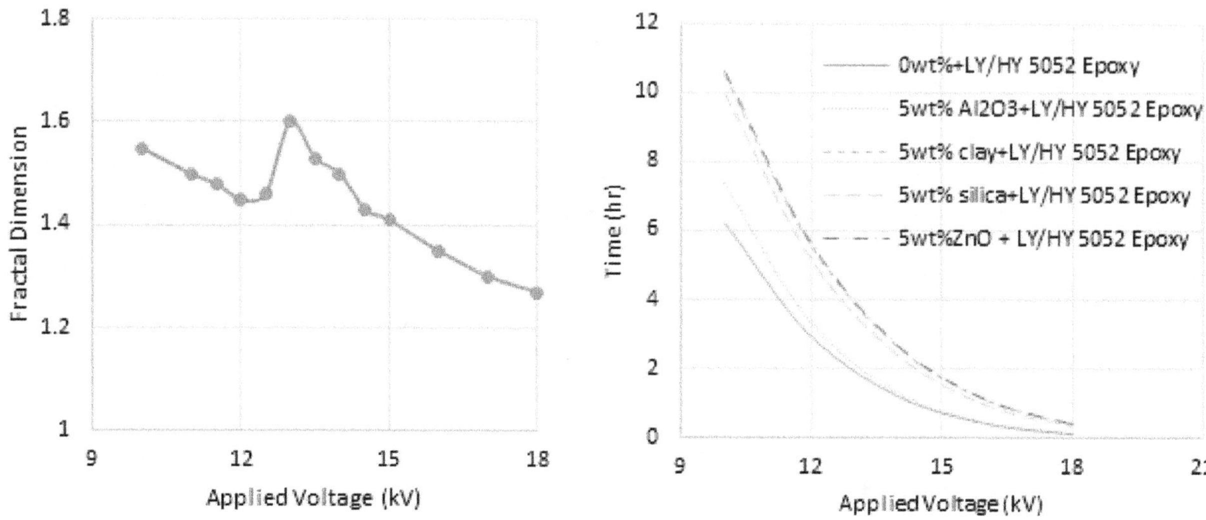

Figure 4. Effect of time on the tree length, with the neat CT200 Epoxy Resin nanocomposites

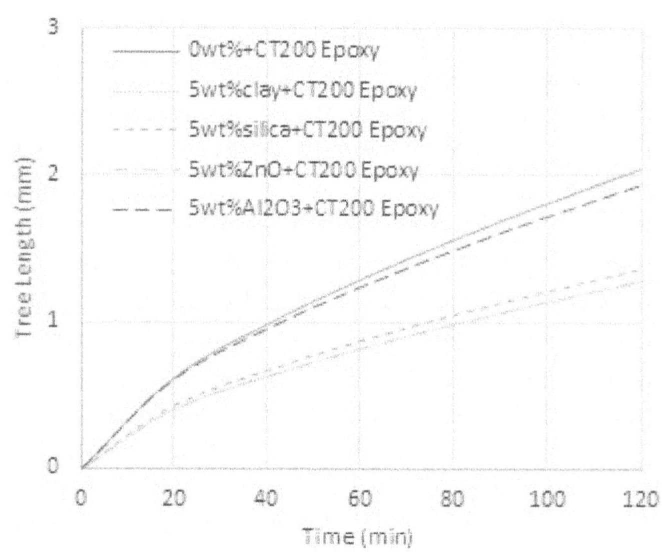

Figure (4) delineates the tree spiral degree with time at 15kV by utilizing different nanocomposite materials by including 5wt% of nanoparticles (Clay, Silica, ZnO, and Al_2O_3) to the CT200 epoxy resin. Tree spiral degree has been expanded with time at 15kV by utilizing different nanocomposite materials of CT200 epoxy resin by including 5wt% of nanoparticles (Clay, Silica, ZnO, and Al_2O_3). It has been seen that the tree length is diminished by expansion of 5wt% nanoparticles (Clay, Silica, ZnO, and Al_2O_3) in CT200 epoxy resin. Al_2O_3 nanoparticles have lower impact on the tree length than the different nanoparticles (Clay, Silica, and ZnO). According to theoretical calculations, it has been deduced that the tree length with time at 15kV for CT200 epoxy resin included 1wt.%, 3wt.%, and 5wt.% nanoparticles (Clay, Silica, ZnO and Al_2O_3) separately. It is seen that the tree length diminishes with time by expanding the rates of mud, silica, and zinc oxide plainly. However, the qualities of tree length diminish marginally with time for CT200 epoxy resin with different rates of Al_2O_3. Figure (5) outlines the tree outspread degree with time at 15kV by utilizing different nanocomposite materials by including 5wt% of nanoparticles (Clay, Silica, ZnO, and Al_2O_3) to the LY/HY5052 epoxy resin. Tree spiral degree has been expanded with time at 15kV by utilizing different nanocomposite materials of LY/HY5052 epoxy rin through including 5wt% of nanoparticles (Clay, Silica, ZnO, and Al_2O_3); however, it is shown that the tree length has diminished by expansion of 5wt% nanoparticles (Clay, Silica, ZnO, and Al_2O_3) in LY/HY5052 epoxy resin. Al_2O_3 nanoparticles have the lower impact on the tree length than the different nanoparticles (Clay, Silica, and ZnO). According to theoretical calculations, it has been deduced that the tree length with time at 15kV for LY/HY5052 epoxy resin included 1, 3, and 5wt% nanoparticles (Clay, Silica, ZnO and Al_2O_3) separately. It is observed that the tree length with time diminishes with expansion of the rates of earth, silica, and zinc oxide plainly; however, the qualities of tree length with time diminishes somewhat for LY/HY5052 epoxy resin with different rates of Al_2O_3.

Figure 5. Effect of time on the tree length in the LY/HY 5052 Epoxy Resin nanocomposites

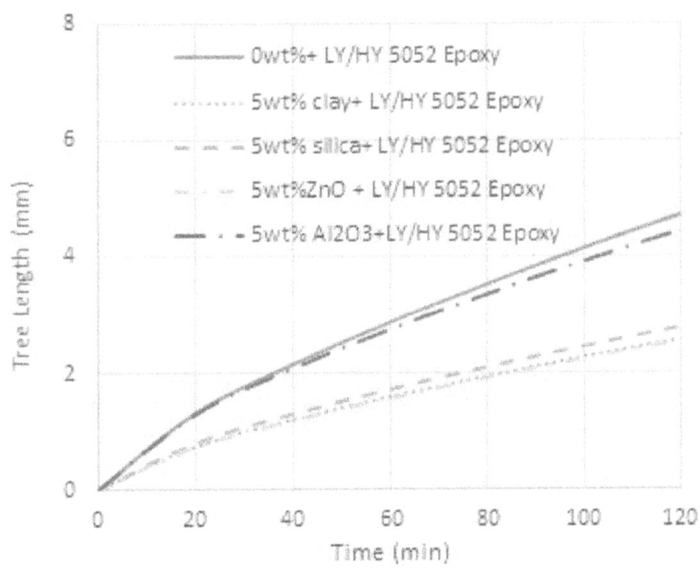

The impact of fractal measurements of electrical tree development relies on the applied-voltage on different nanocomposites (CT200 epoxy resin with 1wt.%, 3wt.%, and 5wt.% of Clay, Silica, ZnO and Al_2O_3 individually). There is a high improvement in the electrical treeing time to breakdown contrasted with perfect CT200 epoxy resin by utilizing Clay, Silica, and ZnO nanoparticles individually. At the same time, there is a slight improvement by utilizing Al_2O_3 nanoparticles. Life-time, an incentive at 14.5 kV, is of central importance. It relies upon the conduct of the fractal measurement in the insulation material and its consequences for electrical tree development. Life-time of CT200 epoxy resin is equivalent to zero at 20kV. Otherwise, the impact of fractal measurements of electrical tree development is reliant on the voltage applied on different nanocomposites (LY/HY5052 epoxy resin with 1wt.%, 3wt.%, and 5wt.% of Clay, Silica, ZnO and Al_2O_3 individually). It is revealed that there is high improvement in the electrical treeing time to breakdown contrasted with the flawless LY/HY5052 epoxy resin by utilizing Clay, Silica, and ZnO nanoparticles. In any case, there is a slight improvement by utilizing Al_2O_3 nanoparticles individually. Life-time, an incentive at 13.5 kV, is of central importance. It relies upon the conduct of the fractal measurement in the insulation material and its consequences for electrical tree development. Taking this into consideration, the impact of fractal measurements of electrical tree development is subject to the applied-voltage on different nanocomposites (CT200 epoxy resin with 1wt.%, 3wt.%, and 5wt% of Clay, Silica, ZnO and Al_2O_3 separately). Regarding the impact of Fig. (6 (a, b)), it

Figure 6. Effect of volume fraction on the life-time of Epoxy Resin types

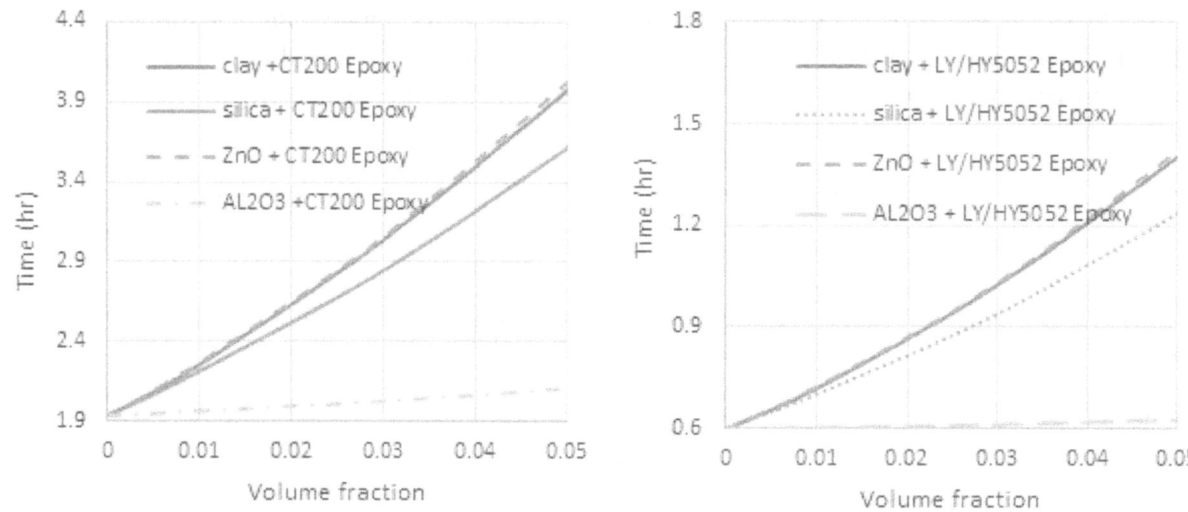

(a) CT200 epoxy resin (b) LY/HY5052 epoxy resin

Figure 7. Effect of nanoparticle volume fraction on fractal dimension with the fractal dimension of Epoxy materials

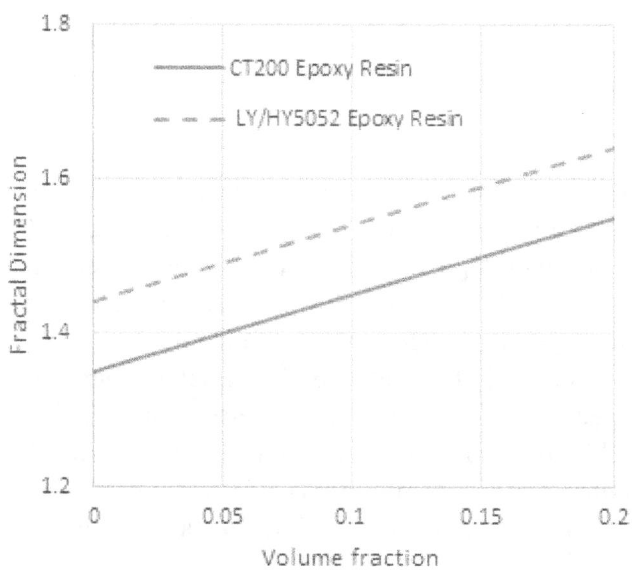

is shown that Clay, ZnO, and Silica are the best nanoparticles for improving the electrical tree time to breakdown. Impact of nanoparticle volume division on fractal measurement with the substance of the adjusted particles in the 0-5wt% territory is shown in Fig. (7). The fractal measurement increments as the nanoparticle volume division increments with the expansion of clay, silica, zinc oxide and aluminum oxide particles in CT 200 epoxy resin, LY/HY5052 epoxy resin.

2 WATER TREEING IN NANO-DIELECTRICS

A method of new nanocomposites to fuse the inhomogeneous model with Vo and Shi's homogeneous model is in this manner proposed (Lombardo, 2007; Vo & Shi, 2002; Yu et al., 2002). Since water treeing has been acknowledged for about 30 years, much research has focused on this issue. It is commonly acknowledged that water tree is a hydrophilic system, and for the age of water trees, the electric field, stickiness and explicit commencement focuses are fundamental. What's more, water tree itself can't cause breakdown; nonetheless, water trees encourage the inception of an electrical tree which gives ascension over the long haul to exorbitant link disappointment (Gulski et al., 2008; Sun et al., 2009; Suzuki et al., 2008). Decay with water tree marvel is a run of the mill issue in the electrical insulation of the high-voltage class power cable. The variables that influence water tree propagation incorporate voltage, recurrence, time, temperature, measure of water, water quality and kind of dissolvable (Gouda et al., 2005; Jiang et al., 2008; Ogiwara et al., 2010; Osada et al., 2009). According to previously known research on the water tree commencement and development, two principal sorts of water tree exist: vented water tree and tie water tree. For the most part, most research is centered on individual vented trees on the grounds that these trees are more hazardous under help maturing conditions than necktie trees, and they become quickly contrasted with tie trees.

A lot of the early research facility work was completed with "water needle" setups, which produce incredibly high electric fields at the tip of a needle-formed water-filled hole. The electrical field at the tip was normally sufficiently high to deliver an electrical tree, if the pit was not loaded up with water. The water tree development in hours to days, as opposed to months to years concerning water treeing, is commonly considered one of the most significant reasons for breakdown in power cables with polymeric insulation (Tamim, 2002). Regarding nanotechnology methods, the use of nanotechnology is guaranteed in the dielectric field, and nanocomposites are accepted to have applications in the cable business. However, the impact of dampness on nanocomposites is a region that has not been completely investigated. Since dampness relocation and electrochemical treeing influence the exhibition of cable insulation in a fundamental manner, it is critical to explore the presentation of nanocomposite dielectrics in a fluid domain before commercialization. Dielectric polymer nanocomposites have attracted much enthusiasm because of their improved electrical properties in correlation with the polymeric materials that don't utilize nanoparticles. Treeing consideration discussed here enhance the past exertion dedicated to create cross-linked polyethylene (XLPE) based nanocomposites with functionalized nanoparticles as the material arrangement of decision for underground cables. The vinyl silane (VS) treated nanocomposite at an enhanced stacking of nanoparticle rate (which is found to have the most elevated AC breakdown quality) is picked. Examination is made with the base pitch, a lighter stacking of nanoparticles rate, and nanocomposites with un-functionalized nano particles at similar two loadings.

At present, just vented water tree development is examined and archived in this report. Each of the four sorts of nanocomposites were found to oppose the development of water trees contrasted with the

XLPE base dielectric. Work is progressing to analyze the united issue of water movement and take-up in nanodielectrics. Likewise, Water tree development in nanocomposites is confined in examination to the base resin. With the expanding nanoparticle stacking and improved holding, this impact is increasingly self-evident. The diverse conduct is credited to the interfacial locale presented by nano particles. The advancement of water trees is restrained through dissipation by the inside interfaces in nanocomposites which prompts a more extensive water tree profile, while in XLPE, the tree profile is thin and moves towards the ground plane. With appropriate treatment of the nanoparticles, an increasingly powerful interfacial area is accomplished which further opposes the advancement of water treeing. It is appropriate to see that this improvement happens in spite of an expanded affinity to ingest water (Crine, 2005; Crine & Jow, 2005; Helles et al., 2008; Hosam-Eldin & Elhadary, 2006; Huang et al., 2010; Hui et al., 2009; Rajagopala et al., 2008). Along these lines, this section portrays the attributes of controlling in water tree by utilizing different nanoparticles in Cross-linked polyethylene. Likewise, electric field strength dissemination on water void, interfacial vitality between water void and dielectric and water treeing length for new nanocomposite insulation materials have been explored. At long last, water tree qualities in unadulterated and nanocomposite insulation materials will be explored,

2.1 Theory and Effective Parameters

Water treeing is a drawn-out debasement marvel in polymeric materials, which brings down the breakdown quality of insulation. Water trees begin in miniaturized scale voids and contaminations in the heft of the material. Water tree inception and development require the nearness of dampness at a state of high-pressure fixation. The fields required for water trees are a lot lower than those tor electrical trees. The impact of high electrical fields on water treeing, electrical treeing, relaxation, conductivity and charge mobility in different polymers is inspected inside the setting of an atomic model. The genuine

Figure 8. Water void in water tree model

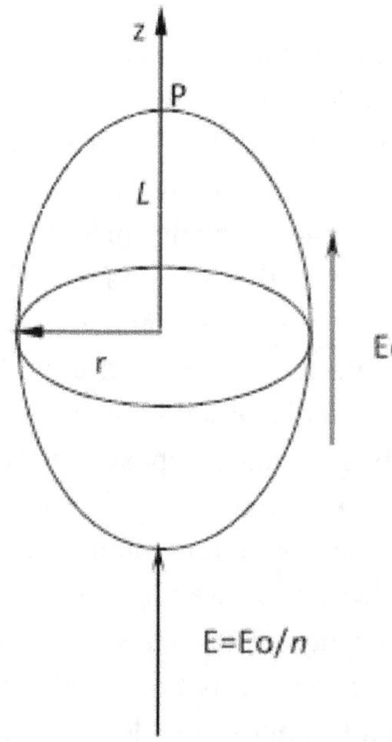

estimation of the air conditioner field following up water trees is addressed, and it is demonstrated that the strain instigated by enormous fields may influence the water tree development in arrangement with high dielectric constant. Power cables are utilized in transmission, organized in urban areas and for abroad power transmission. The polyethylene cables utilized for medium voltage are defenseless against breakdown because of water treeing. Water treeing essentially decreases the electric breakdown quality of the insulation.

The nearness of the water trees brings about the decrease of their dielectric quality. Here, the limited component reproduction strategy is utilized to assess the electric field inside the power cables. Water trees can cause administration breakdown of the cables when they arrive at a basic length. The administration life of links with no metallic water boundary can be drawn out by deferring the water entrance into the cable. The water entrance is reliant on the dispersion rate of water sorption, beginning water content in the cable materials, and the temperature of the link. Let us consider a miniaturized scale void (Tamim, 2002), which is relatively bigger in volume than voids around it. The electrical field near the void is heightened, as shown in Figure (8). The proposed model of water void is developed, and the electric field surface (E_{sur}) on the surface of an ellipsoid can be expressed as follows:

$$E_{sur} = v_x \frac{E_o}{n} \tag{9}$$

Where:

v_x: represents the component of unit length normal vector starting from the surface of the ellipsoid in the direction of the electric field.

$$n = \frac{1-e^2}{2e^3}\left(\ln\frac{1+e}{1-e} - 2e\right) \tag{9.i}$$

$$e = \left[1 - \frac{b^2}{a^2}\right]^{\frac{1}{2}} \tag{9.ii}$$

The value of the electric field inside the micro-void (E_1) is calculated as follows:

$$E_1 = \frac{E_o}{1+\left(\frac{\varepsilon_2^*}{\varepsilon_1} - 1\right).n} \tag{10}$$

$$\varepsilon_2^* = \varepsilon_2 - \frac{J\sigma_w}{w\varepsilon_o} \tag{10.i}$$

$$f_m = \frac{\sigma_w}{2\pi\varepsilon_2\varepsilon_o}\left(\frac{\sqrt{1+a}}{\frac{1}{n}\frac{\varepsilon_1}{\varepsilon_2}+\left(1-\frac{\varepsilon_1}{\varepsilon_2}\right)}\right) \quad (11)$$

$$a = \frac{\frac{q}{n}}{1+q\left(\frac{\varepsilon_2}{\varepsilon_1}-1\right)+\left(\frac{\varepsilon_2^*}{\varepsilon_1}-1\right)n} \quad (11.i)$$

Where,

E_O: electric field strength of insulation (V/m).
ε_1: specific inductive capacity of insulation or Nano-composite material.
E_2*: complex specific inductive capacity of water.
ε_2: specific inductive capacity of water
σ_w: conductivity of water
a: the long axis of ellipsoid.
b: the short axis of ellipsoid.
e: eccentricity of the ellipsoid.
f_m: the frequency which gives the maximum value of the dielectric loss tangent
q: water content ratio to insulation material
The total pressure applied to water is

$$P_{total} = P_e + P_s \quad (12)$$

$$P_e = \left(\frac{1}{2}\varepsilon_o K_o \rho E_1^2 - P_s\right) / \left(1+\frac{3}{2}.\frac{R_o^2\rho^2}{\eta_m D}\right) \quad (12.i)$$

$$P_s = \frac{2\alpha}{R} - \frac{3}{2}\varepsilon_o E_o^2 \quad (12.ii)$$

Viscosity of the dielectric mixture can be calculated with respect to volume fraction of each component in the mixture (main dielectric material and nanoparticle), as follows:

$$\eta_{m=}(1-q)\eta_{insulation} + q\eta_{filler} \quad (13)$$

$$\eta_{insulation} = 10^{\frac{-17.44(T-T_g)}{51.6+T-T_g}} . \eta_{T_g} \tag{13.i}$$

The rate of void radius increase due to the presence of Pe was stated by (Bamji, 1999) as follows:

$$\frac{dR}{dt} = \frac{3}{2}\frac{P_e}{\eta_m} R_o \tag{13.ii}$$

Where,

Pe: pressure of water in micro-void due to mechanical stress of insulation material (N/m^2)
Ps: pressure of water in micro-void due to interfacial tension (N/m^2)
εo: dielectric constant of vacuum
ρ: density of water (kg/m^3)
Ko: constant (=7.9*10-2 m^3/kg)
Ro: the initial radius of water void (m)
$i_{nsulation}$: is the viscosity of the main dielectric of the mixture
$n_{anoparticle}$: is the viscosity of the nanoparticle
D: diffusion constant of water ($1.3*10^{-13}$ $kg.sec/m^3$)
Tg: glass transient temperature of insulation material (Co)
T: temperature (Co)
η (Tg): the viscosity at temperature Tg
dR/dt: is the rate of void radius increase with respect to time
The performance of water treeing length with time for various Nano-composite materials is as follows:

$$L = R_o exp\left(\frac{t}{\tau_o}\right) \tag{14}$$

Where,

t: is the time of water treeing growth (sec)
τo: is the time constant of growth of water treeing

$$\tau_O = \frac{4}{3}\frac{\eta_m}{\rho \varepsilon_O k_O E_1^2} \tag{14.i}$$

2.2 Dielectrics and Nanoparticles

Zinc oxide (ZnO) is a mainstream cross-linker for elastic and for different resins, it is likewise utilized as a UV stabilizer, and it has a generally high refractive record which makes it an effective white color. Zinc oxide is an inorganic compound with the recipe ZnO. It has a high refractive record and high warm conductivity. It is non-poisonous and good with skin, making it a reasonable added substance for materials and surfaces that interact with people. Zinc Oxide is additionally utilized as an impetus for methanol amalgamation. The expansion in the surface region of Nano scale zinc oxide contrasted with bigger powders can possibly improve the effectiveness of these procedures. Aluminum Oxide (Al_2O_3) is the group of inorganic mixes with the synthetic recipe Al_2O_3. It is an amphoteric oxide, and is ordinarily alluded to as alumina. Aluminum oxide is an electrical encasing; however, it has a moderately high warm conductivity. Alumina is a supported nanoparticle for plastics. It is utilized for its hardness and quality and it is generally utilized as a coarse or fine rough, including as a substantially less costly substitute for mechanical precious stones. Cross-linked polyethylene (XLPE), usually condensed PEX or XLPE, is a type of polyethylene with cross-joins. It is framed into tubing, and is utilized prevalently in hydronic brilliant warming frameworks, local water funneling and solid insulation for high strain electrical cables. It is likewise utilized for flammable gas and seaward oil applications, compound transportation, and transportation of sewage and slurries. As of late, it has become a suitable option for polyvinyl chloride (PVC), chlorinated polyvinyl chloride (CPVC), or copper pipe for use as private water pipes. It has been generally utilized as protecting materials for expelled links, since it has great electrical properties along with other alluring properties, such as amazing mechanical quality and adaptability, great solid insulation from synthetic substances, simple handling and minimal effort.

2.3 Effects of Nanoparticles on Nanodielectrics

The electric field strength in smaller scale void and the interfacial vitality among water and solid insulation material have been focused on concerning uniform electric field strength of solid insulation for the proposed model of water treeing in solid insulations (individual, binary-, and multi-mixtures). Additionally, as indicated by the significance of high frequencies on the exhibition of dielectric solid insulations, the electric field strength in small scale void and the interfacial vitality among water and solid insulation material are highlighted regarding uniform electric field strength. In this manner, the dielectric solid insulations and its details which have been utilized in proposed structures are expressed later in Table (3).

Table 3. Characteristics of Industrial Insulation Materials

Nanoparticles and Materials	Dielectric Constant
ZnO	1.7
Al_2O_3	9
XLPE	2.3
5wt%ZnO+XLPE	1.96
10wt%ZnO+XLPE	1.71
15wt%ZnO+XLPE	1.54
5wt%Al_2O_3+XLPE	2.7
10wt%Al_2O_3+XLPE	3.07
15wt%Al_2O_3+XLPE	3.43

Regarding the impact of ZnO nanoparticles on the water tree length with changing field strength solid insulation, it is noticed that expanding the rate of ZnO nanoparticles in the XLPE grid diminishes the water tree length. The impact of ZnO nanoparticles on the water tree length with changing recurrence is also addressed. Additionally, it is noticed that expanding the rate of ZnO nanoparticles in the XLPE network diminishes the water tree length up to 1 kHz; however, expanding ZnO nanoparticles builds the water tree length more than 1 kHz. According to theoretical calculations. Concerning the impact of ZnO nanoparticles on the rate of void radius with shifting field strength solid insulation, it is noticed that expanding the rate of ZnO nanoparticles in the XLPE framework diminishes the rate of void radius. The impact of ZnO nanoparticles on the rate of void radius with fluctuating recurrence is also examined. Similarly, it is noticed that expanding the rate of ZnO nanoparticles in the XLPE lattice doesn't change the rate of void radius up to 2 kHz. However, expanding ZnO nanoparticles builds the rate of void radius in excess of 2 kHz. Furthermore, regarding the impact of ZnO nanoparticles on the weight of water in miniaturized scale void because of mechanical stress in XLPE with changing field strength solid insulation, it is noticed that expanding the rate of ZnO nanoparticles in the XLPE framework diminishes the rate of void radius. According to theoretical calculations, the impact of ZnO nanoparticles on the weight of water in miniaturized scale void because of mechanical stress in XLPE has been inferred. Likewise, it is exhibited that expanding the rate of ZnO nanoparticles in the XLPE network doesn't change the rate of void radius up to 1 kHz. However, expanding ZnO nanoparticles diminishes the weight of water in small scale void because of mechanical stress in XLPE more than 1 kHz.

In theoretical calculations, the impact of ZnO nanoparticles on the time constant of development of smaller scale void in XLPE with differing field strength solid insulation is deduced. It is noticed that expanding the rate of ZnO nanoparticles in the XLPE lattice builds the time constant of the develop-

Figure 9. Effects of ZnO nanoparticles concentration on water tree characteristics in XLPE

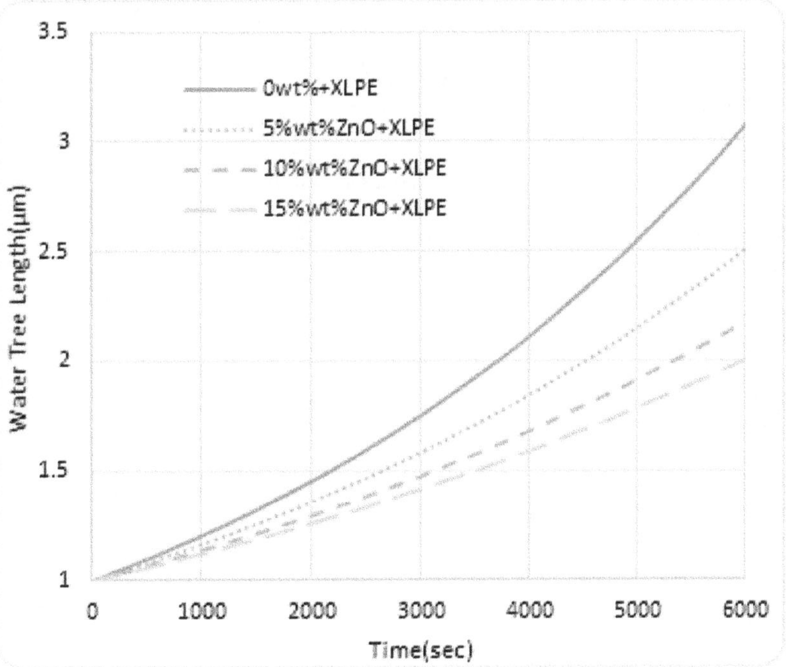

ment of miniaturized scale void. Additionally, the effect of ZnO nanoparticles on the time constant of the development of small-scale void in XLPE is observed. Similarly, it is noted that expanding the rate of ZnO nanoparticles in the XLPE lattice embarks little changes in the time constant of the development of miniaturized scale void in XLPE up to 1 kHz. Nevertheless, no changes in the time constant of the development of small-scale void in XLPE by expanding ZnO nanoparticles more than 1 kHz can be observed. Figure (9) shows the impact of ZnO nanoparticles on the water tree length with time. It is manifested that the water tree length increments with expanding time; however, expanding the level of ZnO nanoparticles in the XLPE grid diminishes the time constant of the development of micro void.

The attributes of water tree length with the ideal opportunity for unadulterated XLPE and different nanocomposites of (5%wt Al_2O_3 + XLPE, 10%wt Al_2O_3 + XLPE, and 15%wt Al_2O_3 + XLPE) are delineated. According to theoretical calculations, the impact of Al_2O_3 nanoparticles on the water tree length with fluctuating field strength solid insulation is examined. It is observed that expanding the rate of Al_2O_3 nanoparticles in the XLPE grid builds the water tree length. Figure (16.b) shows the impact of Al_2O_3 nanoparticles on the water tree length with shifting recurrence. Additionally, it is noticed that expanding the rate of Al_2O_3 nanoparticles in the XLPE grid builds the water tree length up to 10 kHz. Figure (16.f) manifests the impact of Al_2O_3 nanoparticles on the weight of water in small scale void because of mechanical stress in XLPE. Likewise, it is noticed that expanding the rate of Al_2O_3 nanoparticles in the XLPE grid does not change the rate of void radius up to 1 kHz. However, expanding Al_2O_3 nanoparticles diminishes the weight of water in smaller scale void because of mechanical stress in XLPE more than 1 kHz. Regarding the impact of Al_2O_3 nanoparticles on the time constant of the development of miniaturized scale void in XLPE with shifting field strength solid insulation, it is remarked that expanding the rate of Al_2O_3 nanoparticles in the XLPE grid diminishes the time constant of the development of smaller

Figure 10. Effects of Al_2O_3 nanoparticles concentration on water tree characteristics in XLPE

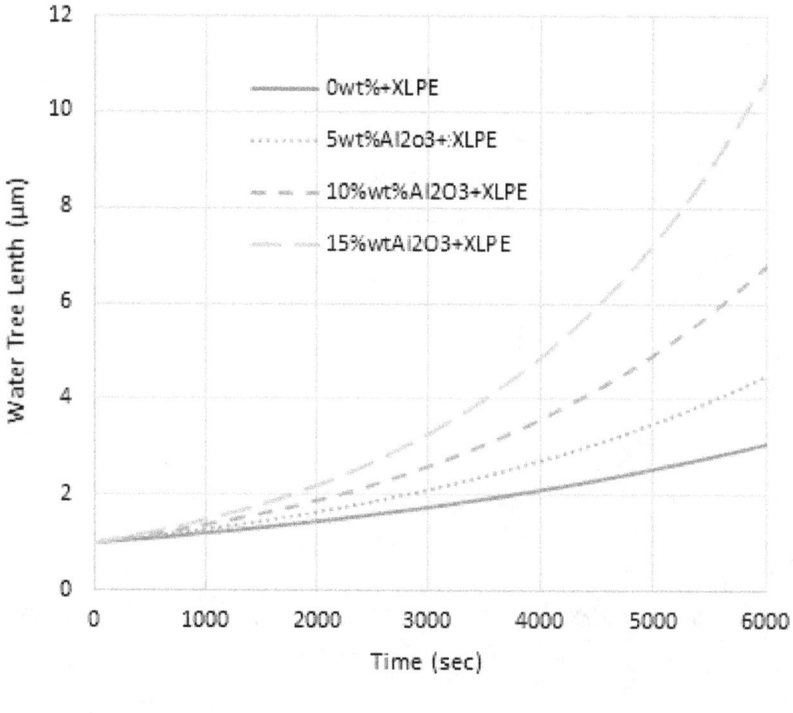

scale void. According to theoretical calculations, the impact of Al_2O_3 nanoparticles on the time constant of the development of smaller scale void in XLPE is addressed. Moreover, it is noticed that expanding the rate of Al_2O_3 nanoparticles in the XLPE grid diminishes the time constant of the development of smaller scale void in XLPE up to 1 kHz. However, there are little changes on the time constant of the development of miniaturized scale void in XLPE by expanding ZnO nanoparticles more than 1 kHz. Figure (10) shows the impact of Al_2O_3 nanoparticles on the water tree length with time. It is shown that the water tree length increments with expanding time; however, expanding the level of Al_2O_3 nanoparticles in the XLPE network builds the time constant of the development of micro void.

3 TRENDS OF NANOPARTICLES ON WATER TREEING

Figure (11) shows the connection between the water tree length and the field strength of solid insulation and recurrence for the different solid insulation materials examined (PURE XLPE, 15%wt ZnO + XLPE, and 15%wt Al_2O_3 + XLPE). Figure (11.a) reveals that the water tree length increments with expanding the field strength of solid insulation. Additionally, it has been demonstrated that including ZnO nanoparticles in the XLPE lattice diminishes the water tree length, but including Al_2O_3 builds the water tree length. Figure (11.b) shows that the water tree length increments with expanding recurrence. Furthermore, it has been indicated that including Al_2O_3 expands the water tree length, while including ZnO nanoparticles in the XLPE network diminishes the water tree length up to 1 kHz of little qualities, at that point, there is a little expansion of more than 1kHz.

Figure (12) manifests the connection between the rate of void radius and the field strength of solid insulation and recurrence for different solid insulation materials contemplated. It reveals that the rate of void radius increments with expanding field strength of solid insulation. Additionally, it has been indicated that including ZnO nanoparticles in the XLPE lattice diminishes the rate of void radius, while including Al_2O_3 expands the rate of void radius. The rate of void radius increments with expanding recurrence up to 1 kHz, after that the rate of void radius will be steady. Moreover, it has been demonstrated that there is high expansion in the rate of void radius by including Al_2O_3 in XLPE, while there are little changes in the rate of void radius by including ZnO nanoparticles in the XLPE network. The weight of water shows the miniaturized scale void because of the mechanical stress in XLPE increments with expanding field strength of solid insulation. Furthermore, it has been indicated that including ZnO nanoparticles in the XLPE grid diminishes the weight of water in small scale void because of the mechanical pressure, while including Al_2O_3 builds the weight of water in miniaturized scale void because of the mechanical stress in XLPE. Moreover, the weight of water displays in smaller scale void because the mechanical stress in XLPE increments with expanding recurrence up to 1 kHz little qualities. At that point, the weight of water in miniaturized scale void is consistent because of mechanical pressure. Added to that, it has been indicated that including Al_2O_3 expands the weight of water in miniaturized scale void because of the mechanical stress in XLPE, while including ZnO nanoparticles in the XLPE lattice diminishes the weight of water in smaller scale void because of the mechanical stress in XLPE. The time constant of the development in miniaturized scale void in XLPE diminishes with expanding field strength of solid insulation. Additionally, it has been demonstrated that including ZnO nanoparticles in the XLPE lattice builds the time constant of the development of smaller scale void, but including Al_2O_3 diminishes the time constant of the development of miniaturized scale void in XLPE. In addition, the time constant of the development of smaller scale void in XLPE diminishes with expanding recurrence up to 1 kHz. At

Figure 11. Effects of nanoparticles on water tree characteristics in XLPE

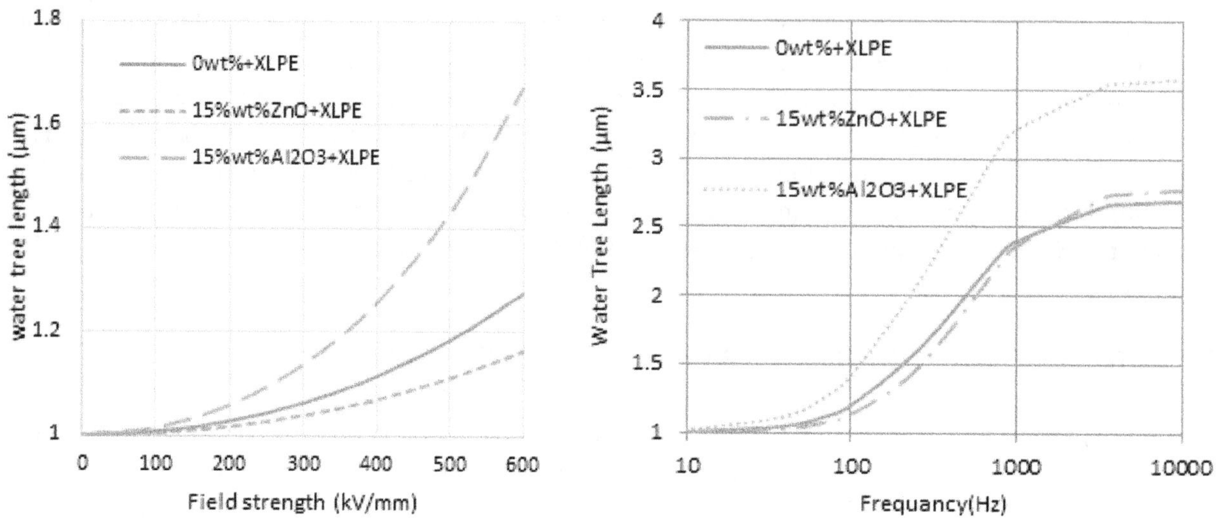

that point, the time constant of the development of miniaturized scale void in XLPE will remain steady for more than 1 kHz. Moreover, it has been demonstrated that including Al_2O_3 diminishes the time constant of the development of small-scale void in XLPE, while including ZnO nanoparticles in the XLPE network expands the time constant of the development of micro void in XLPE.

In case of studying the effect of nanoparticles on the time of water tree length in XLPE, Figure (12) shows the relationship between the water tree length and the time for various studied insulation materi-

Figure 12. Effects of nanoparticles on the time of water tree length in XLPE

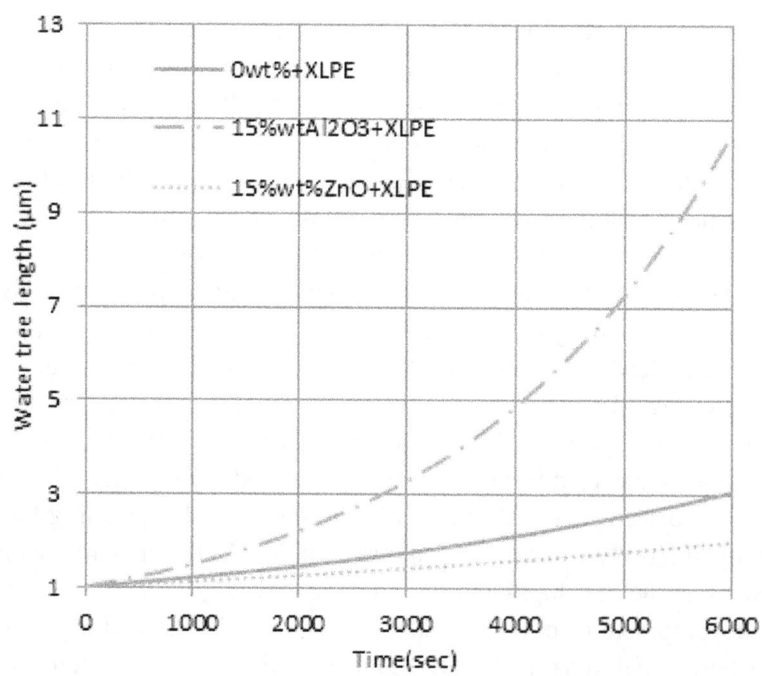

als (PURE XLPE, 15%wt ZnO + XLPE, and 15%wt Al_2O_3 + XLPE). It has been shown that the time of water tree length in XLPE increases and adding ZnO nanoparticles in the XLPE matrix decreases the water tree length, while adding Al_2O_3 increases the water tree length in XLPE.

4 FORECAST AND RECOMMENDATIONS

The electrical treeing recognized halfway releases increments as the principal miniature channels show up at the cathode tip. A diminishing in the release number is then noticed and identified with the presence of an alternate sort of fiber, which proliferates with releases limited at their closures. An abrupt increment of release number and sufficiency is seen when a fiber arrives at the ground terminal. The breakdown will happen a lot later, joined by a general expansion in the release number. This examination renders it conceivable to know the phase of decay of the example by a basic drive estimation. For a counterfeit void, a similar trademark conduct is noticed. Upon applied voltage, releases which influence the entire surface of the hole are noticed; this is trailed by an abatement in their number while they concentrate close to the dividers. Otherwise, the actual premise of water treeing can go from prevalently electromechanical in nature to basically electrochemical. Much of the early research facility work was completed with "water needle" arrangements, which produce amazingly high electric fields at the tip of a needle-molded, water filled hole.

The electric field at the tip was generally and sufficiently high to deliver an electrical tree, if the hole was not loaded up with water, and the water tree development occurred from hours to days, as opposed to months to years concerning a water tree developed under utility working conditions. Accordingly, water treeing is by and large considered as quite possibly the main source of breakdown in power links with polymeric protection. Establishment of water and electric field is fundamental and realizes the prerequisites of commencement, development and proliferation of water treeing, notwithstanding that there are some other irregular components. Thus, water treeing is unimaginable without the electric field and water, while it is conceivable without some help necessities, such as mechanical support focused on areas, unfamiliar considerations or arrangement fixation. Additionally, it is for the most part expected that during creation, water, debasements and leftover items from cross connecting will be made in the nebulous districts of polymer. In the event that these substances are extremity, grouping may occur in existing voids. The game-plan position about nanoparticles (Clay, ZnO, Silica, SiO_2, and TiO_2) inside host framework is a capable parameter for controlling in the successful dielectric constant of multi-nanocomposites. Counting nanoparticles from asserting Clay, ZnO, and silica under CT200 epoxy Resin, LY/HY5052 epoxy sap has caused huge change in the life-time from guaranteeing insulation, the tree Growth rate from guaranteeing electrical tree spread, insulation from electrical tree improvement and the long haul on breakdown and brief extension of fractal degree, as well as the absolute time of electrical tree advancement to disappointment. Counting Al_2O_3 nanoparticles does not cause any significant change in the life-time from asserting insulation and the tree growth rate from guaranteeing electrical tree spread. Filler volume part increases directly the fractal degree from guaranteeing epoxy gum insulation materials. Counting ZnO nanoparticles in the XLPE network decreases the water tree length, rate for void range, and the weight from asserting water on a small-scale void due to mechanical pressure done in the XLPE. In any case, extending the pace of ZnO nanoparticles in the XLPE framework increases the time constant of the growth of miniaturized scale void on XLPE. Anyway, incorporating Al_2O_3 nanoparticles in the XLPE lattice increments the water tree length, rate for void sweep, and the weight about water on a smaller scale void, in view of mechanical worry for XLPE. Nevertheless, extending the pace of Al_2O_3 nanoparticles in the XLPE network reduces the time constant for improvement from asserting miniaturized scale void in XLPE.

Finally, nanoparticles have a huge effect on changing the physical, electrical and dielectric properties. In case of selecting the appropriate nanoparticles, they can reduce the electrical treeing and water treeing. A correlation between the impacts of ZnO and Al_2O_3 nanoparticles on the water treeing qualities on XLPE modern protecting materials can be observed. The examination shows the impacts of ZnO and Al_2O_3 nanoparticles on the water tree length, the rate of void radius and the weight of water in small scale void because of the mechanical stress in XLPE, the time constant of the development of smaller scale void and the water tree length with time.

REFERENCES

Al-Arainy, A. A., Qureshi, M. I., & Malik, N. H. (2006). Investigations of Electrical Trees in XLPE Cable Insulation Samples Using PC Based On-Line Monitoring. in Properties and applications of Dielectric Materials. *IEEE 8th International Conference on Properties and Applications of Dielectric Materials*, 115-118.

Bamji, S. S. (1999). Electroluminescence-A Technique to Detect the Initiation of Degradation in Polymeric Insulation. *IEEE Transactions on Dielectrics and Electrical Insulation*, *15*(3), 9–14. doi:10.1109/57.768531

Crine, J. (2005, August). Influence of Electro-Mechanical Stress on Electrical Properties of Dielectric Polymers. *IEEE Transactions on Dielectrics and Electrical Insulation*, *12*(4), 791–800. doi:10.1109/TDEI.2005.1511104

Crine, J., & Jow, J. (2005, August). A Water Treeing Model. *IEEE Transactions on Dielectrics and Electrical Insulation*, *12*(4), 801–808. doi:10.1109/TDEI.2005.1511105

Crine, J. P. (2007). A Molecular Model for the Electrical Aging of XLPE. *IEEE Annual Report Conference on Electrical Insulation and Dielectric Phenomena (CEIDP)*, 608-610. 10.1109/CEIDP.2007.4451572

Ding, H., & Varlow, B. R. (2004a). Electrical Treeing Studies on the Araldite Lymy 5052 Epoxy Resin Over A Wide Range of Stressing Voltage. *IEEE Annual Report Conference on Electrical Insulation Phenomena (CEIDP)*, 306-309.

Ding, H. Z., & Varlow, B. R. (2004b). Electrical Treeing Studies on the Araldite LY/HY 5052 Epoxy Resin over A Wide Range of Stressing Voltage. *IEEE Annual Report Conference on Electrical Insulation and Dielectric Phenomena (CEIDP)*, 306-309. 10.1109/CEIDP.2004.1364249

Ding, H. Z., & Varlow, B. R. (2005). Thermodynamic Model for Electrical Tree Propagation Kinetics in Combined Electrical and Mechanical Stresses. *IEEE Transactions on Dielectrics and Electrical Insulation*, *12*(1), 81–89. doi:10.1109/TDEI.2005.1394018

El-Zein, A., Talaat, M., & El Bahy, M. M. (2009). A Numerical Model of Electrical Tree Growth in Solid Insulation. *IEEE Transactions on Dielectrics and Electrical Insulation*, *16*(6), 1724–1734. doi:10.1109/TDEI.2009.5361596

Gouda, Thabet, & El-Tamaly. (2005). A New Mathematical Model For Water Treeing In Solid Insulations. *Proceedings of 2005 International Symposium on Electrical Insulating Materials.*

Gulski, E., Putter, H., & Smit, J. J. (2008). Investigation of Water Treeing – Electrical Treeing Transition in Power Cables. *2008 International Conference on Condition Monitoring and Diagnosis.* 10.1109/CMD.2008.4580270

Helles, S. M., Henøen, V. C., & Hvidsten, S. (2008). Simulation of Water Diffusion in Polymeric Cables Using Finite Element Methods. *IEEE Transactions on Dielectrics and Electrical Insulation.*

Hosam-Eldin & Elhadary. (2006). Modeling and Simulation of Water Treeing in Dielectric Samples. *The Eleventh International Middle East power Systems Conference (MEPCON'2006).*

Huang, X., Liu, F., & Jiang, P. (2010, December). Effect of Nanoparticle Surface Treatment on Morphology, Electrical and Water Treeing Behavior of LLDPE Composites. *IEEE Transactions on Dielectrics and Electrical Insulation, 17*(6), 1697–1704. doi:10.1109/TDEI.2010.5658219

Hui, L., Smith, R., Nelson, J. K., & Schadler, L. S. (2009). Electrochemical Treeing in XLPE/silica Nanocomposites. *Annual Report Conference on Electrical Insulation and Dielectric Phenomena.*

Huuva, R., Englund, V., Gubanski, S. M., & Hjertberg, T. (2009). A Versatile Method to Study Electrical Treeing in Polymeric Materials. *IEEE Transactions on Dielectrics and Electrical Insulation, 16*(1), 171–178. doi:10.1109/TDEI.2009.4784565

Jiang, P., Kim, C., Huang, X., & Liu, F. (2008). The Influence of Dielectric Permittivity and It Distribution on Water Tree Initiation and Growth in XLPE Cables. *Proceedings of 2008 International Symposium on Electrical Insulating Materials.* 10.1109/ISEIM.2008.4664433

Kurnianto, R., Kurnianto, R., Murakami, Y., Hozumi, N., Nagao, M., & Murata, Y. (2006). Some Fundamentals on Treeing Breakdown in Inorganic-nanoparticle/LDPE Nanocomposite Material. *IEEE Annual Report Conference on Electrical Insulation and Dielectric Phenomena (CEIDP)*, 373-376.

Kurnianto, R., Murakami, Y., Hozumi, N., & Nagao, M. (2005). Electrical Tree Propagation in Epoxy Resin under Different Characteristics in Electrical Insulating Materials. *International Symposium on Electrical Insulation Materials*, 3, 718 – 721.

Lombardo, N. (2007). A Two-Way Particle Mapping for Calculation of the Effective Dielectric Response of Graded Spherical Composites. *Composites Science and Technology, 67*(3-4), 728–736. doi:10.1016/j.compscitech.2006.04.009

Noskov, M. D., Malinovski, A. S., Sack, M., & Schwab, A. J. (2000). Self-consistent Modeling of Electrical Tree Propagation and PD Activity. *IEEE Transactions on Dielectrics and Electrical Insulation, 7*(6), 725–733. doi:10.1109/94.891982

Ogiwara, Yonaha, Uehara, & Kudo. (2010). Temperature Characteristics of Water Tree Propagation in a Wide Temperature Range Using XLPE Sheets. *Annual Report Conference on Electrical Insulation and Dielectric Phenomena.*

Osada, T., Tohyama, K., Imai, T., & Abe, K. (2009). Dissipation Current Waveforms Monitoring in LDPE Sheet under Water Tree Deterioration Test. *Annual Report Conference on Electrical Insulation and Dielectric Phenomena*. 10.1109/CEIDP.2009.5377883

Pitsa, D., Vardakis, G. E., & Danikas, M. G. (2010). Electrical Tree Growth Simulation in Nanocomposite Polymers: The Role of Nanoparticles and Homocharges. *IEEE International Conference on Solid Dielectrics*, 1-3. 10.1109/ICSD.2010.5568241

Raetzke, S., Ohki, Y., Imai, T., Tanaka, T., & Kindersberger, J. (2009). Tree Initiation Characteristics of Epoxy Resin and Epoxy/Clay Nanocomoposite. *IEEE Transactions on Dielectrics and Electrical Insulation*, *16*(5), 1473–1480. doi:10.1109/TDEI.2009.5293962

Rajagopala, K., Ravindran, M. P., & Kannu, P. D. (2008). Electric Field and Potential Calculation in 10, 15 and 20 kV Water Infested Power Cables. *IEEE Transactions on Dielectrics and Electrical Insulation*.

Sun, Y., Xu, Y., Yang, W., Jin, J., & Cao, X. (2009). A Study of Water Trees in Polymers under Different Conditions. *Proceedings of the 9th International Conference on Properties and Applications of Dielectric Materials*.

Suzuki, M., Itoh, A., & Yoshimura, N. (2008). Three-dimensional equivalent circuit analysis of water tree. *Proceedings of 2008 International Symposium on Electrical Insulating Materials*. 10.1109/ISEIM.2008.4664597

Tamim, O. (2002). *Electric field calculation inside the insulation of three core belted power cables and its effect on the discharge inception voltage and water treeing growth* (Thesis). Cairo University, Egypt.

Tanaka, T., Matsunawa, A., Ohki, Y., Kozako, M., Kohtoh, M., & Okabe, Sh. (2006). Treeing Phenomena in Epoxy/Alumina Nanocomposite and Interpretation by a Multi-Core Model. *IEEJ Trans Fundamentals and Materials*, *126*(11), 1128–1135. doi:10.1541/ieejfms.126.1128

Theiler. (1990). Estimating Fractal Dimension. *IEEE Journal of Optical Society of America, 7*(6).

Vo, H. T., & Shi, F. G. (2002). Towards Model-Based Engineering of Optoelectronic Packaging Materials, Dielectric Constant Modeling. *Microelectronics Journal*, *33*(5-6), 409–415. doi:10.1016/S0026-2692(02)00010-1

Yu, K. W., Gu, G. Q., & Huang, J. P. (2002). *Dielectric Response of Spherical Particles of Graded Materials*. Conducting-Materials.

Chapter 3
Nano-Tech Power Cables

ABSTRACT

This chapter explores the novel nano-metric present-day materials considering power law Profile PLP for redesigning the electrostatic field circulation in the insulation of power cables assessed for scrutinizing charge simulation method (CSM). Moreover, this chapter presents a deep study for using individual and multiple nanodielectrics in power cables manufacturing. An investigation on dielectric strength and partial discharges in the nanodielectrics of power cables is also presented. Furthermore, it offers a detailed theory and effective parameters of partial discharge in nanodielectrics of power cables. Finally, forecast and recommendations are offered for manufacturers to fabricate high quality commercial nano-tech power cables.

1 NANO-TECH VISION

Recently, the electrical insulation properties of new composite materials have been improved toward including explicit qualities of roundabout nanoparticles of earth and seethed silica in a grid about polymer present-day materials. The degree of the roundabout incorporation exhibitions outlines the fortifying time frame and affects indeed the dielectric properties of composite present-day materials. In this manner, with reference to nanoparticles as the supported strengthening stage for the composites, the dielectric interphase has a genuine effect on investigating its dielectric properties. Theoretical and lab-test results have been explored in assessment of regular structure materials and new nanocomposite present day materials for AC and DC power cables. Then, this chapter explains the monetary cost of the new nano-composite insulation materials related to the standard power cables insulation materials. This chapter also presents an examination of the update of electrical insulations for power cables materials using another multi-nanoparticle technique. The effect of guaranteeing includes indicated sorts and focuses on asserting nanoparticles with polymeric materials, for instance XLPE is used for directing as well as for electric and dielectric execution. Expectation of viable dielectric constant has been cultivated for the new nanocomposites considering Interphase power law (IPL) model. The multi-nanoparticles strategy has been prevailing for redesigning electric and dielectric execution for contrasted power cables insula-

DOI: 10.4018/978-1-7998-8536-8.ch003

tion and including novel nanoparticles. Furthermore, electric field distribution in the new prescribed forefront insulations to three-core belted power cables is examined.

Electrical treeing is a pre-breakdown component, and it can be answerable for the long-term degradation of polymer insulation. An efficient and similar examination of the effect of asserting nanoparticles in dielectric insulation is conducted to consider all the pieces of these nanoparticles, Clay, Silica, ZnO, and Al_2O_3 particles in the CT200 epoxy resin and LY/HY-5052 epoxy resin which may cause basic change in the breakdown resistance, life-time for insulation and in lessening the tree advancement rate by guaranteeing electrical tree engendering. Water treeing is the essential purpose behind cable deterioration; thus, for the most part, water treeing might be alluded to as liable for the lessening of the organization term by asserting polymeric cables. This chapter also analyzes the effects of including various sums of nanoparticles, such as ZnO, and Al_2O_3 around water treeing value attributes XLPE smoothed out materials. It further considers the improvement of the nano-composite materials execution for water tree development obstruction superior to the unfilled network, and specifically focuses on the effect of filler volume fraction on the water tree length through various frequencies. This examination renders an accessible, exact and comparable investigation over the effect of how guaranteeing nanoparticles ought to change dielectric insulation properties and will consider the effect of these nanoparticles on the smoothed-out insulation materials which can exasperate essential change to water tree opposition, life-time for insulation and decrease the water tree period in the modern insulation materials. Enormous issues occur in power cables due to the contorting of electrical pressure.

The lab-test issue regarding the nearness of voids in cable insulation which has achieved inadequate discharge and insulation disappointment is further tackled. The electrostatic field dispersal qualities of power cable insulation are addressed through the finite element method (FEM). In this chapter, the effect of nanotechnology procedures for redesigning execution of single core power cable and lessening the electrical stress inside void is examined. In addition, it focuses on the effect of electrical pressure dissemination over nearness of air voids, water voids and cupper contamination voids inside power cable insulation. New systems for nanotechnology techniques are proposed for association with plotting creative polyvinyl chloride insulation materials toward using nanocomposites and multi-nanocomposites. Voids may offer opportunities because of expansion and withdrawal of the power cables. The nearness voids inside the power cable insulation prompts the molding from guaranteeing high electrical pressure and breakdown of the insulation material of the power cable with cable maturing. The electrostatic field movement to control cable insulation is discovered by the finite element method (FEM). In this chapter, the effect of nanotechnology procedures to update the execution of the single core power cable and lessen the electrical stress inside power cable insulation which facilitates multi voids is examined. Additionally, an investigation is conducted on the effect of electrical pressure dispersal over nearness of air voids with differing shapes (cylinder, sphere and ellipse), water voids for assorted shapes (cylinder, sphere and ellipse) and cupper contamination voids with discrete shapes (cylinder, sphere and ellipse) inside power cable insulation. New approaches of guaranteeing nanotechnology techniques for illustrating imaginative polyvinyl chloride insulation materials toward using nanocomposites and multi-nanocomposites are also proposed.

2 NANODIELECTRICS IN POWER CABLES

This chapter researches the electric field distribution on various sorts of dielectrics of AC and Dc power cables toward techniques of precise assurance of the magnitude of electric field inside the cable. It's possible to arrange a cutoff on the voltage rating for an accommodated protecting thickness and choice of best ensuring tape. For AC power cable protecting the electrical field registered, charge simulation method which is a numerical technique for count about electrostatic field is used. A critical number examined in this field reproduces a three-core cable and push ahead charge simulation strategy should have less blunder (Chakradhar Reddy & Ramu, 2008; Hassenzahl et al., 2009). The possibilities about invented charges are taken as explicit outcomes from guaranteeing Laplace's and Poisson's conditions. Genuinely, the scattered surface charges help replace in the long run scrutinizing discrete imaginary charges. These charges are set outside the space for which the field is to be processed. The extent of these charges must be determined with the goal that their incorporated impact fulfills the limit conditions precisely at the chosen number of focuses on the limit. Thus, as the chance in light of focuses satisfy Laplace's and Poisson's examination inside the space viable, the outcome is excellent in this space. Dc cable figuring for field is a more awesome over AC power cables (Nguyen et al., 2010; Takala et al., 2010).

The field distribution in dc cables depends on the insulation permittivity and conductivity, the geometry of the cable and the associated voltage. It is necessary to have the capacity needed to envision the presentation from guaranteeing protecting materials under dc stress, and with information about the best suitability focuses, and pressure scattering qualities of cable insulation. This review discusses such undertakings that shed some light on these issues. This examination is limited to consistent states. The electric field is determined toward using Finite Element Method (FEM) for finding evaluated consequences of Partial Differential Equations PDE for basic examination conditions. The solution approach might be manufactured in whichever manner for dispensing totally with the differential examination, while rendering the PDE under an approximating structure of standard differential Equation's tribute. These conditions would by then reach a numerically planned circuit using standard techniques, such as Runge-Kutta (Nguyen et al., 2010; Pitsa, 2010).

2.1 AC Power Cables

This strategy uses less measurement of charges and accommodates little mistakes and high degrees from guaranteeing rightness (Thabet, 2015a). Assumptions contemplated are expressed as follows: Each computation uses two estimation examinations, at line conductors are seen as round and hollow, on and on long queues equal to each other and corresponding to the earth level. These conductors help suppress their pictures with balanced three period voltage and nothing left to space, and the permittivity will be steady. In the Charge Simulation Method (CSM), the genuine charges on the conductor surfaces are replaced by fictions line, ring or reason charges. These fictions lines would put outside the district a spot; hence a field result will be needed. If the degree of these charges is such, the net chance might be ascending to the known chance during picked measure of cutoff fixations on the conductor surface. The 8 points are shown on the right-hand center of Fig. (1). The estimations of the charges can be discovered as follows:

$$[P]_{nxn}\,[Q]_n=[V]_n \tag{1}$$

Where [P] is a potential coefficient matrix, [Q] is a column vector of values of the unknown charges, [V] is the potential of the boundary points, and n is the number of the simulating charges.

The element of the potential coefficient matrix Pij can be calculated as follows:

$$P_{ij}=\ln (D_i/D_{ij}) \quad (2)$$

Where Di is the distance of charge into origin (cable core), and Dij is the distance of boundary point j to charge.

Finally, the electrostatic field at any point (xk, yk) can be estimated in three belted cable insulation, calculated as the exactness from asserting simulation made in the long run by examining the CSM that depends on the measure of charges and their area, and the region about cutoff reason. The cross territory of a three-core belted cable is delineated in Fig. (1) which shows thickness from guaranteeing belted sheath t, thickness from asserting insulation cable T, and cable conductor measurement d.

Figure 1. Cable configuration, and its coordinates axis

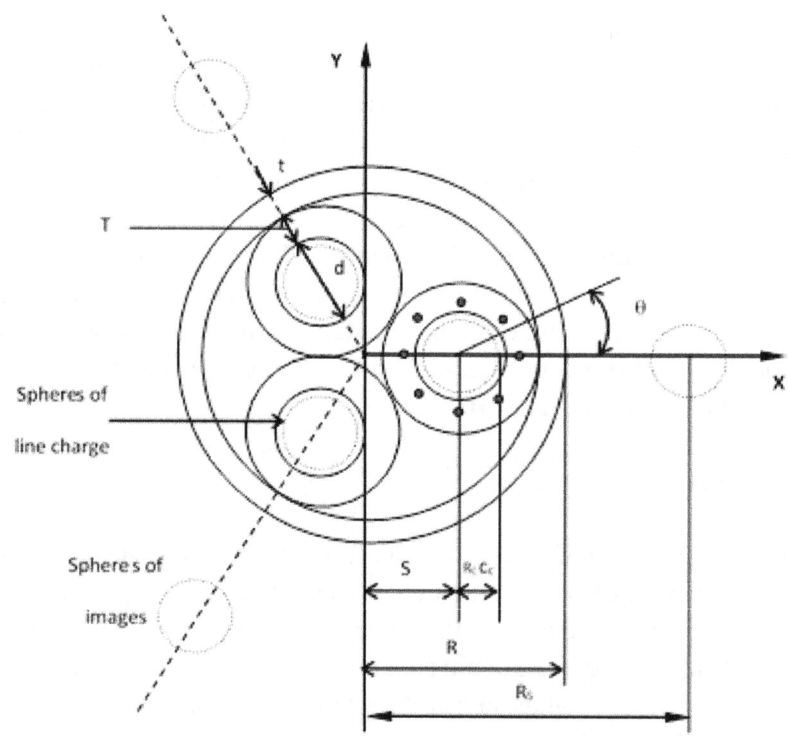

$$[P]_{m\times n}[Q]=[V]_n \quad (3)$$

Where [P] is the potential coefficient matrix, [Q] is the column vector of values of the unknown charges, [V] is the potential of the boundary points, and n is the number of simulating charges.

It might be customary that the chance and field don't move qualities z-heading. What's all the more along these lines is that the boundless understanding charges has help used for Recreation. The electrostatic field in any side of the point (xk, yk) inside solid insulation materials (unadulterated, or nanocomposites, or multi-nanocomposites) and around the cable conductor may be surveyed follows:

$$E_x = \sum_{i=1}^{N_T} \left(\left(\frac{q_i}{2\pi\varepsilon_o \tilde{\varepsilon}} \right) \cdot \left(\frac{x_k - x_i}{|D_{ij}|^2} \right) \right) \tag{4}$$

$$E_y = \sum_{i=1}^{N_T} \left(\left(\frac{q_i}{2\pi\varepsilon_o \tilde{\varepsilon}} \right) \cdot \left(\frac{y_k - y_i}{|D_{ij}|^2} \right) \right) \tag{5}$$

Where qi are the charges stated in the conductors and its images

This sort of the cable built ordinarily is an insulated cable for voltage examinations below 25kV. Polymer insulated cables have help essentially of the sheathed sort with particular centers screened, achieving winding field distribution for each center. It's traditional that regarding cables, the chance and field do not move with Z-direction and appropriately boundless understanding charge would be used inside simulation. A total of nc line charge would be used to represent each conductor, and a similar number about line charges are utilized to address the sheath Ns=Nc. The conductor charges are discovered evenly on the border of a sphere from guaranteeing range (Rc > d/2) whose middle matches with the conductor's inside. The number for charge pictures Ns has help found evenly around circumduction of a sphere about range Rs=Rc, and its inside lies in question going through start and conductor center, and investigating detachment Rs=R+T+t+d/2 beginning with inception, and t=0. 155T. Nowadays, theoretical models study the insulation materials PVC and XLPE with various nanocomposites, such as Clay, Glass Beads, and $BaTiO_3$. Glass spot 15wt% composite has a high electric field distribution inside insulation in light of its low dielectric constant. At the same time, PVC-$BaTiO_3$ 15wt % has low electric field distribution due to its high dielectric constant and diverse composite electric appropriation reviewed at their center.

2.2 DC Power Cables

On displaying these physical systems, we should consider both perfect insulation and imperfect insulation. Figure (2) demonstrates the 30kV cable that is used inside this chapter. In this application, FEM is associated determination cables mode. In essence, the FEM reduces the problem to that of solving a number of simultaneous algebraic equations. The steady state of the field in the first layer can be computed by insulation heating, caused by losses due to friction of the molecular polarization process in dielectric materials. A polluted insulation has a constrained opposition, with the objective that the spillage present in the insulation will warn the dielectric for the emulating connection:

$$E = \frac{\sigma_2}{r\left(\sigma_2 ln \frac{r_2}{r_1} + \sigma_1 ln \frac{r_3}{r_4} \right)} \tag{6}$$

In which r1, r2 and r3 is a radius of cable core and insulation layers, respectively and $\sigma 1, \sigma 2$ is a conductivity of insulation layers. Moreover, the conductivity and thermal equation used in simulation is given by:

$$J_1 = \sigma_1 \cdot E_1, \text{ and } J_2 = \sigma_2 \cdot E_2 \quad (7)$$

Figure 2. DC power cable models with insulation layers

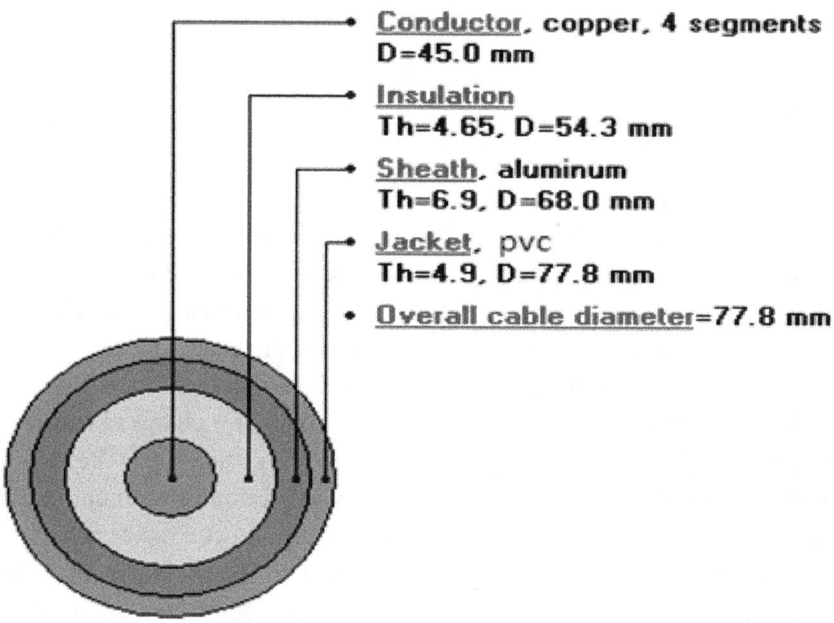

The presented current materials in this part are PVC and XLPE which are used inside protecting structures (AC and dc cable insulation). Although premium here is essentially the dielectric properties from asserting this new class from guaranteeing material, it might be likely that a noteworthy number of the arrangements will additionally exploit expert movements in various qualities, particularly warm conductivity, coefficient of the warm turn of events and warm tirelessness (Nguyen et al., 2010). Then the Electric field distribution in dc cable insulation comes to fruition and assessment. Recently, theoretical models have outlined the chance and electric field transport through the thickness for the insulation layer model. Figuring is accomplished using different nanocomposites insulation materials in the primary layer about models (XLPE with $BaTiO_3$, Al_2O_3 and mud nanocomposites (15% wt, at 0. 3T). Likewise, XLPE-Clay 15%wt nanocomposite has high electric field distribution and low chance dispersion in view of its low dielectric constant. XLPE-Al_2O_3 has low electric field movement and high chance transport in view of its high dielectric constant, as well as other composite electric dispersal assessments between them.

3 MULTIPLE NANODIELECTRICS IN POWER CABLES

Nowadays, electric and dielectric properties of power cables insulation materials could be directed using nanotechnology techniques under various warm states (Ebnalwaled & Thabet, 2016; Ebnalwaled et al., 2016; Thabet, 2015a; Thabet, 2015b; Thabet, 2015c; Thabet, 2016; Thabet, 2017a; Thabet & Mubarak, 2015; Thabet & Mubarak, 2016; Thabet & Mubarak, 2017a; Thabet & Mubarak, 2017b). This section displays novel modern materials for improved dielectric angles from guaranteeing new multi-nanocomposites smoothed out materials toward Interphase Power Law (IPL) model. The suggested model takes under record the relationship between the portions from asserting multi-nanocomposites system in the structure about interphase districts. The dielectric estimations of the interphase region have been researched around new present-day materials considering particular and multi-nanocomposites. Likewise, this section looks at the electrostatic field scattering in the new propelled dielectric insulations from asserting three-core belted power cables considering charge reproduction strategy (CSM). This investigation also strives to determine perfect plans of variation sorts of multi-nanoparticles for overhauling polymeric power cables insulations. Maxwell-Garnett debilitate obsession result for the dielectric constant has been exhibited for an inhomogeneous interphase zone enveloping every thought. Power law associations are used inside dielectric showing composite structures (Gouda et al., 2010; Karkkainen et al., 2000; Todd & Shi, 2003a). The assurance of the interphase thickness to remarkable polymeric nanocomposites will be additionally depicted for focal point (Polizos et al., 2010; Tagami et al., 2010; Todd & Shi, 2003b; Yi & Sastry, 2002). In any case, the interphase zone for new polymeric multi-nanocomposites is represented in Fig. (3). In multi-nanoparticles strategy, decline in the viable interphase volume fraction has been recognized in numerous occasions of close-by from guaranteeing multi-nanoparticles together and the spread of interphase territories enveloping each nanoparticle. Approximations are being framed to evaluate the spread from asserting such particles using logical answers for penetration models (Brosseau et al., 2001; Jing et al., 2000).

Figure 3. Interphase region surrounding the filler particles in multi-nanocomposites system

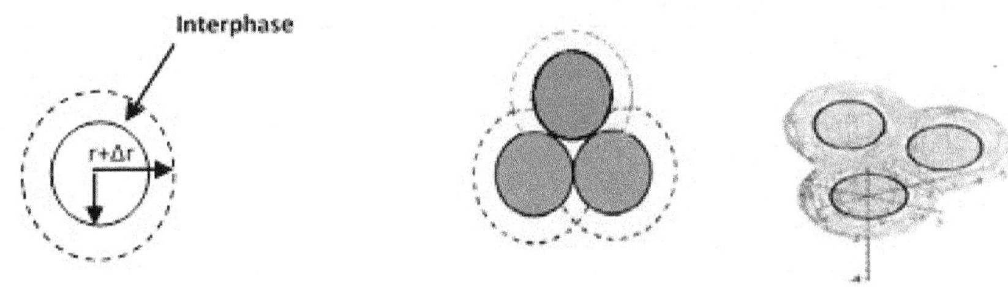

The distribution of the electric field to a three-center cable will be uncommonly basic for the right design and the insulated activity from guaranteeing power cables. Theoretical models have recognized most outrageous high voltage tensions for new insulated three-center pelted power cables. From asserting by the use of a single one or different nanoparticles techniques inside polymer lattice, power hypothesis affiliations used inside dielectric are exhibited from guaranteeing composite structures. Hence, the composite system has three components (matrix, interphase region and nanoparticles). There are various techniques conveyed that can disturb the computations of the electric field in the three-core belted power cables for choosing the pressure cables (Chakravorti & Mukherjee, 1992; Hefny, 2003; Mahdy, 1988; Malik & Al-Arainy, 1985; Nosseir & Zaky, 1979; Salama et al., 1984). This part associated the counts of the electric field in the three-core belted cables toward using the best blame generation framework which jobs less numbers for charges and accommodates little mistakes and high degrees of exactness achieved (Hefny, 2003). In modern Measurement Devices, HIOKI 3522-50 LCR Hi-tester device measures the characterization of the nanocomposite insulation industrial materials. All fabricated polyvinyl chloride nanocomposites and multi-nanocomposites have been characterized in this research, as shown in Table 1.

Table 1. Dielectric Constant of Nanoparticles, Polyvinyl Chloride Nanocomposites and Multi-Nano-composites Materials

Nanoparticles and Materials	Dielectric Constant
Pure PVC	3
Clay	2
$BaTiO_3$	3.8
MgO	9
TiO_2	48
Al_2O_3	9.5
Fe_2O_3	25
10wt.% Clay/ PVC	2.625
(10wt.%Clay+15wt.%$BaTiO_3$)/ PVC	2.802
(10wt.%Clay+15wt.%MgO)/ PVC	3.913
(10wt.%Clay+15wt.% Fe_2O_3)/ PVC	45.2
(10wt.%Clay+15wt.% TiO_2)/ PVC	4.469
10wt.% $BaTiO_3$/PVC	3.084
(10wt.% $BaTiO_3$+15wt.% Clay)/ PVC	2.507
(5wt.% $BaTiO_3$+15wt.% MgO)/ PVC	4.336
(5wt.% $BaTiO_3$+15wt Fe_2O_3)/ PVC	44.01
(5wt.% $BaTiO_3$+15wt.% TiO_2)/ PVC	4.888
10wt.% Al_2O_3/ PVC	6.041
(10wt.% Al_2O_3 +15wt.% Clay)/ PVC	3.669
(10wt.% Al_2O_3 +15wt.% $BaTiO_3$)/ PVC	5.616
(10wt.% Al_2O_3 +15wt.% MgO)/ PVC	6.793
(10wt.% Al_2O_3+15wt.% Fe_2O_3)/ PVC	38.58

Clay nanoparticles are utilized to decrease the thickness of result (Thabet & Mobarak, 2010). Aluminum oxide is a strong high warmth obstruction, particularly steady with high hardness, high mechanical quality, and electrical insulation. Zinc oxide (ZnO) is used inside paints and coatings, cross-linked for elastic and sealants. Magnesium oxide has high warm conductivity and low electrical conductivity, and barium titanate is a dielectric clay used for capacitors. Cross-linked polyethylene (XLPE) will for the larger part smooth out the generally perceived structure of polymeric insulation for creation of power cables insulations (Danikas & Tanaka, 2009; Mauseth et al., 2012). Table 1 portrays inconspicuous components for numerical investigation for first new polymeric nanocomposites materials. This section discusses the effects from guaranteeing multi-nanoparticles ahead of modern materials. The viable dielectric constant of the composite system will be analyzed in terms of the exertion for multi-nanoparticles and power cables insulations, with the end goal of XLPE. The representation showed in Fig. (4,5) allude to the lead method for convincing dielectric constant for XLPE nanocomposites with different volume fractions of nanoparticles. It might act naturally, since earth, ZnO and silica nanoparticles lessen the dielectric constant of XLPE insulation materials. In any case, TiO_2, and alumina nanoparticles increase the dielectric constant property from guaranteeing XLPE insulation materials. Using the multi-nanoparticle technique may be an opportunity to advance the arrangement from asserting nanoparticles; henceforth, controlling dielectric insulation might be just a glimpse of something larger, straightforward and gainful as demonstrated graphically. Figure 6 shows the effects of different nanoparticles on developing the electrostatic field circulation around centers of three-core belted power cables. It is noteworthy MgO and SiO_2 nanoparticles grow the electric field distribution, while TiO_2 nanoparticles reduce the electric field dissemination. The strategy from asserting nanoparticles inside multi-nanocomposites will be an extra compelling procedure for managing the electric field spread over insulation for power cables. In addition, Multi-nanoparticles technobabble has been examined as the best new examples for updating the incredible dielectric constant with reference to using individual nanoparticles; accordingly, the new nanocomposites should make Hosting multi-nanoparticles to overhauling the convincing dielectric constant about base framework dielectric material. Strategy and fixations from guaranteeing nanoparticles

Figure 4. Effect of Clay, SiO_2 and Silica on effective dielectric constant of XLPE

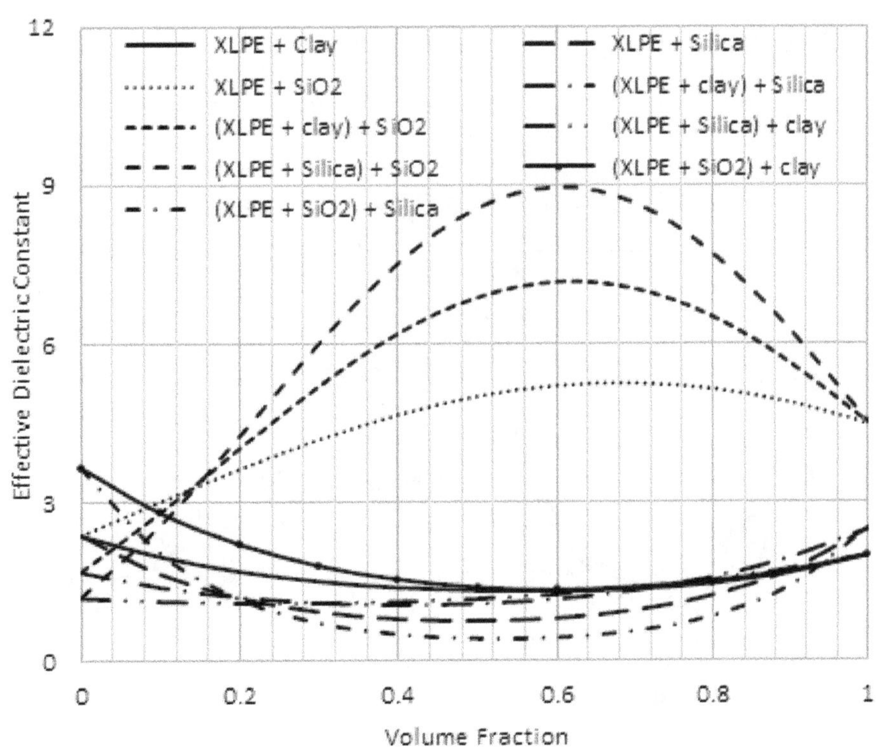

Figure 5. Effect of ZnO, SiO_2 and TiO_2 on effective dielectric constant of XLPE

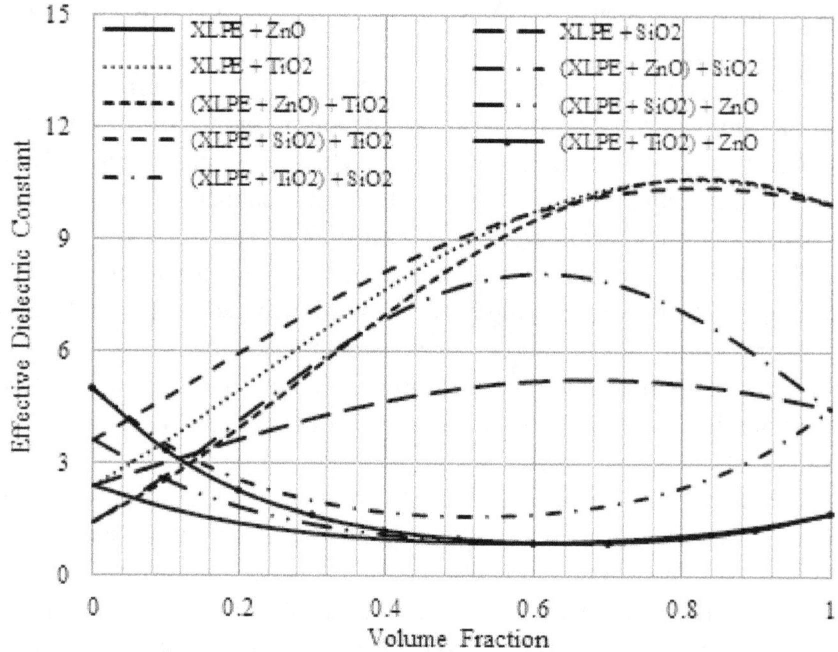

in polymeric form grid are controlling in dielectric portrayal. The suggested examination of the electric field transport in the new forward-thinking insulations for three-core belted power cables demonstrates

Figure 6. Effect of MgO, SiO_2 and TiO_2 on electric field distribution in 3-phase core belted power cables

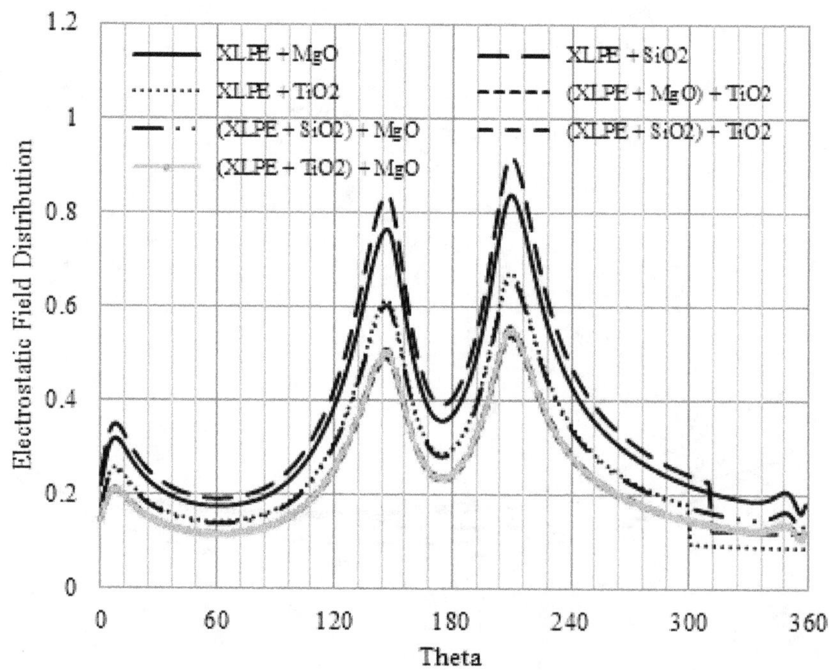

that multi-nanoparticles technique might be extra gainful for directing electric field appropriation in insulation from asserting three-core belted power cables.

4 DIELECTRIC STRENGTH OF POWER CABLES NANODIELECTRICS

In modern Measurement Devices, HIOKI 3522-50 LCR Hi-tester device measures characterization of nanocomposite insulation industrial materials. All fabricated polyvinyl chloride nanocomposites and multi-nanocomposites have been characterized in this research, as shown in Table 1. Spherical nanoparticles shape (Dia. Less than 50nm) have been used in our research and in the most polymer applications. Penetration of nanoparticles polyvinyl chloride in PVC nanocomposites has been detected by using Scanner of Electron Microscopic "SEM" images, as shown in Fig. (7).

Figure 7. SEM measurements for nanoparticles distribution in PVC

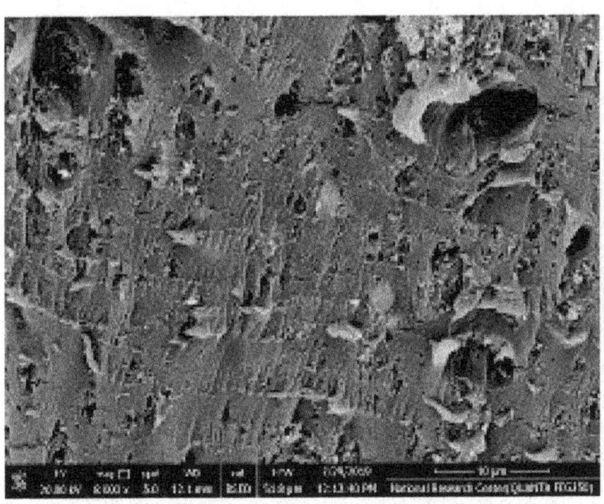

(a) Clay/PVC nanocomposites

(b) ZnO/PVC nanocomposites

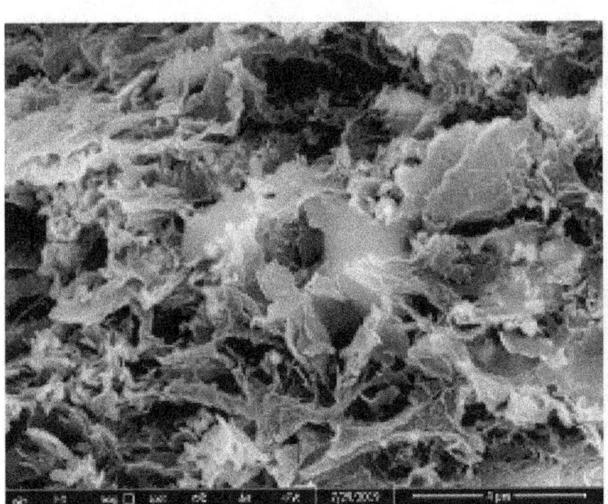

(c) (Clay+BaTio3)/PVC nanocomposites

(d) (ZnO+BaTio3)/PVC nanocomposites

HI-POT TESTER Model ZC2674 device has been used to experiment uniform and non-uniform electric field distribution through the thickness of insulation layer with different nanocomposite materials. The Hi-tester device has been used for measuring breakdown characterization of pure and nanocomposite insulation industrial materials under uniform electric fields, as shown in Fig (8.a-8.c). Recently, it has been observed that adding individual nanoparticles (Clay, ZnO and $BTiO_3$) raise maximum dielectric strength of polyvinyl chloride, especially, in case of adding two different types of nanoparticles together in polyvinyl chloride base matrix materials (Thabet & Fouad, 2021). When nanoparticles increase the

Figure 8a. Dielectric strength of nanocomposite insulation material: Clay and $BaTiO_3$ nanoparticles under uniform electric fields

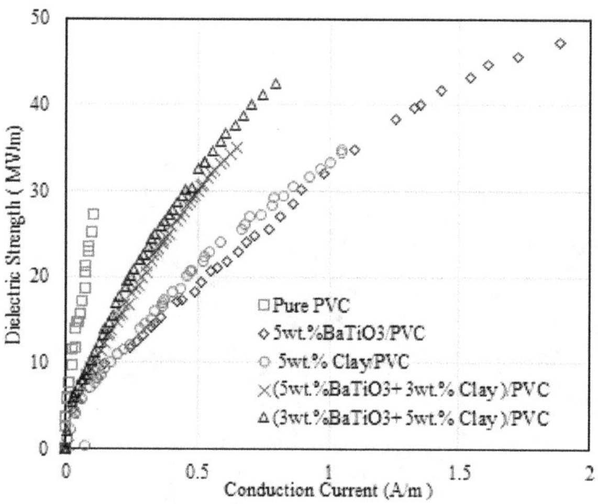

Figure 8b. Dielectric strength of nanocomposite insulation material: Clay and ZnO nanoparticles under uniform electric fields

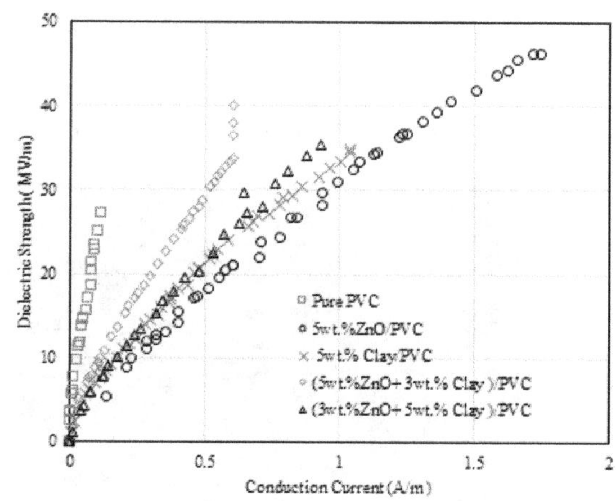

Figure 8c. Dielectric strength of nanocomposite insulation material: ZnO and $BaTiO_3$ nanoparticles under uniform electric fields

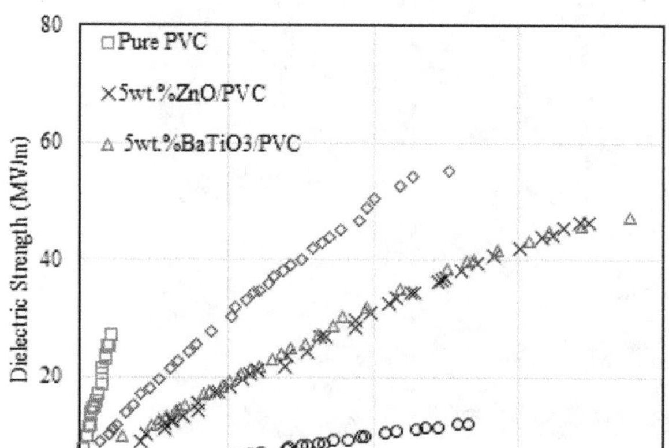

Figure 8d. Dielectric strength of nanocomposite insulation material: Clay and $BaTiO_3$ nanoparticles under non-uniform electric fields

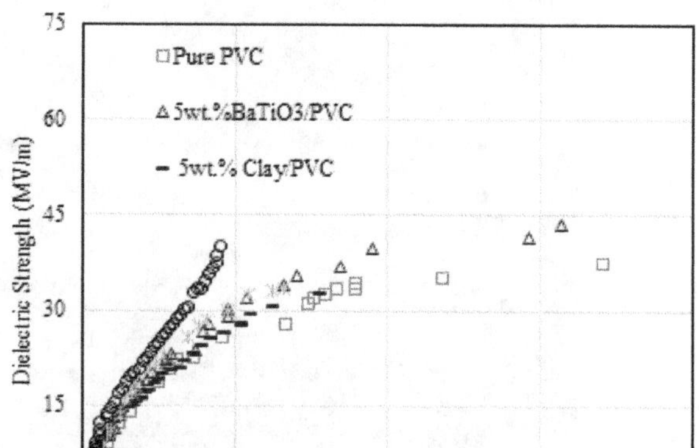

Figure 8e. Dielectric strength of nanocomposite insulation material: ZnO and $BaTiO_3$ nanoparticles under non-uniform electric fields

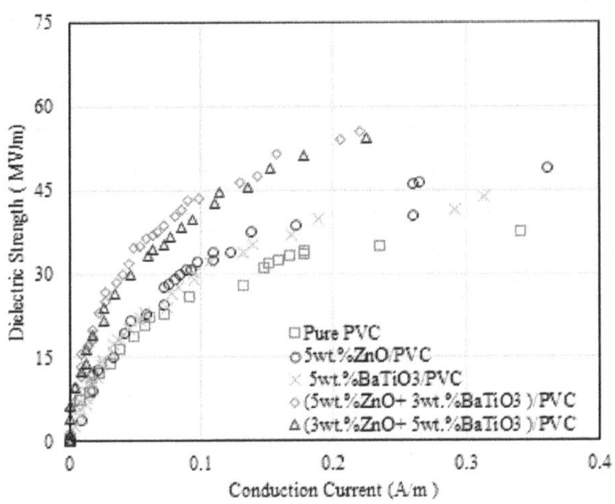

Figure 8f. Dielectric strength of nanocomposite insulation material: Clay and ZnO nanoparticles under non-uniform electric fields

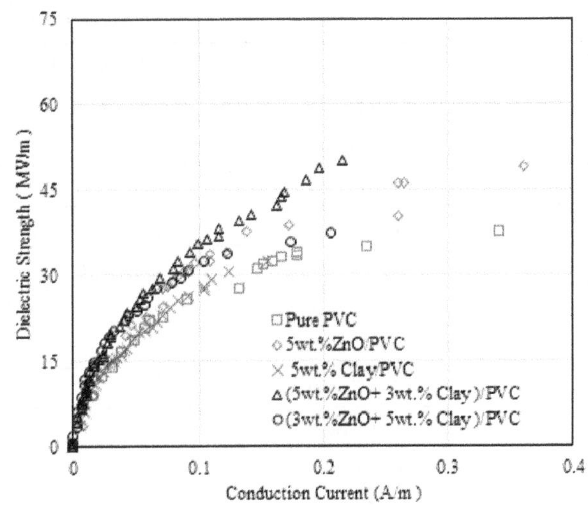

dielectric strength of polyvinyl chloride, it is noticed that the conduction current increases in nanocomposites specimens compared with the traditional polyvinyl chloride material. Moreover, the breakdown characterization of pure and nanocomposite insulation industrial materials under non-uniform electric fields is measured, as shown in Fig (8.d - 8.f). It has been observed that adding individual nanoparticles (Clay, ZnO and $BTiO_3$) raise the dielectric strength of polyvinyl chloride, especially in case of adding two different types of nanoparticles together in polyvinyl chloride base matrix materials. When the

Table 2. Max. Dielectric Strength of Nanocomposites Insulation Materials (20 °C)

Materials	MAX Dielectric Strength (MV/m)	
	Uniform Electric Field	Non-Uniform Electric Field
Individual Nanoparticles Technique		
PURE PVC	27.23	37.61
5%wt. Clay/PVC	34.90	32.45
5wt.%ZnO/PVC	46.34	48.95
5%wt. $BaTiO_3$/PVC	47.33	43.70
Multiple Nanoparticles Technique		
(3wt.%ZnO+5wt.%Clay)/PVC	35.35	37.37
(5wt.%ZnO+3wt.%Clay)/PVC	40.00	50.13
(3wt.%ZnO+5wt.%$BaTiO_3$)/PVC	55.08	54.19
(5wt.%ZnO+3wt.%$BaTiO_3$)/PVC	54.30	55.43
(5wt.%$BaTiO_3$+3wt.%Clay)/PVC	35.11	33.26
(3wt.%$BaTiO_3$+5wt.%Clay)/PVC	42.38	40.00

Table 3. Max. Conduction Current in Nanocomposites insulation materials (20 °C)

Materials	MAX Conduction Current (A/m)	
	Uniform Electric Field	Non-Uniform Electric Field
Individual Nanoparticles Technique		
PURE PVC	106.1	340.9
5%wt. Clay/PVC	1043.2	155.6
5wt.%ZnO/PVC	1743.7	361.3
5%wt. $BaTiO_3$/PVC	1884.8	310
Multiple Nanoparticles Technique		
(3wt.%ZnO+5wt.%Clay)/PVC	930.4	205.7
(5wt.%ZnO+3wt.%Clay)/PVC	604.7	215.9
(3wt.%ZnO+5wt.%$BaTiO_3$)/PVC	1258.5	224.6
(5wt.%ZnO+3wt.%$BaTiO_3$)/PVC	1327.4	219.7
(5wt.%$BaTiO_3$+3wt.%Clay)/PVC	644.7	131.6
(3wt.%$BaTiO_3$+5wt.%Clay)/PVC	792.3	89.7

nanoparticles increase the dielectric strength of polyvinyl chloride, it is noticed that the conduction current decreases in nanocomposites specimens compared with the traditional polyvinyl chloride material.

HI-POT TESTER Model ZC2674 device has been used to experiment uniform and non-uniform electric field distribution through the thickness of insulation layer with different nanocomposite materials. The Hi-tester device has been used for measuring breakdown characterization of pure and nanocomposite insulation industrial materials, as shown in Table 2. and Table 3.

5 PARTIAL DISCHARGES IN NANODIELECTRICS OF LOW VOLTAGE POWER CABLES

Partial discharge happens for attributes whose electric field power level is higher than the recognized level for insulation of a power cable, Nowadays, Finite Element Method (FEM) has been utilized to examine air pockets sway in high voltage power cable insulations as a fundamental methodology. The constrained part methodology is an opportunity that can undoubtedly be used to figure out the pre-lab-test stress inside the void (Patel et al., 2014). This finally prompts the right appraisal about electric field movement for cable insulation system. In the long run by examining the use of 2-measurement Maxwell programming, the electric field esteems for each breaking point and zone can be discovered (Lachini et al., 2010). With the tally of the degree of the electric field power level as its spatial dispersal inside the cable, this could sensibly and normally set a breaking point to the voltage rating for an accommodated insulation thickness, or otherwise (Thabet & Mubarak, 2017a). The major breaking down outcomes are seen as an opportunity, since hydrogen, carbon monoxide, methane and carbon dioxide would produce the power cable upon exposure to PD development (Avinash & Rajagopala, 2013; Nosseir, 1979). Thus, this section focuses on overhauling insulation execution about single core power cable that is exposed to collecting abandons like voids and metal polluting influences. Therefore, this book concentrates on the effect of electrical pressure appropriation in nearness of discrete voids, by using unmistakable and unique nanoparticle techniques on cable insulation.

5.1 Theory and Effective Parameters

Ionization and incomplete releases inside the voids may cause scraped spot and nearby rot of the insulation, which prompts a total disappointment of the power cable, thus the essentialness is accommodated for the thought of the conductor voltage in which Partial discharges take spot inside the void (Avinash & Rajagopala, 2013; Elayyan & Abderrazzaq, 2005; Nosseir, 1979; Thabet, 2017b). Fractional discharge happens in spots whose electric field power level is higher than the recognized level from guaranteeing insulin in a power cable. These days, Finite Element Method (FEM) has been utilized to consider air pockets sway in high voltage power cable insulations as a basic method. In recent years, the progression of nanotechnology science has allowed for the advancement of new insulation materials that have impelled physical and electrical properties (Thabet & Ebnalwaled, 2017; Thabet & Salem, 2017; Thabet et al., 2018). Accordingly, this section focuses on overhauling insulation execution for the single core power cable that is discovered to create surrenders like voids, polluting influences, etc. Hence, the effect from guaranteeing electrical pressure dispersion attributes of nearness of various voids shapes (cylinder-sphere-ellipse) has been examined, eventually scrutinizing by the use of particular and distinctive nanoparticles techniques for cable insulation. It is known theoretically that a solitary center cable Hosting (r) range from asserting transmitter and (r) internal range from guaranteeing insulation, the chance inclination (g) toward a partition (x) beginning with the point of convergence of the channel inside the dielectric material is

$$g = \frac{q}{2\pi\varepsilon x} = \xi x \tag{8}$$

Where ξx is the electric field intensity, q is the charge per unit length, ε is the permittivity of the dielectric material. Therefore, the potential of the conductor will be,

$$V = -\int_{R}^{r} \xi x . dx \tag{9}$$

$$\begin{aligned} &= \int_{R}^{r} \frac{q}{2\pi\varepsilon x} . dx \\ &= \frac{q}{2\pi\varepsilon} . \log_e \frac{R}{r} \end{aligned} \tag{9.i}$$

Since, $\frac{q}{2\pi\varepsilon x} = \xi x$ (From eq. (8)). So,

$$\xi x = \frac{V}{x \log_e \frac{R}{r}} \tag{10}$$

Here, “x”’ is the only variable in the equation, the maximum stress in dielectric material occurs at the minimum value of the radius (here, x = r) (Thabet, 2017b). Thus,

$$\xi x = \frac{V}{r \log_e \frac{R}{r}} \tag{10.i}$$

In any case, each and every of the flighty issues of geometry about void might be accordingly jumbled, so the logical outcome will be along these lines troublesome. In light of this explanation, several endeavors have attempted to find new determining methodologies about the electric field. FEM might be picked for computation use. Electric field conditions result in the long run scrutinizing. This technique also takes into consideration Maxwell conditions for limit states.

The point behind guaranteeing this fill in might be to have new dielectric materials that are necessary for redesigning electric demands. The emulating impacts show sway of (air – contamination – water) voids with discrete shapes (cylinder – sphere – ellipse) for nanocomposites and multi nanocomposites, as shown in Fig. (9). Therefore, the least electric stress inside power cable insulation is obtained from what’s progressively inside the sort of asserting voids. Most extraordinary electric pressures will find

Figure 9. Cross chapter of Single Core Power Cable with (cylinder – sphere – ellipse) voids

attributes of cylinder void due to its territory near the channel. Accordingly, the lesser electric pressure will be found on sphere void in light of its region much away from the conveyor.

The ground-breaking dielectric constant of the incorporation and interphase might be an opportunity to be imparted for single individual nanocomposite model that holds an interphase zone as expressed by (Todd & Shi, 2003a) underneath:

$$\varepsilon_{effi}^{\beta} = \varphi_i \varepsilon_i^{\beta} + \varphi_{phi} \varepsilon_{phi}^{\beta} + \varphi_m \varepsilon_m^{\beta} \tag{11}$$

In any case, the effective dielectric constant of the inclusion and interphase could be expressed for a multiple Nanocomposite model that contains an interphase region.

$$\varepsilon_{effj}^{\beta} = \varphi_j \varepsilon_j^{\beta} + \varphi_{phij} \varepsilon_{phij}^{\beta} + \varphi_{effi} \varepsilon_{effi}^{\beta} \tag{12}$$

where, φ_j is the volume fraction of second nanoparticle component of the multi-composite system, φ_{phij} is the volume fraction of the interphase region component of the multi-composite system, φ_{effi} is the volume fraction of the initial matrix component of the multi-composite, ε_{effj} is the dielectric permittivity of the multi-composite system. ε_j is the second nanoparticle permittivity of the multi-composite system, ε_{phij} is the interphase permittivity of the multi-composite system, ε_{effi} is the dielectric permittivity of the individual nanocomposites system. The second nanoparticle volume fraction, φ_j is directly measured for a given composite system. The matrix volume fraction is given by $\varphi_{effi} = (1 - \varphi_j - \varphi_{phij})$.

The interphase volume fraction of multi-composite system, φ_{phij}, is estimated as follows:

$$\varphi_{phij} = (1\text{-}F).\ (S_j.\ \Delta r).\ \rho_j\ .\ \varphi_j \tag{12.i}$$

where, S_j is the specific surface area of second nanoparticle (measured in m^2/g), ρ_j is the density of second nanoparticle (measured in g/m^3), Δr is the thickness of the interphase region, F is an overlap probability function.

5.2 Dielectrics and Nanoparticles

The cable plan in this part includes inward conductor for cross sectional area of 8.55 cm^2 (1. 65 cm range), and the outer insulation from asserting thickness from guaranteeing is perceived to be 2. 85cm. The material polyvinyl chloride (PVC) is regarded as the insulation material for relative permittivity around 3. The cable arrangement of a conductor is toward 11 kV, and the insulation material is held at ground plausibility. The Power cable has three voids (cylinder-sphere-ellipse) with the same cross-sectional domain of 0.2826 cm^2. Table 3 depicts the specifications of case study single core power cable. The voids will detect the center of the conductor surface and the insulation outside surface. Finite Element Method Magnetics (FEMM) item will be used for this proliferation. Finite Element Method Magnetics (FEMM) programming will be a constrained part one group. This programming might be used for fathoming 2-dimensional planar and axisymmetric issues in electrostatics, current stream, high temperature stream and low repeat magnetics. Polyvinyl chloride is for the larger part by and large used for cables, thermoplastics, polymerized vinyl chloride which are prepared from ethylene and anhydrous

Table 4. Specifications of Single Core Power Cable

Specification	Rating
Rated voltage (V)	11kV
Number Core	1
Copper conductor radius (r)	1.65 cm
Insulation radius (R)	4.5 cm
Insulation dielectric constant (PVC)	3
Water permittivity	80
Air permittivity	1
Copper impurity	5.6

Table 5. Dielectric Constant of Nanocomposite materials

Nanocomposite and Materials	Dielectric Constant
Pure PVC	3
5wt.% Clay/PVC	2.801
(5wt.%Clay+7wt.%ZnO)/PVC	2.608
(5wt.%Clay+12wt.%ZnO)/PVC	2.484
(5wt.%Clay+7wt.%$BaTiO_3$)/PVC	2.872
(5wt.%Clay+12wt.%$BaTiO_3$)/PVC	2.924
(5wt.%Clay+7wt.%Al_2O_3)/PVC	4.871
(5wt.%Clay+12wt.%$Al2O_3$)/PVC	6.62
5wt.% ZnO/PVC	2.831
(5wt.% ZnO+7wt.% Clay)/PVC	2.608
(5wt.% ZnO+12wt.% Clay)/PVC	2.47
(5wt.% ZnO+7wt.% $BaTiO_3$)/PVC	2.901
(5wt.% ZnO+12wt.% $BaTiO_3$)/PVC	2.95
(5wt.% ZnO+7wt.% Al_2O_3)/PVC	4.898
(5wt.% ZnO+12wt.% Al_2O_3)/PVC	6.638
5wt.% Al_2O_3/PVC	4.413
(5wt.% Al_2O_3+7wt.% Clay)/PVC	3.697
(5wt.% Al_2O_3+12wt.% Clay)/PVC	3.277
(5wt.% Al_2O_3+7wt.% ZnO)/PVC	3.882
(5wt.% Al_2O_3+12wt.% ZnO)/PVC	3.554
(5wt.% Al_2O_3+7wt.% Fe_2O_3)/PVC	16.7
(5wt.% Al_2O_3+12wt.% Fe_2O_3)/PVC	30.84

hydrochloric destructive. Polyvinyl Chloride (PVC) might be more grounded and even more rigid over various general purposes behind existing thermoplastic materials. PVCs are in a general sense exceptional and solid, in contrast to water and scraped spot, and are incredible electrical separators (Gouda et al., 2010). Mud nanoparticles can be used to diminish the thickness of outcome (Karkkainen et al., 2000). Aluminum oxide (Al_2O_3) might be of strong high warmth opposition, truly steady, with high hardness and high mechanical quality. Electrical insulin and Zinc oxide (ZnO) will be used inside paints, coatings, cross cable for flexibles and sealants. Magnesium oxide (MgO) has high thermal conductivity and low electrical conductivity. Barium titanate ($BaTiO_3$) is a dielectric earthenware used for capacitors. Table 4 delineates focuses on numerical dielectric constant qualities for the first and new polymeric nanocomposite materials (Thabet & Mubarak, 2017b).

5.3 Nanoparticles Trends for Partial Discharges

In our study, Fig. (10-12) outlines the lead strategy for electric stress inside variation polyvinyl chloride insulation materials and nanocomposites from guaranteeing nearness air void. It has been observed that including Al_2O_3 nanoparticles polyvinyl chloride extends the electric field appropriation inside (cylinder and ellipse) air voids, but reduces the electric field distribution inside sphere air void and decreases the electric field dissemination inside power cable insulation in view of the effect of high dielectric constant of Al_2O_3 nanoparticles. Clearly, the extension from guaranteeing centralization of Al_2O_3 nanoparticles

Figure 10. Stress distribution in cable insulation with different air voids (cylinder- sphere- ellipse) compare 5wt.% multiple nanoparticles/PVC

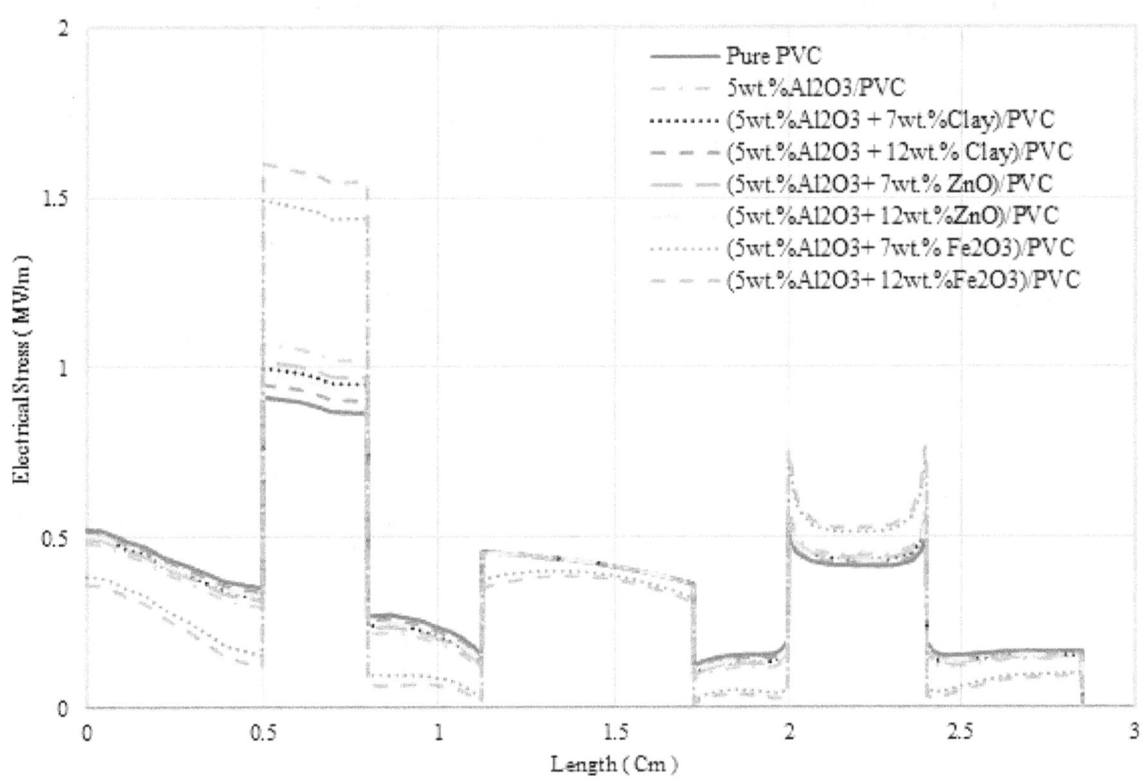

Figure 11. Stress distribution in cable insulation with different air voids (cylinder- sphere- ellipse) compare 5wt. % multiple nanoparticles/PVC

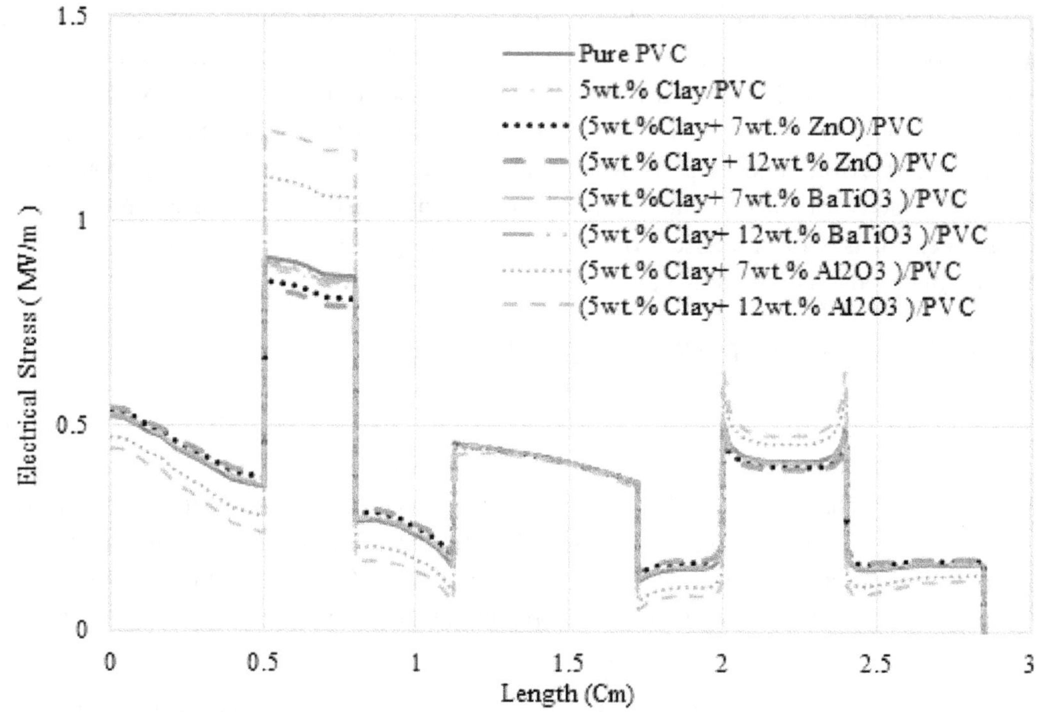

Figure 12. Stress distribution in cable insulation with different air voids (cylinder- sphere- ellipse) compare 5wt. % multiple nanoparticles/PVC

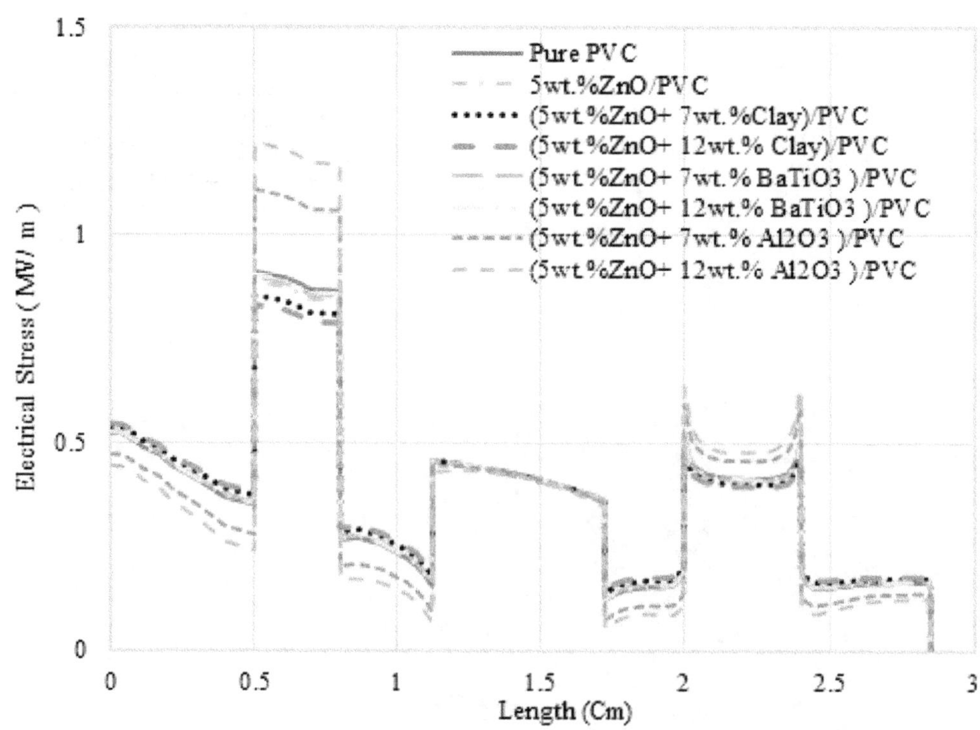

extends the electric field movement inside (cylinder and sphere) air voids. In any case, the electric field transport inside (cylinder and ellipse) air void in the insulation for single core power cable has been decreased and the electric field movement inside power cable insulation has been extended by including mud and ZnO nanoparticles as a result of low dielectric constant. Thus, the growing obsession about asserting clay nanoparticles is because it is just a glimpse of something larger with profitable effect on lessening electric field movement inside (cylinder and ellipse) air voids; however, it may unmistakably lead to electric pressure execution in case of including diverse nanoparticles inside polyvinyl chloride. The strategy for different nanoparticles will be a fundamental parameter that is impacted for extending or lessening the electric field transport inside (cylinder, sphere and ellipse) air voids and others. It becomes clear that using distinctive nanoparticles technique favors the individual nanoparticles methodology for reducing the electrostatic field inside air voids. The reason is that mud and zinc oxide are the best nanoparticles that cause lessening in the electric transport inside (cylinder, sphere and ellipse) air voids, while Al_2O_3 nanoparticles endeavor to raise the electric appropriation inside the equivalent geometric air voids. More details are shown for the best nanoparticles that can cause a reduction in the electric movement inside power cable insulation.

Figures (13-15) shows the cause behind impeccable electric stress inside variation polyvinyl chloride insulation materials and nanocomposites in the occasion of asserting nearness water void. It is observed that including Al_2O_3 nanoparticles polyvinyl chloride reduces the electric field circulation inside power cable insulation and extends the electric field dissemination inside (cylinder, sphere and ellipse) water

Figure 13. Stress distribution in cable insulation with different water voids (cylinder – sphere - ellipse), compare 5wt. % multiple nanoparticles/PVC

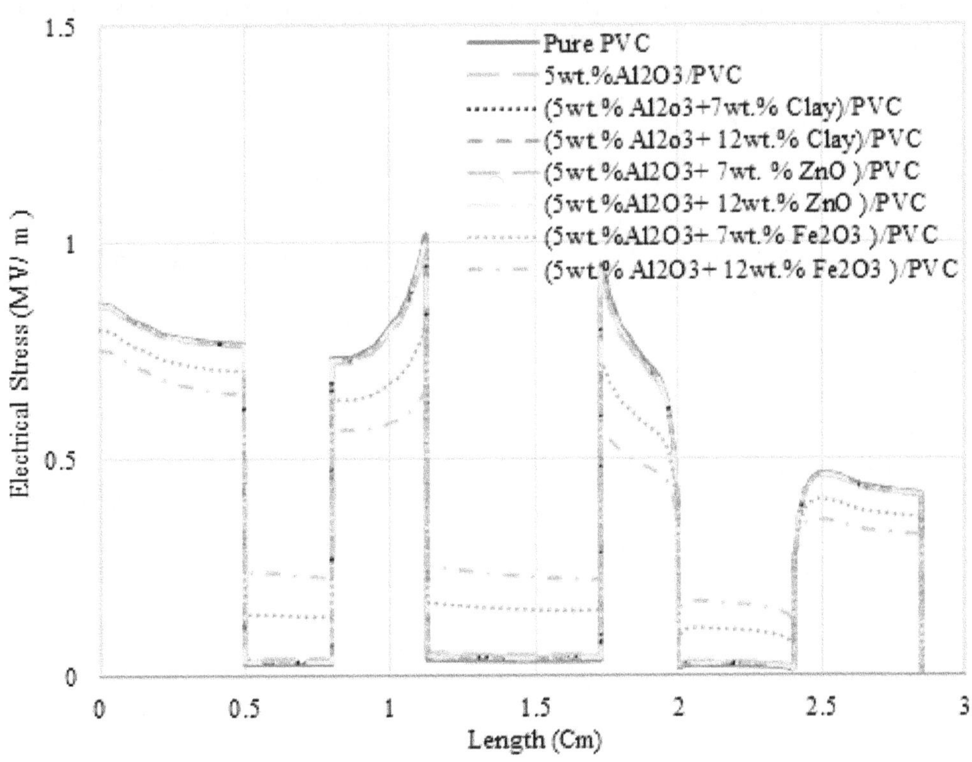

Figure 14. Stress distribution in cable insulation with different water voids (cylinder - sphere - ellipse), compare 5wt. % multiple nanoparticles/PVC

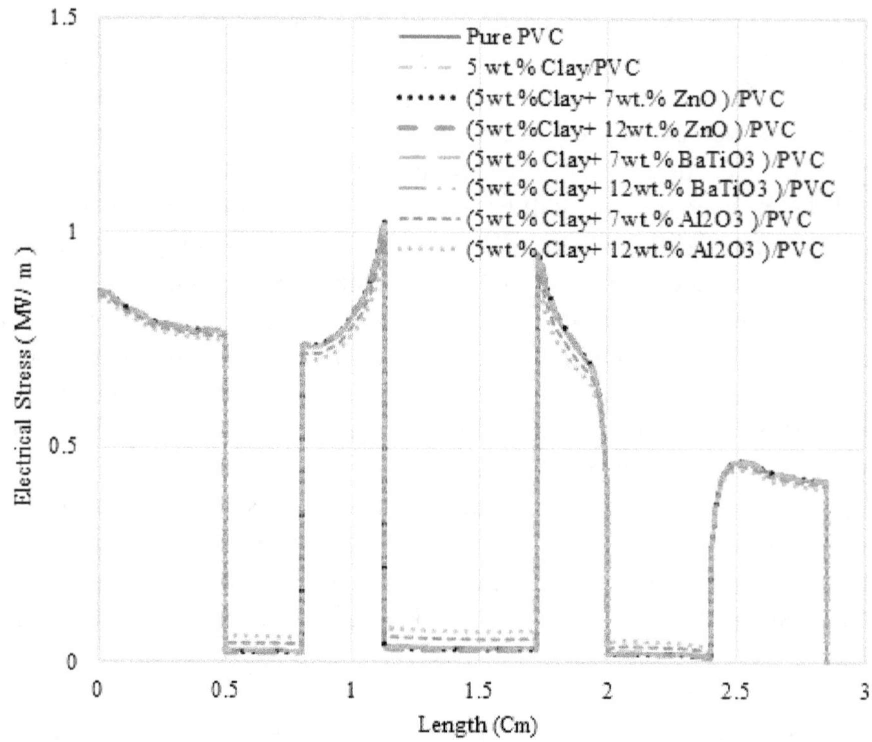

Figure 15. Stress distribution in cable insulation with different water voids (cylinder- sphere- ellipse), compare 5wt. % multiple nanoparticles/PVC

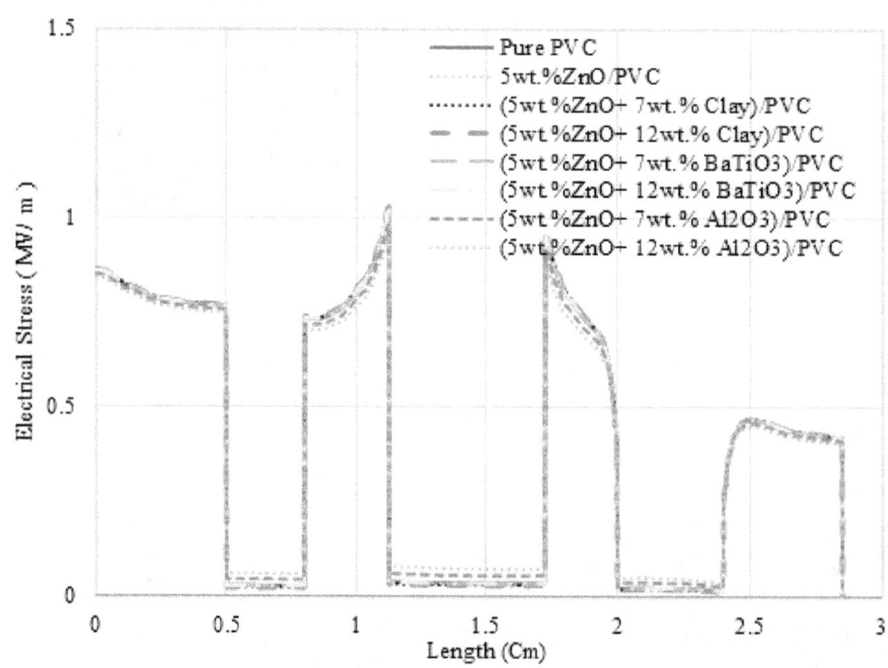

voids on account of the effect of high dielectric constant for Al_2O_3 nanoparticles. Clearly, the development in focus from asserting Al_2O_3 nanoparticles lessens the electric field circulation inside power cable insulation. In any case, the electric field dissemination in (cylinder, sphere and ellipse) water voids inside the insulation from guaranteeing single core power cable has been decreased, and the electric field appropriation inside power cable insulation has been extended toward including clay and ZnO nanoparticles due to low dielectric constant. Thus, the growing interest in mud nanoparticles is because of its more prominent effect on decreasing the electric field distribution inside (cylinder, sphere and ellipse) water voids, since it can unmistakably affect the electric pressure execution if asserting includes various nanoparticles inside polyvinyl chloride. The game-plan about different nanoparticles is a basic parameter that is affected by extending, while the electric field circulation inside water voids is decreased. Hence, it is clear that using distinctive nanoparticles technique is predominant in the individual nanoparticles framework for reducing electrostatic field inside (cylinder, sphere and ellipse) water voids. It is found that clay and zinc oxide are the best nanoparticles that will cause an abatement in the electric dispersal inside water void. Anyway, Al_2O_3 nanoparticles endeavor to raise the electric dissemination inside the equivalent geometric water voids. Furthermore, it helps the nanoparticles which will cause reduction in the electric scattering inside power cable insulation.

Figure 16. Stress distribution in cable insulation with different solid impurities voids (cylinder- sphere-ellipse), compare 5wt. % multiple nanoparticles/PVC

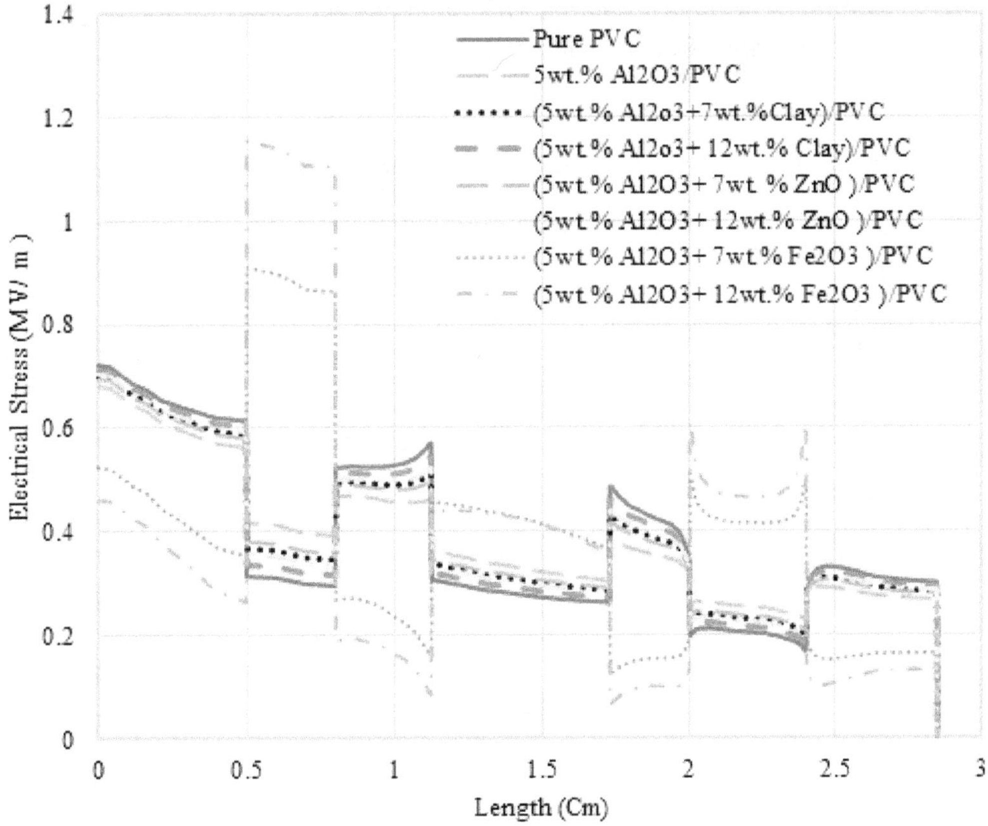

Figure 17. Stress distribution in cable insulation with different solid impurities voids (cylinder- sphere-ellipse), compare 5wt. % multiple nanoparticles/PVC

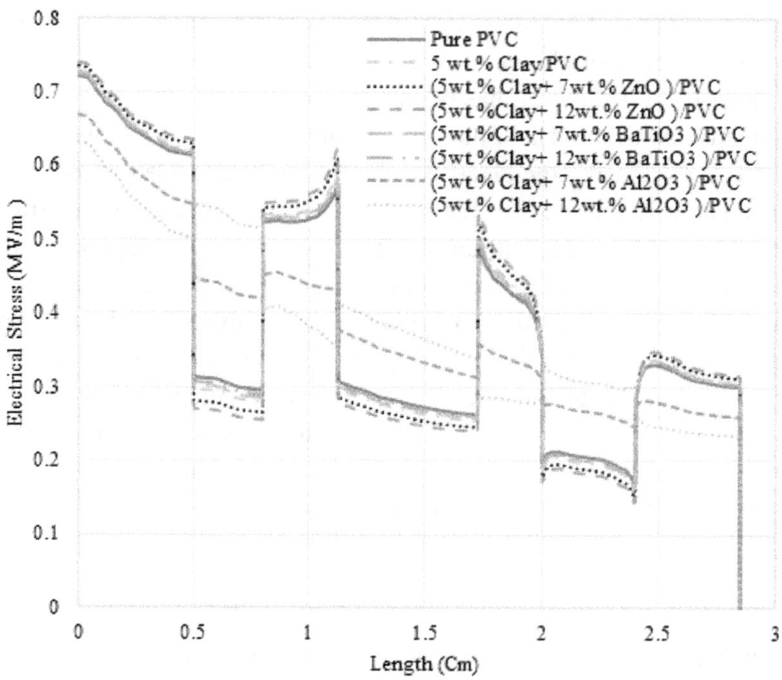

Figure 18. Stress distribution in cable insulation with different solid impurities voids (cylinder- sphere-ellipse), compare 5wt. % multiple nanoparticles/PVC

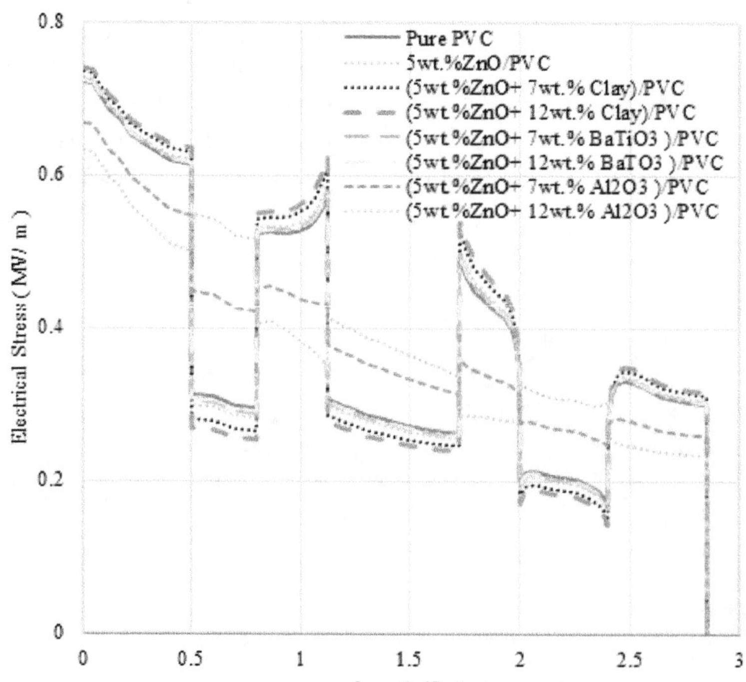

Figures (16-18) delineates that the impeccable direction of electric stress inside variation polyvinyl chloride insulation materials and nanocomposites in the occasion nearness cupper contamination void. It is observed that including Al_2O_3 nanoparticles polyvinyl chloride extend the electric field scattering inside (cylinder, sphere and ellipse) cupper contamination voids and reduce the electric field transport inside power cable insulation in view of the effect from guaranteeing high dielectric constant of Al_2O_3 nanoparticles. Clearly, the development about the focus of Al_2O_3 nanoparticles decreases the electric field appropriation inside power cable insulation. In any case, the electric field transport inside (cylinder, sphere and ellipse) cupper contamination voids inside the insulation from guaranteeing single core power cable has been lessened, and the electric field scattering inside power cable insulation has been extended in the long run by including mud and ZnO nanoparticles in light of the low dielectric constant. Thus, mud nanoparticles have received great attention due to its effect on decreasing the electric field scattering inside (cylinder, sphere and ellipse) cupper contamination voids. Nevertheless, a diverse direct method on electric pressure execution by including various nanoparticles inside polyvinyl chloride has been observed. The game-plan for various nanoparticles might be a principal parameter that cause increase, while the electric field appropriation inside contamination voids decreases. Therefore, it is evident that using a different nanoparticles framework might be better for an unmistakable nanoparticles methodology that reduces the electrostatic field inside contamination voids. This is because clay and zinc oxide are considered the best nanoparticles that would exasperate a diminishing in the electric dissemination inside cupper contamination voids, while Al_2O_3 nanoparticles seek to raise the electric dispersion inside the equivalent geometric contamination voids. Likewise, it is the best nanoparticle that would cause a diminishing in the electric dissemination inside power cable insulation.

6 PARTIAL DISCHARGES IN NANODIELECTRICS OF HIGH VOLTAGE POWER CABLES

This chapter investigates the electrical field stress in new cross-linked polyethylene (XLPE) nanocomposites in case of containing voids under high voltages. Moreover, the effect of nanoparticles on decreasing the electrical stress inside three-core power cables insulation is investigated, especially, in case of the presence of multiple voids of air, water and copper impurities with different shapes (cylinder, sphere and ellipse) inside insulation of three-core power cables. The electrostatic field distribution in power cable insulation has been calculated by finite element method (FEM). This research has succeeded in designing innovative patterns of Cross-linked polyethylene insulation materials for enhancing the electrical stress within void up to 74.23% and inside insulator up to 85.67%, according to the type and concentration of nanoparticles which are added to pure materials while applying new nanocomposites and multi-nanocomposite. Partial discharge (PD) in electrical cables is one of the main causes for insulation degradation and localized electric discharges, as well as defects within the insulation material of electrical cables. Partial discharges degrade the insulation of cable to a failed point in the cable. PD usually occurs due to cavities or voids within the insulation material of the electrical cables. Degradation in insulation materials is caused by electrical, chemical and mechanical stress that causes partial discharges within the cable and leads to cable failure. Power cable plays the key role in state power transmission, especially cross-linked polyethylene (XLPE) cable. The defects in cable joints, induced in the process of manufacturing or assembling, can alter the distribution of the electrical field seriously. XLPE insulation has much better electrical, thermal, and mechanical performances than that of oil paper insulation. The manufacturing of XLPE power cables needs to avoid defects of insulation and joints of power cables which are the vital components in power cable manufacturing. Due to the existing defects in the insulation and joints of power cables, they may lead to a system fault, such as partial discharge cable joints, the cable joints are considered the weakest points in power cables transmission lines (Thabet & Fouad, 2021). The partial discharge study provides a useful tool to obtain information about discharging defects in high-voltage equipment.

6.1 Insulation Materials Design

Nanocomposites are formed by the homogenous dispersion of nano-sized inorganic particles into a polymeric matrix, and the special properties come about because of the strong interaction between the organic and inorganic phases, which is attributed to the very large active surface of the nano-sized particles (Thabet, 2015a; Thabet, 2015c; Thabet, 2016; Thabet & Ebnalwaled, 2017; Thabet & Mubarak, 2017a; Thabet & Mubarak, 2017b). Theoretical models are capable of predicting the life and reliability of HV cable systems, including accessories, as functions of the applied stresses, and this would be of interest to cable manufacturers and transmission system operators. The service lifetime of an insulation is a mathematical relationship between time to failure and the level of one or more of the applied stresses. The life-time model becomes a reliable model, if it is also capable of predicting the reliability of the insulation after a certain time under the applied stresses (Thabet, 2017a). This section focuses on analysing the electric fields distribution in new multi-nanocomposites and studying the effects of different types of multi-nanocomposites for industrial progress power cables applications. Furthermore, it examines the effect of multi-nanoparticles arrangement on the new suggested electrical nanocomposites materials. The electric field distribution has been simulated to explain the ability of new nanocomposites and multiple nanocomposites materials to enhance three core belted power cables insulation.

The aim of this chapter is presenting recent enhancement of insulation performance of AC and DC power cables that are exposed to manufacture defects like air voids, water voids and impurities, etc. Thus, this paper examines electrical stress distribution in presence of different shapes of voids (cylinder - sphere - ellipse), in case of using individual and multiple nanoparticles techniques for insulation of three-core power cables. Power law relationships are used in dielectric modelling of composite systems by using individual nanoparticles; thus, the composite system has three components as follows: Matrix, nanoparticle and interphase region (Karkkainen et al., 2000; Todd & Shi, 2003a). Furthermore, the interphase region is dependent upon the nanoparticle volume fraction, the nanoparticle surface area and the thickness of the interphase region surrounding each nanoparticle. Anyway, a multi-nanoparticles technique has been proposed for developing the electric and dielectric properties of polymer (). Estimation of interphase thickness is further developed based on ref. (Polizos et al., 2010; Tagami et al., 2010).

6.2 High Voltage Power Cable Simulation Model

It is known by theory that in a three-core cable having (r) radius of conductor and (R) inner radius of insulation, the potential gradient (g) at a distance (x) from the center of the conductor within the dielectric material is (Avinash & Rajagopala, 2013; Elayyan & Abderrazzaq, 2005; Jing et al., 2000; Lachini et al., 2010; Nosseir, 1979; Patel et al., 2014; Thabet, 2017b). However, at all times, an irregular problem in geometry for void can be so complicated that an analytical solution is so hard. Due to this reason, researches have sought to find new calculating methods to obtain electric field. FEM is chosen

Figure 19. Cross section of three core cables with Multi-voids

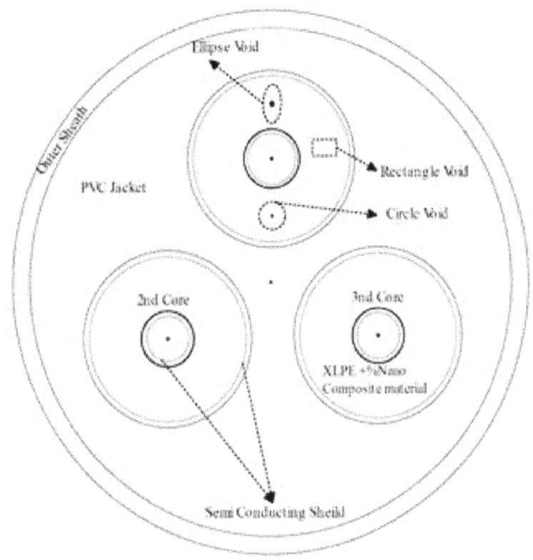

Table 6. Configuration of Three-Core Power Cable (Brosseau et al., 2001)- (Salama et al., 1984)

Specifications	Rating
Rated voltage (V)	33kV
Number core	3
Nominal cross-sectional area	150 mm
Diameter over conductor	14.3 mm
Approximate diameter over insulation	26.5 mm
Approximate overall diameter	83 mm
Insulation dielectric constant (XLPE)	2.4
Water permittivity	81
Air permittivity	1
Copper impurity permittivity	5.6
Radius of bubble	0.5 mm

Table 7. Dielectric constant of Cross-linked polyethylene (XLPE) Nanocomposites and Multi-Nano-composite Materials.

Materials	Dielectric Constant
Pure XLPE	2.4
Clay	2.8
BaTiO3	3.8
MgO	9.7
TiO2	11
Al2O3	9.1
Fe2O3	25
10wt.% Clay/ XLPE	2.44
(10wt.%Clay+15wt.%$BaTiO_3$)/ XLPE	2.616
(10wt.%Clay+15wt.%MgO)/ XLPE	27.42
(10wt.%Clay+15wt.% Fe_2O_3)/ XLPE	45.81
(10wt.%Clay+15wt.% TiO_2)/ XLPE	3.57
10wt.% $BaTiO_3$/ XLPE	2.536
(10wt.% $BaTiO_3$+15wt.% Clay)/ XLPE	2.495
(5wt.% $BaTiO_3$+15wt.% MgO)/ XLPE	27.26
(5wt.% BaTiO3+15wt Fe_2O_3)/ XLPE	45.73
(5wt.% $BaTiO_3$+15wt.% TiO_2)/ XLPE	3.6
10wt.% Al_2O_3/ XLPE	4.38
(10wt.% Al2O3 +15wt.% Clay)/ XLPE	2.921
(10wt.% Al_2O_3 +15wt.% $BaTiO_3$)/ XLPE	3.077
(10wt.% Al_2O_3 +15wt.% MgO)/ XLPE	24.7
(10wt.% Al_2O_3+15wt.% Fe_2O_3)/ XLPE	44.36

Figure 20. FEMM program steps flow chart

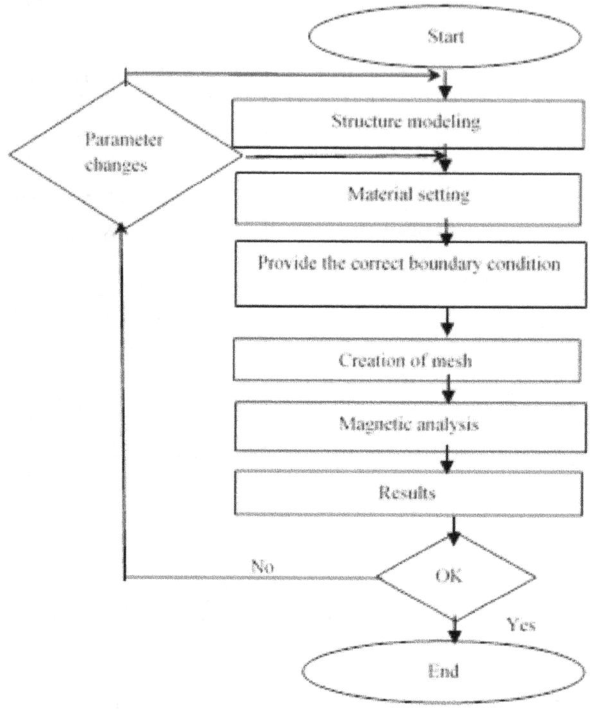

for computation use. Electric field equations solutions by this method are based on Maxwell equations with boundary conditions. In this section, the three-core power cable configuration that is shown in Fig. 19 has been investigated. The model configuration setup is based on the standard dimensions designed by NEXANS Energy Networks Company, Design Standards 6622 - BS 7835(). Table 1 shows the model configuration of three-core power cables. The void diameter is considered 1mm, located between the conductor surface and the insulation inner surface. Finite Element Methods Magnetics (FEMM) software is used for this simulation. Finite Element Method Magnetics (FEMM) software is a finite element package (). This software is used for solving 2-dimensional planar and axisymmetric problems in electrostatics, current flow, heat flow and low frequency magnetics (Patel et al., 2014). The FEMM used method flow chart is shown in Fig. 20 and the dielectric constant of new cross-linked polyethylene nanocomposites and multi-nanocomposite materials is shown in Table 6.

6.3 Trends of Nanoparticles in Partial Discharges

The electric field stress can be controlled by changing the relative permittivity of insulator material through adding individual nanocomposites or multi-nanocomposite materials to insulator material. Furthermore, obtaining new dielectric materials is due to our needs for enhancing electric applications (Thabet & Fouad, 2021). In this study, the behaviour of electric field stress inside insulation material of power cable that have (air – impurity – water) voids and have different shapes (cylinder – sphere – ellipse) by adding nanocomposites and multi nanocomposites are shown. Therefore, the minimum electric stress within power cable insulation and inside any type of voids is achieved.

The following figures illustrate the behaviour of electric stress within variant pure new Cross-linked polyethylene insulation materials and nanocomposites in case of the presence of different air void shapes in three-core belted power cable. Electric stress distribution is drawn around the first core axis passed through air voids. It has been observed that adding concentration of (10wt.%Clay+15wt.% Fe_2O_3) nanoparticle to XLPE decreases the electric field distribution about 74.23% inside air sphere void, decreases the electric field distribution about 66.13% inside air cylinder void, decreases the electric field distribution about 82.39% inside air ellipse void and decreases the electric field distribution inside power cable insulation about 85.37% due to the effect of high dielectric constant of Fe_2O_3 nanoparticles. On the other side, it has been observed that adding the concentration of (10wt.%Clay +15wt.% $BaTiO_3$) nanoparticle to XLPE increases the electric field distribution about 0.26% inside air sphere void, increases the electric field distribution about 2.09% inside air cylinder void, decreases the electric field distribution about 1.66% inside air ellipse void and decreases the electric field distribution inside power cable insulation about 3.06% due to the effect of low dielectric constant of $BaTiO_3$ and Clay nanoparticles. Furthermore, adding the concentration of (10wt.% Clay) nanoparticle to XLPE electric stress inside cable insulator is of higher value than adding the concentration of (10wt.% Clay+15wt.% $BaTiO_3$) nanoparticles to XLPE and other composite electric distribution graded between them, as shown in Fig. (21). It noticed that increasing the concentration of Fe_2O_3 nanoparticle causes more efficient effect for decreasing electric field distribution inside air voids. The different behaviours of electric stress performance in case of adding multiple nanoparticles inside XLPE have been observed.

Figure 21. Electric field distribution comparing pure XLPE with % Clay nanoparticles insulation material containing different shapes of air voids

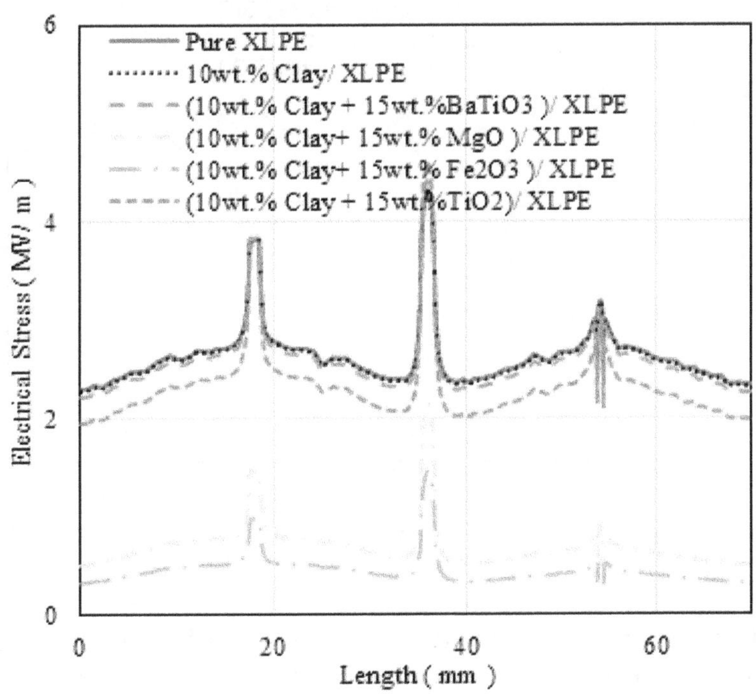

Moreover, it has been observed that adding the concentration of (10wt.%Clay+15wt.%Fe_2O_3) nanoparticle to XLPE decreases the electric field distribution about 56.67% inside water sphere void, decreases the electric field distribution about 60.84% inside water cylinder void, decreases the electric field distribution about 33.45% inside water ellipse void and decreases the electric field distribution inside power cable insulation about 85.41% due to the effect of high dielectric constant of Fe_2O_3 nanoparticle. On the other side, it has been observed that adding the concentration of (10wt.%Clay +15wt.% $BaTiO_3$) nanoparticle to XLPE decreases the electric field distribution about 6.28% inside water sphere void, decreases the electric field distribution about 5.54% inside water cylinder void, decreases the electric field distribution about 5.64% inside water ellipse void and decreases the electric field distribution inside power cable insulation about 3.06% due to the effect of low dielectric constant of $BaTiO_3$ and Clay nanoparticles. Moreover, adding the concentration of (10wt.% Clay) nanoparticle to XLPE, electric stress inside cable insulator is of higher value than adding the concentration of (10wt.% Clay+15wt.% $BaTiO_3$) nanoparticle to XLPE and other composite electric distribution graded between them, as shown in Fig.22. It is noticed that increasing the concentration of Fe_2O_3 nanoparticle causes more efficient effect for decreasing electric field distribution inside water voids. Additionally, the different behaviours of electric stress performance in case of adding multiple nanoparticles inside XLPE have also been observed.

Figure 22. Electric field distribution comparing pure XLPE with % Clay nanoparticles insulation material containing different shapes of water voids

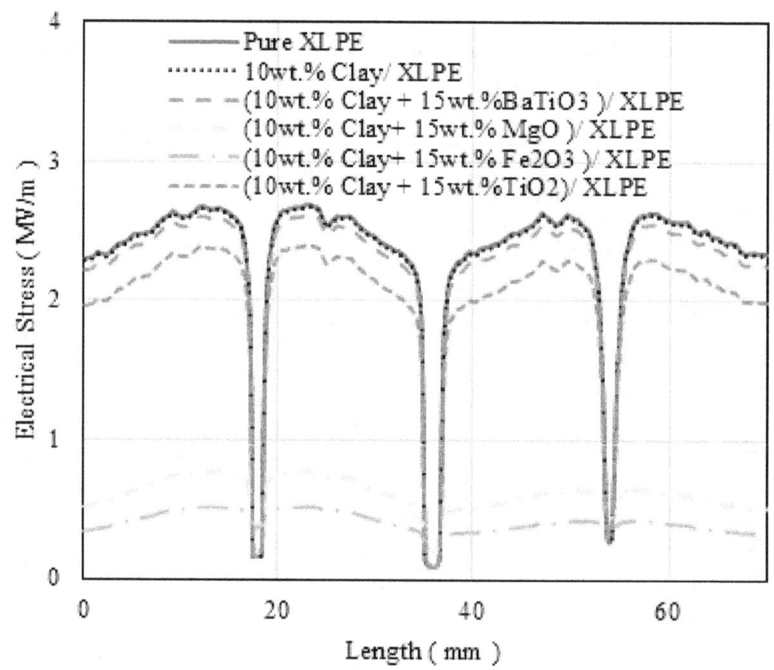

The behaviour of electric stress within the variant of New Cross-linked pure polyethylene insulation materials and nanocomposites in case of the presence of different copper impurity shapes in three core belted power cable has been investigated as follows:

Adding the concentration of (10wt.%$BaTiO_3$+15wt.%Fe_2O_3) nanoparticles to XLPE decreases the electric field distribution about 45.39% inside impurity sphere void, decreases the electric field distribution about 1.78% inside impurity cylinder void, decreases the electric field distribution about 76.30% inside impurity ellipse void and decreases the electric field distribution inside power cable insulation about 85.37% due to the effect of high dielectric constant of Fe_2O_3 nanoparticle. On the other side, it has been observed that adding the concentration of (10wt.% $BaTiO_3$+15wt.%Clay) nanoparticle to XLPE increases the electric field distribution about 1.21% inside impurity sphere void and increases the electric field distribution about 2.67% inside impurity cylinder void, does not cause any effect on the electric field distribution inside impurity ellipse void and decreases the electric field distribution inside power cable insulation about 1.31% due to the effect of low dielectric constant of $BaTiO_3$ and Clay nanoparticles and other composite electric distribution graded between them, as shown in Fig. (23).

Figure 23. Electric field distribution comparing pure XLPE with % $BaTiO_3$ nanoparticles insulation material containing different shapes of copper impurities voids

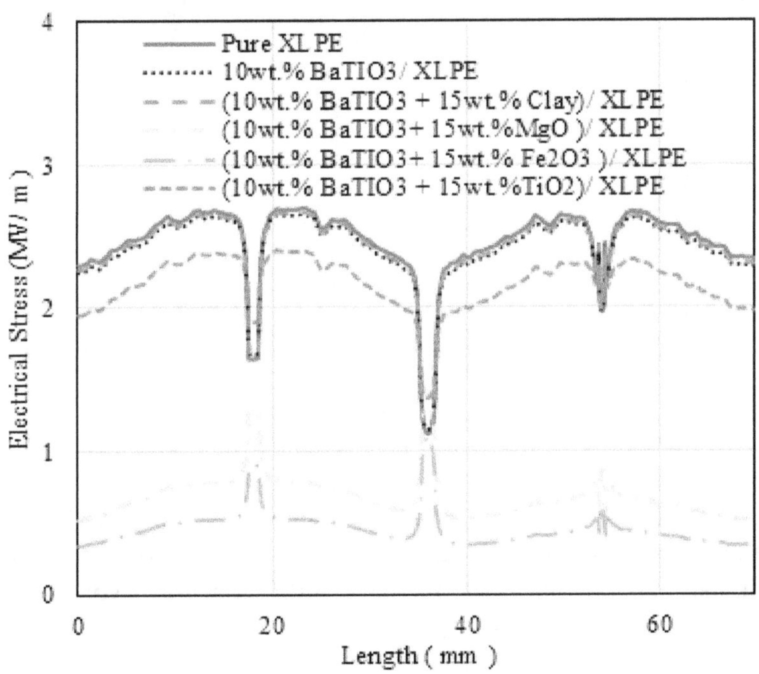

Moreover, it has been observed that adding the concentration of (10wt.%Clay+15wt.%Fe_2O_3) nanoparticle to XLPE decreases the electric field distribution about 45.45% inside impurity sphere void, decreases the electric field distribution about 1.78% inside impurity cylinder void, decreases the electric field distribution about 76.34% inside impurity ellipse void and decreases the electric field distribution inside power cable insulation about 85.37% due to the effect of high dielectric constant of Fe_2O_3 nanoparticle. On the other side, it has been observed that adding the concentration of (10wt.%Clay +15wt.% $BaTiO_3$) nanoparticle to XLPE increases the electric field distribution about 3.63% inside impurity sphere void, increases the electric field distribution about 5.35% inside impurity cylinder void, does not cause any effect on the electric field distribution inside impurity ellipse void and decreases the electric field distribution inside power cable insulation about 3.06% due to the effect of low dielectric constant of $BaTiO_3$ and Clay nanoparticles and other composite electric distribution graded between them, as shown in Fig. 24. It is noticed that increasing the concentration of Fe_2O_3 nanoparticle causes a more efficient effect on decreasing electric field distribution inside impurity void. Hence, different behaviours of electric stress performance in case of adding multiple nanoparticles inside XLPE have been observed.

Figure 24. Electric field distribution comparing pure XLPE with % Clay nanoparticles insulation material containing different shapes of copper impurities voids

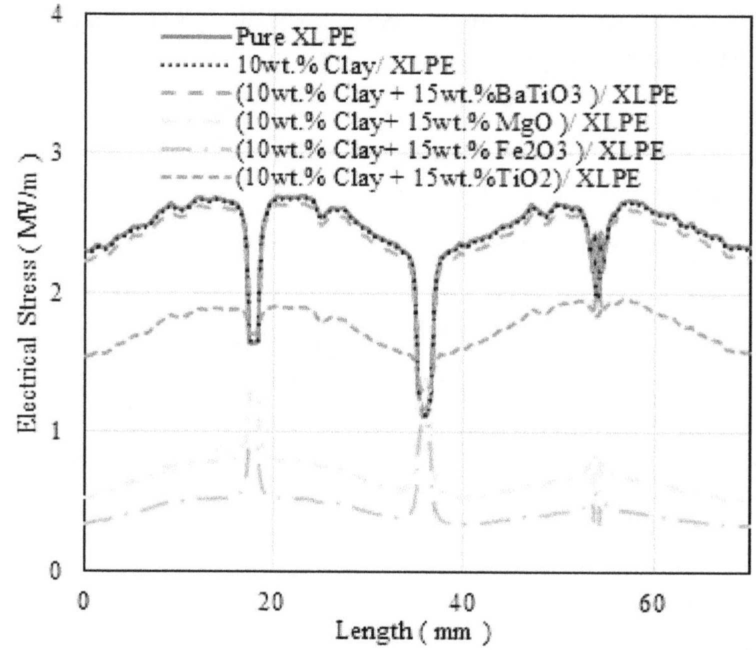

Nowadays, new theories have been proposed for decreasing the electric stress inside voids due to different shapes of air voids through comparing highly effective nanocomposite and multiple nanocomposites for insulation materials. In the case of the present different water void shapes, highly effective

Table 8. Electric Stress in Three Core Power Cables with Air Voids

Modern Nanocomposites Insulation Material	Air Voids Shapes			
	Sphere void	Cylinder void	Ellipse void	Inside insulation
(10wt.% Al_2O_3+15wt.% Fe_2O_3)/ XLPE	¯up to 73.6%	¯up to 65.2%	¯up to 81.89%	¯up to 84.97%
(10wt.%Al_2O_3+15wt.%Clay)/XLPE	—	up to 4.2%	¯up to 4.32%	¯up to 6.98%
(10wt.%Clay+15wt.%Fe_2O_3)/XLPE	¯up to 74.2%	¯up to 66.13%	¯up to 82.39%	¯up to 85.37%
(10wt.% Clay+15wt.% $BaTiO_3$)/XLPE	up to 0.26%	up to 2.09%	¯up to 1.66%	¯up to 3.06%

Table 9. Electric Stress in Three Core Power Cables with Water Voids

Modern Nanocomposites Insulation Material	Water Voids Shapes			
	Sphere void	Cylinder void	Ellipse void	Inside insulation
(10wt.% Al_2O_3+15wt.% Fe_2O_3)/ XLPE	¯up to 33.3%	¯up to 80.07%	¯up to 19.92%	¯up to 85.67%
(10wt.%Al_2O_3+15wt.%Clay)/XLPE	¯up to 19.5%	¯up to 15.42%	¯up to 18.04%	¯up to 24.02%
(10wt.%Clay+15wt.%Fe_2O_3)/XLPE	¯up to 56.7%	¯up to 60.84%	¯up to 33.45%	¯up to 85.41%
(10wt.% Clay+15wt.% $BaTiO_3$)/XLPE	¯up to 6.28%	¯up to 5.54%	¯up to 5.64%	¯up to 3.06%

Table 10. Electric Stress in three core cables with Impurities

Modern Nanocomposites Insulation Material	Copper Impurities Voids Shapes			
	Sphere void	Cylinder void	Ellipse void	Inside insulation
(10wt.% Al_2O_3+15wt.% Fe_2O_3)/ XLPE	¯up to 44.12%	—	¯up to 75.70%	¯up to 84.97%
(10wt.%Al_2O_3+15wt.%Clay)/XLPE	up to 7.87%	up to 10.71%	¯up to 0.45%	¯up to 6.98%
(10wt.%Clay+15wt.%Fe_2O_3)/XLPE	¯up to 45.5%	¯up to 1.78%	¯up to 76.34%	¯up to 85.37%
(10wt.% Clay+15wt.% $BaTiO_3$)/XLPE	up to 3.63%	up to 5.35%	—	¯up to 3.06%

nanocomposite material in decreasing electric stress inside void and low effective nanocomposite material in decreasing electric stress inside void are compared. Moreover, in case of the present different copper impurity shapes, highly effective nanocomposite materials in decreasing electric stress inside void and low effective nanocomposite material in decreasing electric stress inside the void have been compared.

7 FORECAST AND RECOMMENDATIONS

Counting nanoparticles (Clay, $BaTiO_3$, and Al_2O_3) with polymer mechanical materials has improved their electrical properties and their electrostatic field withstand quality. Counting costive nanoparticles for clay has raised electric field quality for nanocomposite materials regarding the different nanoparticles of (seethed silica and Al_2O_3). Thickness from asserting insulation power cables has depended upon electric field dissemination withstand through nanocomposite materials which rely on the sort and pace of included nanoparticles. Solitary nanoparticles like Clay, ZnO, and silica offer help by decreasing the ground-breaking dielectric constant of polymeric insulation. In any case, SiO_2, MgO, TiO_2, and alumina are important for growing dielectric portrayal. The usage of various nanoparticles will be a hint of something larger controlling both dielectric properties and physical properties of Power cable materials. Electric field distribution inside insulation materials from asserting three-core belted Power cables has been improved by individual use and is eventually scrutinizing by using multi-nanoparticles system. It can be controlled in the dielectric properties up or down, as expressed by sorts and pace of nanoparticles which add to the flawless smoothed out insulation materials. Counting individual nanoparticles (Clay and ZnO) of polyvinyl chloride has defiled the electrical insulation properties and extended the electric field spread inside Power cable insulation. In this manner, the dielectric quality is incapacitating that development of particular nanoparticles (Al_2O_3 and Fe_2O_3). It can thus be concluded that they have an ability of overhauling the electrical properties and the electrostatic field distribution for growing withstand quality. It can successfully manage the electrical properties and the electrostatic field scattering inside the insulation of single core Power Cables, eventually scrutinizing by the use of multi-nanoparticles technique. Accordingly, including single individual nanoparticle (Clay and ZnO) cause a lessening in the electrostatic field dispersal inside voids (air, water, pollution) snap for polyvinyl chloride, thus, the development exceptional nanoparticle (Al_2O_3 and Fe_2O_3) raises the electrostatic field appropriation. A Multi-nanoparticles system has a huge capacity to direct electrical properties and the electrostatic field appropriation inside the void snap for insulation about single core Power cables.

Electrical stress inside sphere void has a higher quality, since it might be spotted near to the high voltage conduit, nonetheless. Electrical stress inside bended void has cut down worth, since it is found out of sight. Air void makes electrical stress at a higher quality in light of its low permittivity. However, water void makes electrical stress in a simpler manner as a result of its high permittivity. Counting particular nanoparticles (Clay, and ZnO) of polyvinyl chloride has adulterated the electrical insulation properties and extended the electric field distribution inside Power cable insulation. Subsequently, the dielectric quality might be crippling the development in any case from asserting extraordinary nanoparticles (Al_2O_3, and Fe_2O_3), concluding that they have an ability for overhauling the electrical properties and the electrostatic field movement for growing withstand quality. Without doubt, it could manage the electrical properties and the electrostatic field transport inside the insulation from asserting single core Power cables by using multi-nanoparticles technobabble. In any case, including unmistakable nanoparticle (Clay and ZnO) disturbed a lessening in the electrostatic field circulation inside voids (air, water, debasement) snap on polyvinyl chloride. In any case, the extension of the exceptional nanoparticle (Al_2O_3, and Fe_2O_3) raises the electrostatic field appropriation.

Multi-nanoparticles technique accommodates an extra ability for controlling the electrical properties and the electrostatic field transport inside the void snap for insulation from asserting total center Power cables. In this chapter, new nanocomposites have been created and tentatively tried under high voltages and at that point, they have been analyzed between dielectric strength of new and customary materials. Likewise, the impact of nanoparticles on diminishing the electrical pressure inside three-center power cables insulation, particularly, in the event of quality numerous voids has been researched. Addition of nanoparticles impacts the dielectric constant of insulation material of power cables that can change the self-capacitance of power cables, thus impacting the electric field dissemination. Electrical pressure inside the rectangular void has higher qualities in three core belted power cables than different shapes link due to its mathematical shape. Air void renders higher electrical pressure esteems than water voids, regarding the permittivity property. Using individual nanocomposite materials and multi-nanocomposite materials rather than unadulterated PVC insulation material declines the electric pressure circulation from 3.55% to 89.4%, as indicated by the type and centralization of the nanocomposite material. Using individual nanoparticles (Al_2O_3 and Fe_2O_3) for polyvinyl chloride is discovered for their capacity for diminishing the electrostatic field dispersion inside insulation of three-core belted power cables. In any case, adding the individual nanoparticle (Clay and $BaTiO_3$) expands the electric field appropriation inside power cable insulation. It tends to be more controlling in electrical properties and the electrostatic field circulation inside the insulation of three core belted power cables by utilizing multi-nanoparticles strategy.

REFERENCES

Avinash & Rajagopala. (2013). Some Aspects of Stress Distribution and Effect of Voids Having Different Gases in MV Power Cables. *IOSR Journal of Electrical and Electronics Engineering, 5*(6), 16-22.

Brosseau, C., Queffelec, P., & Talbot, P. (2001). Microwave Characterization of Filled Polymers. *Journal of Applied Physics, 89*(8), 4532–4540. doi:10.1063/1.1343521

Chakradhar Reddy, Ch., & Ramu, T. S. (2008). Polymer Nanocomposites as Insulation for HVDC Cables – Investigations on the Thermal Breakdown. *IEEE Transactions on Dielectrics and Electrical Insulation, 15*(1), 221–227. doi:10.1109/T-DEI.2008.4446754

Chakravorti, S., & Mukherjee, P. K. (1992, February). Efficient field calculation in three-core belted cable by charging simulation using complex charges. *IEEE Transactions on Electrical Insulation, 27*(6), pp1208–pp1212. doi:10.1109/14.204873

Danikas, M. G., & Tanaka, T. (2009). Nanocomposites-a review of electrical treeing and breakdown. IEEE Electrical Insulation Magazine, 25(19). doi:10.1109/MEI.2009.5191413

Ebnalwaled, A. A., & Thabet, A. (2016, July). Controlling the optical constants of PVC nanocomposite films for optoelectronic applications. *Synthetic Metals Journal, 220*, 374–383. doi:10.1016/j.synthmet.2016.07.006

Ebnalwaled, A. A., Yousef, A., Gerges, M. K., & Thabet, A. (2016). Synthesis of Nano-Polyimide for Microelectronic Applications. *Journal of Applied Chemical Science International, 6*(1), 18–30.

Elayyan, H., & Abderrazzaq, M. (2005, December). Electric Field Computation in Wet Cable Insulation Using Finite Element Approach. *IEEE Transactions on Dielectrics and Electrical Insulation, 12*(6), 1125–1133. doi:10.1109/TDEI.2005.1561792

Gouda, O., Mobarak, Y. A., & Samir, M. (2010). A Simulation Model for Calculating the Dielectric properties of Nano-Composite Materials and Comprehensive Interphase Approach. *14th International Middle East Power Systems Conference (MEPCON)*, 151-156.

Hassenzahl, W. V., Eckroad, S., Grant, P. M., Gregory, B., & Nilsson, S. (2009). A High-Power Superconducting DC Cable. *IEEE Transactions on Applied Superconductivity, 19*(3), 1756–1761. doi:10.1109/TASC.2009.2017844

Hefny, A. F. (2003). *Novel technique to improve the calculation of the electrostatic field and the discharge inception voltage inside the three-core belted power cables* (MSc. thesis). High Institute of Energy.

Jing, X., Zhao, W., & Lan, L. (2000). The Effect of Particle Size on Electric Conducting Percolation Threshold in Polymerr Conducting Particle Composites. *Journal of Materials Science Letters, 19*(5), 377–379. doi:10.1023/A:1006774318019

Karkkainen, K. K., Sihvola, A. H., & Nikoskinen, K. I. (2000). Effective Permittivity of Mixtures: Numerical Validation by the FDTD Metho. *IEEE Transactions on Geoscience and Remote Sensing*, *38*(3), 1303–1308. doi:10.1109/36.843023

Lachini, S., Gholami, A., & Mirzaie, M. (2010). *The Simulation of Electric Field Distribution on Cable under the Presence of Moisture and Air Voids*. In The 4th International Power Engineering and Optimization Conf. (PEOCO2010), Shah Alam, Selangor, Malaysia. 10.1109/PEOCO.2010.5559239

Mahdy, A. M. (1988). *Evaluation of human exposure to 50Hz electric field* (PhD thesis). Cairo University.

Malik & Al-Arainy. (1985). Charge simulation modeling of three-core belted cables. *IEEE Transactions on Electrical Insulation*, *20*(3), 499–503.

Mauseth, F., Amundsen, M., Lind, A., & Faremo, H. (2012). Electrical Insulation and Dielectric Phenomena (CEIDP). *Annual Report Conference, Montreal, QC Montreal, 692.*

McCalley, J. D., & Krishnan, V. (2014). A survey of transmission technologies for planning long distance bulk transmission overlay in US. *International Journal of Electrical Power & Energy Systems*, *54*, 559–568. doi:10.1016/j.ijepes.2013.08.008

Nguyen, M. Q., Malec, D., Mary, D., Werynski, P., Gornicka, B., Therese, L., & Guillot, P. (2010). Silica Nanofilled Varnish Designed for Electrical Insulation of Low Voltage Inverter-fed Motors. *IEEE Transactions on Dielectrics and Electrical Insulation*, *17*(5), 1349–1356. doi:10.1109/TDEI.2010.5595535

Nosseir, A. (1979, April). Calculation of Discharge Inception Voltage Due to the Presence of Voids in Power Cables. *IEEE Transactions on Electrical Insulation*, *EI-14*(2), 117–120. doi:10.1109/TEI.1979.298165

Nosseir, A., & Zaky, A. A. (1979, August). Representation of three-core cables using line charge simulation. *IEEE Transactions on Electrical Insulation*, *14*(4), 207–210. doi:10.1109/TEI.1979.298174

Patel, Chaudhari, & Patel. (2014). Analysis of Electric Stress in High Voltage Cables Containing Voids. *International Journal of Engineering Research & Technology, 3*(3).

Pitsa, D. (2010). Electrical Tree Growth Simulation in Nanocomposite Polymers: The role of nanoparticles and homocharges. *International Conference on Solid Dielectrics*, 1-3. 10.1109/ICSD.2010.5568241

Polizos, G. T., Sauers, E., More, I., & Karren, L. (2010). Properties of a Nanodielectric Cryogenic Resin. *Applied Physics Letters*, *96*(15), 152903–152903, 3. doi:10.1063/1.3394011

Salama, Nosseir, Hackam, Soliman, & El-Shieekh. (1984). Methods of calculations of field stresses in a three-core power cable. *IEEE Transactions on PAS, 103*(12).

Tagami, N., Hyuga, M., Ohki, Y., Tanaka, T., Imai, T., Harada, M., & Ochi, M. (2010). Comparison of Dielectric Properties between Epoxy Composites with Nanosized Clay Fillers Modified by Primary Amine and Tertiary Amine. *IEEE Dielectrics and Electrical Insulation Transactions*, *17*(1), 214–220. doi:10.1109/TDEI.2010.5412020

Takala, M., Ranta, H., Nevalainen, P., Pakonen, P., Pelto, J., Karttunen, M., Virtanen, S., Koivu, V., Pettersson, M., Sonerud, B., & Kannus, K. (2010). Dielectric Properties and Partial Discharge Endurance of Polypropylene-Silica Nanocomposite. *IEEE Transactions on Dielectrics and Electrical Insulation*, *17*(4), 1259–1267. doi:10.1109/TDEI.2010.5539698

Thabet, A. (2015a, January). Experimental enhancement for dielectric strength of polyethylene insulation materials using cost-fewer nanoparticles. *International Journal of Electrical Power & Energy Systems*, *64*, 469–475. doi:10.1016/j.ijepes.2014.06.075

Thabet, A. (2015b, March). Experimental Study of Space Charge Characteristics in Thin Films of Polyvinyl Chloride Nanocomposites. *International Journal on Electrical Engineering and Informatics*, *7*(1), 1–11. doi:10.15676/ijeei.2015.7.1.1

Thabet, A. (2015c, June). Experimental Verification for Improving Dielectric Strength of Polymers by Using Clay Nanoparticles. *Advances in Electrical and Electronic Engineering Journal*, *13*(2), 182–190.

Thabet, A. (2016). Thermal experimental verification on effects of nanoparticles for enhancing electric and dielectric performance of polyvinyl chloride. *Journal of the International Measurement Confederation*, *89*, 28–33. doi:10.1016/j.measurement.2016.04.002

Thabet, A. (2017a, June). Theoretical Analysis for effects of nanoparticles on dielectric characterization of electrical industrial materials. *Electrical Engineering Journal*, *99*(2), 487–493.

Thabet, A. (2017b). Experimental Control of Dielectric Loss Behavior of Polyvinyl Chloride Nanocomposites under Thermal Conditions. *IEEE International Middle East Power System Conference "MEPCON"*, 12-17.

Thabet & Ebnalwaled. (2017). Improvement of surface energy properties of PVC nanocomposites for enhancing electrical applications. *Journal of the International Measurement Confederation, 110*, 78–83.

Thabet & Fouad. (2021). Dielectric Strength and Patterns of Partial Discharges in Nanocomposites Insulation of Three-Core Belted Power Cables. *Transactions on Electrical and Electronic Materials Journal, 22*(3).

Thabet, A., & Mobarak, Y. A. (2010). Dielectric Characteristics of New Nano-Composite Industrial Materials. *International Conference on High Voltage Engineering and Application "ICHVE"*, 568 – 571.

Thabet, A., & Mubarak, Y. A. (2015, February). Experimental Dielectric Measurements for Cost-fewer Polyvinyl Chloride Nanocomposites. *Iranian Journal of Electrical and Computer Engineering*, *5*(1), 13–22. doi:10.11591/ijece.v5i1.pp13-22

Thabet, A., & Mubarak, Y. A. (2016, September). Thermal Experiment Analysis for Dielectric Characterization of High-Density Polyethylene Nanocomposites. *Advances in Electrical and Electronic Engineering Journal*, *14*(3), 295–303.

Thabet, A., & Mubarak, Y. A. (2017a, March). Experimental Enhancement for Electric Properties of Polyethylene Nanocomposites under Thermal Conditions. *Advances in Electrical and Electronic Engineering Journal*, *15*(1), 55–62. doi:10.15598/aeee.v15i1.1727

Thabet, A., & Mubarak, Y. A. (2017b, June). The Effect of Cost-Fewer Nanoparticles on the Electrical Properties of Polyvinyl Chloride. *Electrical Engineering Journal, Springer*, *99*(2), 625–631. doi:10.100700202-016-0392-3

Thabet, A., & Salem, N. (2017). Optimizing Dielectric Characteristics of Electrical Materials Using Multi-Nanoparticles Technique. *IEEE International Middle East Power System Conference MEPCON*, 716-721.

Thabet, A., Samir, M., & Mountasser, M. M. (2018). New Design for polypropylene thin film of HV metallized power capacitor using multiple nanoparticles technique. *IEEE International Middle East Power System Conference MEPCON.*

Todd, M., & Shi, F. (2003a). Molecular Basis of the Interphase Dielectric Properties of Microelectronic and Optoelectronic Packaging Materials. *IEEE Transactions on Components and Packaging Technologies, 26*(3), 667–672. doi:10.1109/TCAPT.2003.817862

Todd, M., & Shi, F. (2003b). Characterizing the Interphase Dielectric Constant of Polymer Composite Materials: Effect of Chemical Coupling Agents. *Journal of Applied Physics, 94*(7), 4551–4557. doi:10.1063/1.1604961

Yi, Y., & Sastry, M. (2002). Analytical Approximation of the Two-Dimensional Percolation Threshold for Fields of Overlapping Ellipses. Phys. Rev. E, 66. doi:10.1103/PhysRevE.66.066130

Chapter 4
Nano-Tech Power Capacitors

ABSTRACT

This chapter sheds light on AC and DC power capacitors and the recent impacts of nanoparticles on enhancing capacitive charge in power capacitors. In addition, this chapter displays theories and effective parameters of nano-tech power capacitors. Lab-test measurements have been carried out for variant sorts and concentrations of organic and inorganic nanoparticles during different frequencies (10mHz – 10MHz) and temperatures (20°C-80°C). Moreover, this chapter reviews the ideal sorts and concentrations of nanoparticles for upgrading the capacitive charging on insulation of power capacitors. Thus, the novel nanodielectrics have been designed and fabricated for upgrading the electrical execution of multi-section metalized film capacitors.

1 NANO-TECH VISION

Charging capacity of insulation materials is a fundamental parameter for choice of materials on electrical applications. It has been investigated on the compelling nanoparticles that have been specified for upgrading electric industrial power capacitors applications. This chapter clarifies the upgrade of the execution of routine high voltage metalized film capacitor doped with nanoparticles; the model under analysis reveals how aspects of high voltage metalized film capacitor have been improved contingent upon the concentration of the doped nanoparticles and additionally their electric and dielectric properties. Furthermore, this chapter investigates novel Nano-metric industrial materials for upgrading the electrical execution of multi section metalized film capacitors which satisfy a large variety of ac and electronic applications and upgrading the aspects of metalized film capacitors for particular applications by selecting suitable nanocomposite materials. Including nanoparticles of the traditional materials of multi-segment metalized film capacitors has improved the dielectric constant, dielectric strength, self-healing properties, temperature solidness and volumetric efficiency that permit accomplishing the highest capacitance of every unit volume for this kind of capacitors. Simulation results have been investigated to for considering routine structure materials and new nanocomposite industrial materials about multi-segment metalized film capacitor. The current study has computed equivalent series inductance (ESI), equivalent series resistance (ESR), volumetric energy density, peak discharge current …etc. For multi

DOI: 10.4018/978-1-7998-8536-8.ch004

section, high voltage windings, including restrictions brought about by different new nanocomposite materials are addressed. The target of the available study will be to discover ESL, ESR, volumetric energy density, peak discharge current, and so on. For multiple sections, high voltage windings, including restrictions created toward different new nanocomposite materials, are also examined.

This chapter clarifies the upgrade on the routine high voltage metalized film capacitor doped for the Nanoparticles. The recent propelled model addresses how the aspects of high voltage metalized film capacitor have been improved contingent upon the concentration of the doped nanoparticles and their electric and dielectric properties. Furthermore, this chapter investigates novel nano-metric industrial materials for upgrading the electrical execution of multi section metalized film capacitors which satisfy a large variety of ac applications, electronic applications and upgrading the aspects of metalized film capacitors for particular applications by selecting suitable nanocomposite materials. Including nanoparticles of the traditional materials of multi section metalized film capacitors improves the dielectric constant, dielectric strength, self-healing properties, temperature Strength and volumetric efficiency. Hence, permitting will accomplish the highest capacitance per unit volume for this type of capacitors. Simulation outcomes have been investigated for analyzing traditional structure materials and new nanocomposite streamlined materials of multi section metalized film capacitor. Engineering these days expanded significantly to have all stronger devices compact and fit to withstand more stringent conditions of use; polypropylene slim films are greatly suitable for many significant applications. This chapter displays another plan for polypropylene slim film of high voltage metallized film capacitor. It has been deduced that the multi-nanoparticles technique has the capacity to modify the electrical series resistance ESR, electrical series inductance ESI, energy density and current rise time. Moreover, the breaking points that power the construction of polypropylene slim film of HV metallized power capacitor by utilizing unique nanoparticles and new multi-nanoparticles system have been delineated.

2 TRADITIONAL POWER CAPACITORS

Capacitors are devices for storing electric charges; every Capacitor differs in state and size; however, the essential setup will be two conductors carrying equivalent but inverse charges. Capacitors have many vital applications in hardware. Some cases incorporate storing electric potential energy, delaying voltage transforms coupled with resistors, filtering out unwanted recurrence signals, framing full circuits and making frequency-dependent, in addition to the free voltage dividers at the point of joining with resistors. Throughout the charging process, a charge Q may be moved starting with a particular case conductor of a different one, providing for one conductor a charge Q+, and the other particular case a charge. A possibility contrast may be created, for the positively charged conductor at a higher possibility over the contrarily charged conductor (Baker, 2002; Coursey, 1927; Longland, 1984; Longland et al., 1984; Natarajan, 2019; Shi et al., 1991).

2.1 Capacitor Construction

The parallel plate capacitor is the simplest type of capacitors, as shown in Fig 1. It could be constructed utilizing two metals at a separation parallel between every other, with its capacitance worth for Farads, continuously altered by the surface zone of the conductive plates and the separation of detachment between them. Adjusting any two of these values alters the characteristic of its capacitance and this manifestation on the support of operation of the variable capacitors. Moreover, in view capacitors store the energy of the electrons in a type of electrical charge on the plates. This will be the charge that the capacitor holds

Figure 1. Shows connection of the capacitor

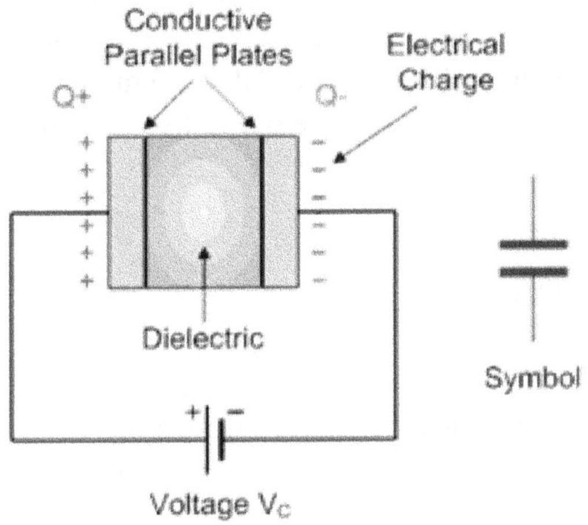

for any given voltage across its plates, in other words, larger plates mean more diminutive distance and more capacitance (Shi et al., 1991).

2.2 Capacitance

The capacitor comprises two directing plates divided by a layer of protecting tape called the dielectric. Hence, the leading surfaces have a chance to be in the manifestation of hardware or rectangular plates. The motivation of the capacitor may be on storing electric energy (Se'vigny et al., 2000). The capacitance of a capacitor is characterized as the measure of charge required for making a unit possibility distinction between the plates. Assume Q coulombs of charge is provided for a standout amongst the two plates of a capacitor and assume that a possibility distinction of v volts is made at the middle of the plates at the point of the capacitance. Therefore, the capacitance is the charge required for every unit possibility Contrast. The unit of capacitance will be the coulomb/volt, which may be known as the farad (Andrei, 1999; Phillips, 1994). Moreover, the traditional power capacitors have aid accessible in fundamental innovations and are utilized over an expansive range for applications. The current economically accessible advances recorded are Ceramic, Aluminum electrolytic, Tantalum electrolytic, Film (polymeric) and Film (mica). In spite of the five innovations indicated above, the rundown is generally about the three fundamental innovations of electrolytic, film, and ceramic capacitors. The qualification at the middle of aluminum and tantalum electrolytic is the vicinity of a fluid impregnate (aluminum) versus dry (tantalum). The refinement at the middle of polymeric and mica or chapter film is engineered versus regular or film-like material (Porter & McCall, 1990).

2.3 Dielectric of a Capacitor

The conductive plates are for the most part settled on a metal foil or alternately a metal film, but the dielectric material is an insulation. The different insulating materials utilized as the dielectric in a capacitor vary regarding their capability on piece alternately pasquinading an electrical charge. This dielectric material is a chance to be made starting with a number of insulating materials of alternate combinations of these materials, with the greater part as relatable point sorts utilized being: air, chapter, polyester, polypropylene, Mylar, ceramic, glass, oil, or an assortment of other materials. The variable by which the dielectric material, alternately insulator, expands the capacitance of the capacitor contrasted with air. The Dielectric constant will be a dimensionless material at the middle of the plates may be at the result of permittivity of nothing space (ε_0) and the relative permittivity (ε_r) of the material constantly utilized concerning illustration of the dielectric (Baker, 2002).

3 AC POWER CAPACITORS

Capacitors utilized within energy framework applications incorporate electrolytic capacitors, metalized-chapter capacitors, dry metalized-film capacitors, ultra-capacitors, and high back capacitors. Few characteristics of these capacitors are discussed underneath (Andrei, 1999; Baker, 2002; Longland et al., 1984; Se'vigny et al., 2000).

3.1 Electrolytic Capacitors

Electrolytic capacitors utilize aluminum foil that approach anodic partner to process an oxide film. One electrode is the aluminum itself, and to a unidirectional application, a plain aluminum foil may be utilized. For exchanging current operation, a second anodized foil may be utilized as the other electrode. The foils are differentiated by paper insulation that is doused over electrolyte on gatherings to give contact to the middle of the oxide layers. The aluminum could be approached by carving to base the surface zone. This may be known as carved foil. In some other capacitors, the aluminum electrode might give a chance to be sprayed on the fabric preceding anodizing to expand the surface range. This will be known as sprayed foil. In this way, the capacitance of every unit is expanded fundamentally. The electrolytic capacitors would be just for the long run applications. For example, the electrolytic capacitors for engine begin applications in the voltage go of 125–350V rams which are suitable for a 30-second operation. The beginning capacitors are utilized within single-phase motors. Electrolytic capacitors are not suitable for nonstop obligation.

3.2 Metalized-Chapter Capacitors

In order to decrease the size of the capacitor, dainty electrodes and hence decreased insulation thickness will be utilized. By utilizing metal deposition techniques, it is conceivable to decrease the thickness of the electrodes. In this sort of capacitor, a particular case layer for insulation will be utilized as opposed to the two layers utilized within customary capacitors. The paper insulation has a chance to be impregnated with the remaining unreasonable voltage stress. These capacitors are likewise utilized for engine begin and run applications and would short the long run rated. They are a great deal heavier than the electrolytic capacitors but have a tendency to furnish more extended administration term to have down losses. These capacitors might be worked ceaselessly in order to provide suitableness for engine begin and run applications. The capacitor might have a chance to be altered on the engine frames.

3.3 Dry Metallized-Film Capacitors

The power factor of the fluorescent and discharge lighting circuits is around 0.4 and necessities are to be remunerated. These capacitor units ought to be low cost and profoundly dependable. Since these capacitors utilized within lighting fittings would be mounted in counterweight assembly, the working temperatures could have a chance to be high. The lighting capacitor comprises a dainty turned polypropylene film, metalized on each particular case, and two films wound together as a barrel-shaped component. The dry metalized-film kind capacitors would be accessible dependent upon 440 V.

3.4 Ultra-Capacitors

An ultra-capacitor may be an electro compound device comprising two porous electrodes on an electrolyte result that saves charges electrostatically. An ordinary view of an ultra-capacitor is indicated in figure 5. The device comprises two porous electrodes, as a rule committed to and dependent upon initiated carbon drenched in an electrolyte result those streams under and around the electrode plates. The electrolytic result is normally potassium hydroxide or sulfurized. This structure successfully makes two capacitors toward every carbon electrode, joined in arrangement. The ultra-capacitors allotment is of the same battery chemistry, but the approach will be to work them in a cell voltage reach that prompts electrostatic stockpiling of charge. The resistance in the middle of the units because of the electrolyte will be a great deal more modest than that of a battery. Therefore, ultra-capacitors can accomplish a great part higher energy density than batteries, by an element about 10 alternately more. Figure 2 shows the different outlines for ultra-capacitor with distinctive electrodes utilized within energy capacity applications of a really short span (Baker, 2002).

Figure 2. an ultra-capacitor (Baker, 2002)

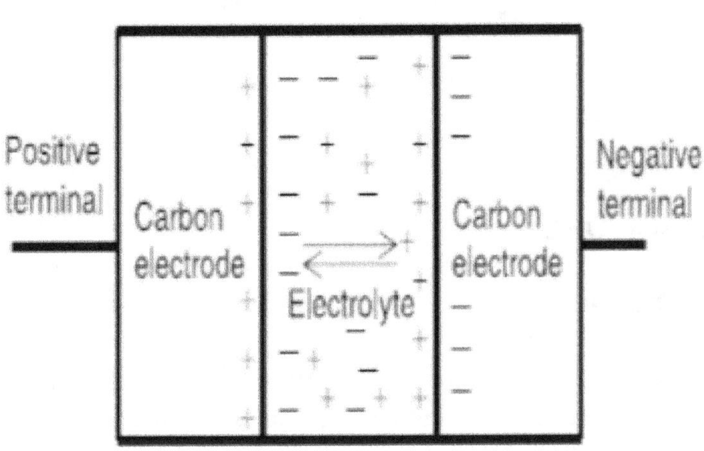

3.5 Capacitors use for Commutation Networks

Particular case requisition for capacitors will be done in thyristor powered chopper and inverter circuits for substitution. When the thyristor starts with conduct, it proceeds to direct until transformed off (Commutated) by interference of the supply or by the application on a contradicting current supplied by a commutating capacitor. The capacitors accessible to this end goal are the individuals with impregnated chapter dielectric, metalized chapter, and plastic film dielectric/mixed dielectric. The operating voltages might make 200–1,200 v dc. These capacitors have aid liable to charging and discharging in supply recurrence alternately in higher levels. Therefore, fitting forethought ought to be taken for picking the capacitors for the application (Se'vigny et al., 2000).

3.6 High Frequency Capacitors

In high back applications, the measure of high temperature produced in a capacitor may be more amazing and obliges exceptional considerations. The high back capacitors are utilized within incitement heating, pulse, commutation, show transmission, drives, sidestep equipment, recurrence converters, filters, high voltage power supplies, shock absorber, couplers, voltage dividers, flash generators, and symphonious filters. The voltage reach might make 1–300 kV and capacitance measure might have a chance to be 100 pF–5000 mf. Commonplace high back capacitors are indicated in Fig. (3). These capacitors are oil filled. The oil serves like Insulation and acts similarly as a cooling agenize. A sample rating of a high back capacitor will be 800 kVAR, 800 V, 10,000 Hz, 1000 A, and 19. 19 MFD (Andrei, 1999).

Figure 3. High frequency capacitors with several taps (Andrei, 1999)

4 DC POWER CAPACITORS

4.1 Electrolytic Capacitor

Electrolytic capacitors are by and large utilized when thick, as vast capacitance qualities would have obliged, as shown in Fig. (4). Here, instead of utilizing a dainty metallic film layer to a standout amongst the electrodes, a semi-liquid electrolyte result in the manifestation of a jam or pasta sauce will be utilized, which serves as the second electrode (usually the electrode). The dielectric is a slim layer of oxide which is developed electro-chemically in handling the thickness of the film being under ten microns (Bhargava & Haas, 2002; Butler & Concordia, 1937; Hammond, 1988; Kimbark, 1966; Phillips, 1994; Porter & McCall, 1990).

Figure 4. Electrolytic capacitor (Bhargava & Haas, 2002; Butler & Concordia, 1937; Hammond, 1988; Kimbark, 1966; Phillips, 1994; Porter & McCall, 1990)

4.2 Polyester Film

Polypropylene capacitors of similar determinations might utilize plates of foil, metalized film, or a mix. Pet film capacitors are just about totally swapped chapter capacitors for the vast majority of dc electronic applications. Working voltages are up to 60kV dc and operating temperatures are dependent upon 125 c's. Low dampness absorption is utilized on the dc applications, such as filtering, bypassing, coupling, spherical segment concealment (Bhargava & Haas, 2002; Butler & Concordia, 1937; Hammond, 1988; Kimbark, 1966; Maini, 1998; Porter & McCall, 1990).

5 POWER CAPACITORS APPLICATIONS

5.1 Timing (Charging---Discharging)

When a capacitor is associated with the terminals of a battery, each plate of the capacitor gets charged. The plate associated with the sure terminal loses electrons. If this plate has an absence of electrons, it accepts a sure charge. The plate is associated with the negative terminal additions electrons. On the plate, it has an overabundance of electrons, so it accepts a negative charge. This transform proceeds until the charge over the plates equals the connected voltage. Toward this perspective, current ceases should stream in the circlet. Concerning illustration in length similar to any progressions in the circuit, the capacitor will hold its charge and there will a chance to for only the circlet. Whether the heads of the capacitor are right away shorted together, the current once more streams in the circlet. The current will proceed to stream until the charges on the two plates are turned equivalent. At this point, the current ceases to stream. For a DC voltage source, the current will stream in the main long circlet sufficiently on charge (or discharge) of the capacitor. Thus, a capacitor doesn't permit dc current with stream ceaselessly on and out (Christie, 1938; Ramo et al., 1994). When utilized within an immediate current or DC circuit, a capacitor squares the stream of the current in light that the dielectric of a capacitor may be non-conductive. However, when a capacitor is associated with an exchanging current or AC circuit, the stream of the current gives the idea of pasquinade straight through it with little or alternately no resistance.

Dc voltage will be connected to capacitors' conductive plates, a current may be unabated to stream through the capacitor itself because of the dielectric insulation, and an electrical charge assembles up on the capacitors plates with electrons generating a certain charge ahead particular case and an equivalent and inverse negative charge on the other plate. The immediate current i and the charge q of the plates will once have more exponential capacities of the run through. The time constant of a resistance – capacitance as shown in Fig. (5) and the characteristic are defined according to the equation (Masuda et al., 1986; Ue et al., 1996):

$$T = RC \tag{1}$$

T = time constant (in second s)
R = resistance of the resistor (in ohms)
C = capacitance of the capacitor (in farads)

Figure 5. Charging-discharging capacitor in DC circuit

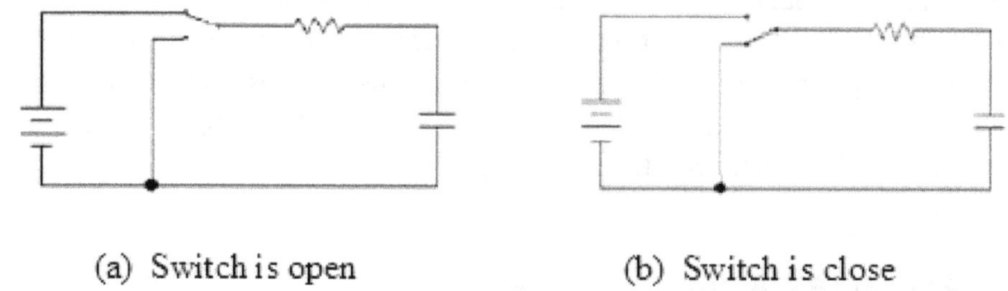

When the switch is open, the capacitor has a charge q_0. When the switch is closed, charges leave the capacitor and flow through the resistor.

5.2 HVDC Design of DC Voltage and DC Capacitors (Filter)

The plan about dc side capacitor is a critical issue for the outline of an HVDC framework. Because of PWM exchanging movement in the VSC-HVDC, the current streaming of the dc side of a converter holds sounds which will have about a ripple on the dc side voltage. The size of the ripple relies on the dc side capacitor size and the exchanging recurrence. Here, a little DC capacitor CDC could be used, which ought to have simulation about a quicker converter reaction. It can likewise give an energy capacity in place to power the power stream. The DC capacitor size is described concerning the illustration of a time constant, characterized similarly as the pro-fraction between the saved energy in the rated dc voltage UDCN and the ostensible evident power of the converter SN: The time constant is equivalent to the long haul necessary to charge the capacitor from zero to rated voltage UDCN, assuming that the converter will be supplied with a constant dynamic energy equivalent to nibble the time constant - can be chosen to fulfill the little ripple and little transient through voltages on the dc voltage. This will be checked in the simulations. The generally little time constant permits quick power about animated and sensitive power (Masuda et al., n.d.).

5.3 Capacitor Coupling (CR-Coupling)

The reason for coupling circuits in sound units will be to separate the unwanted dc voltage from the off-service AC indicator which we have to let fly through to the following venture for indicator transforming alternately of the yield device. A coupling out might be envisioned as a straightforward C-R differentiator with ostensible capacitance c's about coupling capacitor and information resistance r so that following a unit like amplifier, a sign processor alternately yields device in. Headphones are alternately amplifiers. Designers picking the straight worth of the capacitor for the coupling circlet think about the information impedance of the taking after circlet versus low pasquinade recurrence, which will be administered toward the human ear capability to listen to bass tones, and it's as a rule that it is in the range of 20 to 50 Hz.

In coupling for circuits for high information impedance similar to operational amplifiers inputs, low esteem for coupling capacitor can make additions for a flawless bass tone arm exchange (Watanabe et al., 1999). Smoothing is performed by an extensive strength electrolytic capacitor associated with the dc supply to go about as a reservoir, supplying current of the yield to the point at which the changing dc voltage starting with the rectifier may be falling. The outline indicates the unsmoothed changing dc (dotted line) and the smoothen dc (solid line), as shown in Fig. (6). The capacitor charges rapidly close to the top of the fluctuating DC and discharges as it supplies current of the yield. Note that smoothing fundamentally bases the normal dc voltage to nearly the top esteem smooth birch dc. Smoothing is not flawless because of the capacitor voltage tumbling a minimal concerning illustration it discharges, providing for a little ripple voltage as shown in Fig. (7). To a significant number of circuits, a ripple which may be 10% of the supply voltage will be acceptable and the comparison beneath provides for the obliged strength of the smoothing capacitor. A larger capacitor will provide for less ripple. The capacitor esteem must be multiplied when smoothing half-wave dc (Watanabe et al., 1999).

Figure 6. Charging-discharging capacitor

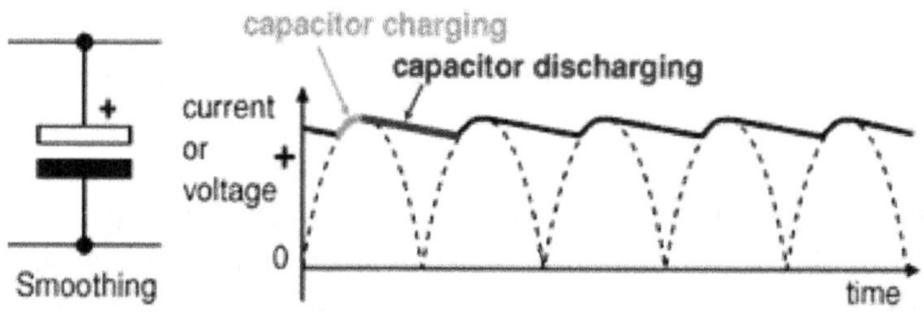

Figure 7. the regulated DC output is very smooth with no ripple. It is suitable for all electronic circuits (Watanabe et al., 1999)

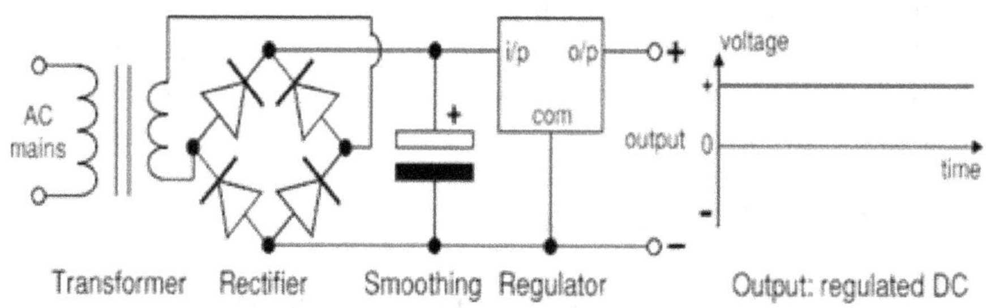

6 CAPACITIVE CHARGE IN NANODIELECTRICS

Over recent years, polymer nanocomposites have received great attention with respect to upgrading polymer properties and extending their utility. Therefore, the abilities to moving forward dielectric properties have been reported. However, there is a close relationship for the interfacial conduct technique between the nanoparticles and the polymer matrix in such nanocomposites. The physical properties of the new nanocomposite materials made of nanoparticles are settled inside a polymeric dielectric matrix. The addition of nanoparticles under polymers structures new engineering materials to moving forward different characterizations for streamlined items. (Bois et al., 2009; Chatterjee et al., 2013; Kamalapur & Udaykumar, 2011; Sarathi & Umamaheswari, 2013; Thabet, 2013a; Thabet, 2013b; Thabet & Mubarak, 2011; Thabet et al., 2010). Furthermore, the nanotechnology science is concerned with upgrading electrical applications. Therefore, the primary target for pertinent nanotechnology science may be to orchestrate dielectric nanocomposites and improve the physical properties of electrical applications. Extraordinary desires have concentrated on the impact of nanoparticles sorts on the electrical properties of polymeric nanocomposites. The electric and optical properties for these materials have been exhibited to be exceedingly reliant on the size, structure and concentration of the nanoparticles, and additionally on the kind of polymeric matrix (Aman et al., 2013; Ebnalwaled & Thabet, 2016; Gouda et al., 2014; Thabet, 2015a; Thabet, 2015b; Thabet, 2015c; Thabet, 2016a; Thabet, 2016b; Thabet & Mubarak, 2015; Thabet & Mubarak, 2016).

The pattern of nanotechnology science should create dielectric materials for upgrading their possibility applications on future energy storage/transfer devices that have association with the interfacial conduct technique between the nanoparticles and the polymer matrix of such nanocomposites. Thus, the reason for these investigations is measuring the capacitive charge of specific new nanodielectrics in view of polyvinyl chloride in variant temperatures in place to augment the minimal information accessible for their electrical properties. Polyvinyl chloride will be picked similarly as a base for polymer matrix in structuring new nanocomposites due to establishing its industrial essentialness on energy link insulations, high voltage applications, engineered thermoplastic polymers and so on. The chosen nanoparticles like clay, fumed silica (SiO_2), zinc oxide (ZnO), and titanium dioxide (TiO_2) have been contemplated in terms of their impacts on electric and dielectric properties of polyvinyl chloride. Therefore, the impacts on the sorts and concentrations of nanoparticles (organic, and inorganic) will focus on variant frequencies and temperatures with respect to capacitive charge for new polyvinyl chloride nanocomposites; the recurrence and temperature ranges utilized have been aggravated as expansive as practicable.

6.1 Characterization of Nanoparticles in Dielectrics

All selected nanoparticles, in this chapter, have been in the round shape and measured as 10nm diameter to each grain size, they have impetuses to become the best filler "around base matrix industrial materials". Additionally, polyvinyl chloride is the base matrix polymer for the vast majority, generally utilized for any of the thermoplastics. Polymerized vinyl chloride is stronger and is characterized by

Figure 8. SEM images for polyvinyl chloride nanocomposites

(a) Clay/ PVC nanocomposite

(b) SiO_2 / PVC nanocomposite

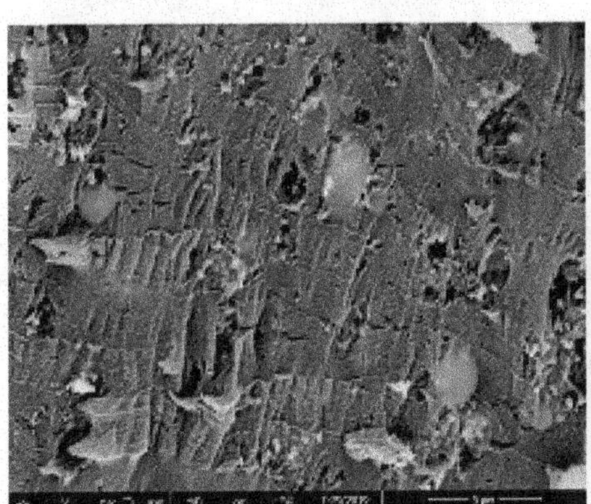

(c) ZnO /PVC nanocomposite

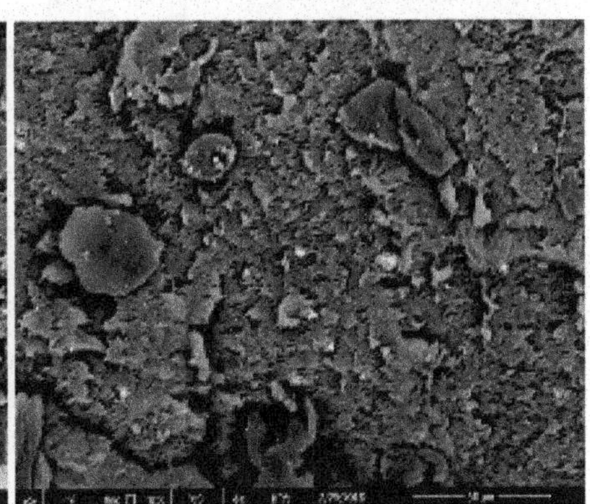

(d) TiO_2/ PVC nanocomposite

additional unbending over all different motivation thermoplastic materials. It has a high elasticity and modulus of flexibility. Preparation of the mulled over polyvinyl chloride nanocomposites has utilized SOL-GEL strategy towards additives of nanoparticles of the base industrial polymers that have been created by utilizing mixing, ultrasonic, and thermal methods (Bois et al., 2009; Ma et al., 2008; Xiu et al., 2007). Figure (8) indicates Scanning Electron Microscope (SEM) pictures that reveal infiltration of nanoparticles scattered uniformly in the polymer matrix of polyvinyl chloride and depict the flakes in morphological tenet with high surface range. Dielectric spectroscopy measures the dielectric properties of a medium as a capacity for recurrence by system of measuring the impedance of a framework in a reach for frequencies. It is based on the interaction of an external field with the electric dipole moment of the sample and it has been regularly communicated toward permittivity.

This chapter investigates the impacts of nanoparticles with respect to permittivities of polyvinyl chloride nanocomposites. It is measured as a capacity of recurrence in the go for 0. 01 Hz with 1MHz at variant temperatures (20°C - 80°C). SOLARTRON dielectric reaction is a method dependent upon the estimation of the dielectric properties concerning the illustration of a capacity of recurrence of a Lab-test sandwiched between two electrodes. The dielectric constant might have been measured concerning the illustration of a work of recurrence in the extent of 0. 01 Hz with 1MHz at variant temperatures for every last one of the Lab-test specimens. It is paramount to master the characterization of nanocomposite materials under variant thermal conditions.

Determination of the measured SOLARTRON device may be energy supply, switch selectable: 90 on 126V, 198252V ac, 48 to 65Hz. Utilization 200VA, Earth temperature operating: 0 to 50 °C, Storage: -30 with 70 °C, Determination limits: 10 to 30 °C, Humidity, non-condensing: 95% in 40 °C, Vibration: tried for understanding of IEC68 (BS2011), Resistance: intended to go along for IEC348 (BS4743). Dielectric spectroscopy is a capable Lab-test strategy which explores the dynamical conduct technique of an example through the dissection of its recurrence reliant dielectric response. Thus, two parallel plates have been utilized as a layer nanocomposite example for dispersing uniform electric fields distribution.

Figure 9. Measured capacitive charge of pure polyvinyl chloride

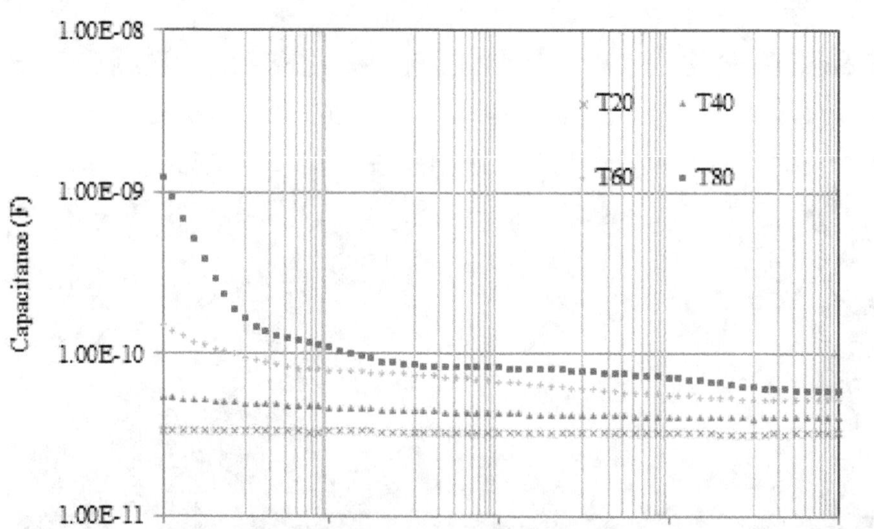

This method is dependent upon the estimation of the capacitance concerning illustration work of recurrence of an example sandwiched between two electrodes. Capacitance (C) might have been measured as a capacity of recurrence in the extent of 10^{-2}Hz with 10^{6}Hz during variant temperatures for every last one of Lab-tests. Lab-test capacitive charge measurements have been carried out for polyvinyl chloride nanocomposite materials under variant thermal ecological conditions. Figure (9) indicates the measured capacitive charge of pure polyvinyl chloride specimens as a capacity of recurrence toward variant temperatures (20°C - 80°C). The measured capacitive charge contrasts upon increasing the capacitive charge for increasing natural temperatures (20°C - 80°C). In any case, the capacitive charge may be decreasing for increasing power frequencies, especially during high frequencies.

Recently, it has been detected that the capacitive charge decreases in 1wt. % clay/PVC nanocomposites with respect to pure polyvinyl chloride specimens. Thus, the measured capacitive charge expands for increasing natural temperatures (20°C - 80°C). There are semi-matching outcomes during variant temperatures (20°C - 40°C) and there may be a decreasing in the capacitive charge for increasing energy frequencies, especially during high frequencies. Nevertheless, Fig. (10) contrasts the measured capacitance which expand with the increasing rate of clay nanoparticles in the nanocomposites, compared with 1wt. % clay/PVC nanocomposites Lab-tests. Consequently, the capacitive charge outcomes are about 5 wt. % clay/ PVC to shut toward variant temperatures (20°C - 40°C) with respect to 1wt. % of clay/ PVC nanocomposites as a capacity for recurrence. Moreover, decreasing the capacitive charge on SiO_2/ PVC nanocomposites is detected inside the centralization extent (1 wt. % - 5 wt. %) for fumed silica nanoparticles with respect to pure polyvinyl chloride specimens. Hence, the measured capacitive charge expands for increasing ecological temperatures (20°C - 80°C). In addition, the Lab-test outcomes indicate a divergence towards the the middle of variant temperatures (20°C - 40°C), uncommonly in high energy frequencies as shown in Fig. (11). On the other hand, Fig. (12) indicates the measured capacitive charge for 5wt. % ZnO/ PVC contrasts upon decreasing the capacitive charge with respect to pure polyvinyl chloride specimens. In any case, the measured capacitive charge increments for increasing natural temperatures (20°C - 80°C). Moreover, Lab-test Outcomes indicate a divergence between outcomes during

Figure 10. Measured capacitive charge of 5wt. % clay/PVC nanocomposites

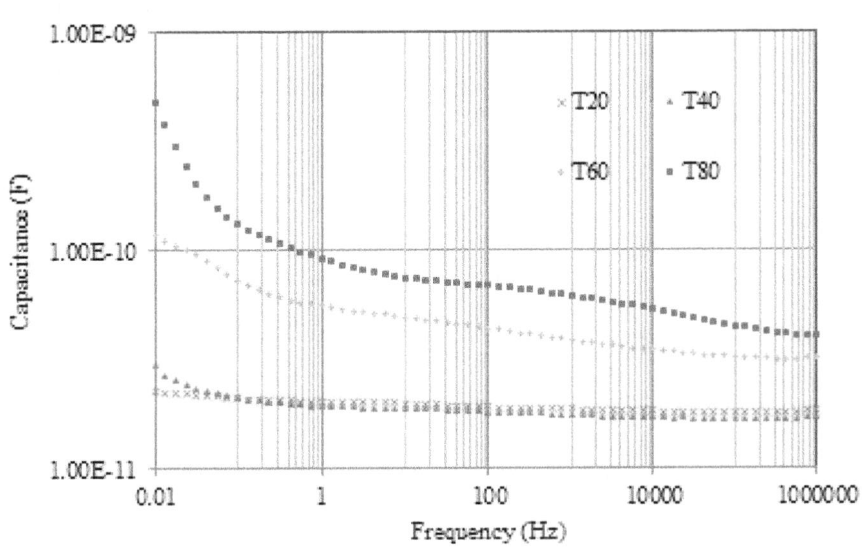

Figure 11. Measured capacitive charge of 5wt. % SiO_2/PVC nanocomposites

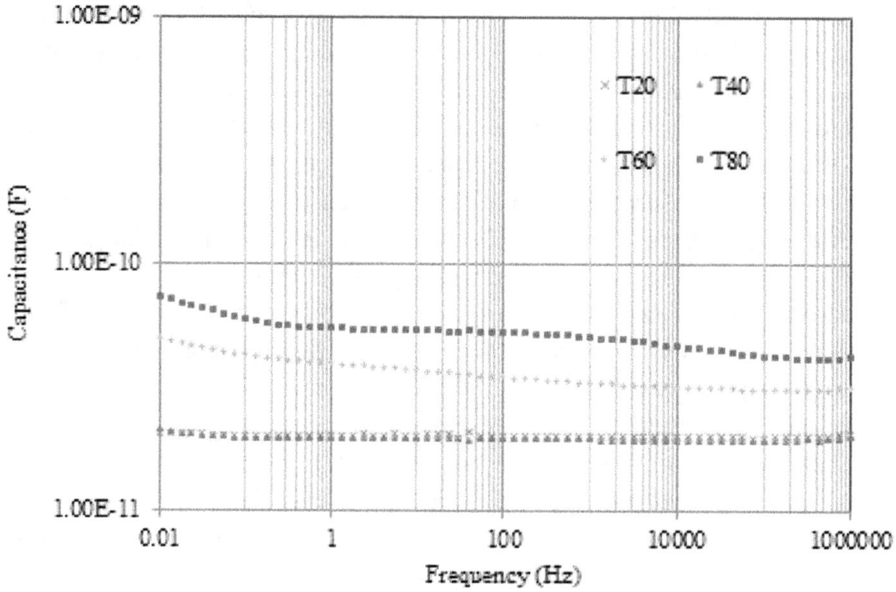

Figure 12. Measured capacitive charge of 5wt. % ZnO/PVC nanocomposites

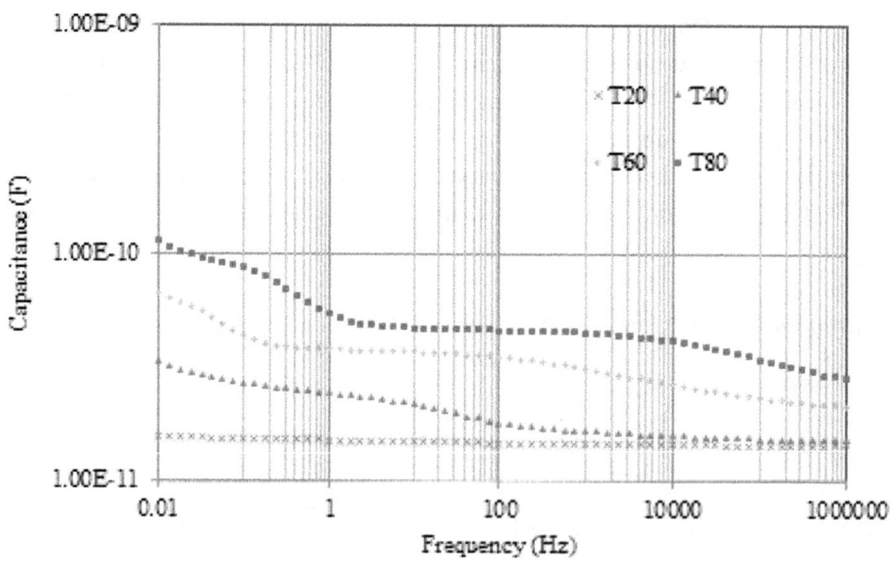

Figure 13. Measured capacitive charge of 5wt. % TiO_2/PVC nanocomposites

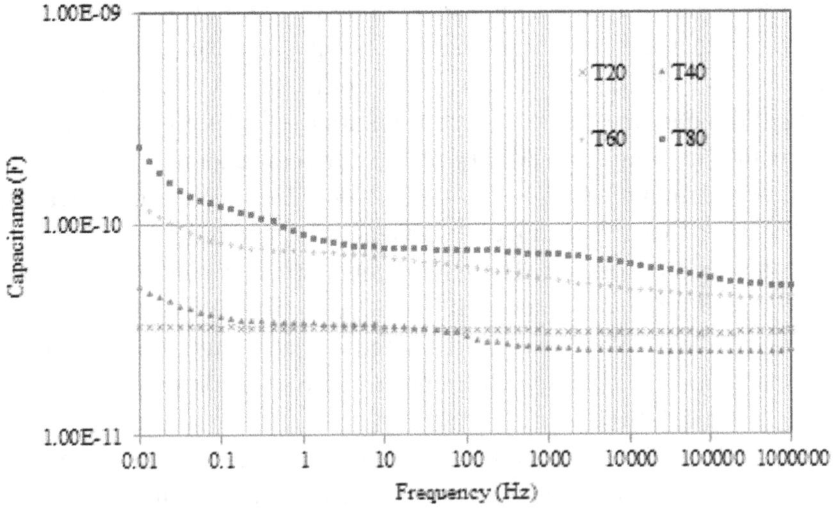

variant temperatures (20°C - 40°C), extraordinarily in high power frequencies. Fig. (13) contrasts the measured capacitance of 5wt. % TiO_2/PVC nanocomposites that base for increasing concentration about titanium dioxide nanoparticles in polyvinyl chloride nanocomposites compared with 1wt. % TiO_2/PVC nanocomposite specimens.

6.2 Trends of Organic and Inorganic Nanoparticles on PVC Dielectrics

Impact of clay nanoparticles has aid pointed out in the previous figures that self-evident the measured capacitive charge is decreasing the capacitive charge toward increasing clay nanoparticles up to 1wt. % with respect to pure polyvinyl chloride specimens; but the measured capacitive charge increments with increasing concentration of clay nanoparticles dependent upon 5 wt. % in clay/PVC nanocomposites. There is a decreasing in the capacitive charge for SiO_2/PVC nanocomposites with fluctuating fumed silica concentration up to 5wt. %. The divergence has been occurred them the middle of outcomes in variant temperatures (20°C - 40°C), extraordinarily in high power frequencies. Zinc oxide nanoparticles have a solid impact inside polyvinyl chloride. This means that the capacitive charge declines with bases zinc oxide nanoparticles dependent upon 5wt. %. Titanium dioxide may be the best nanoparticles during concentration (5 wt. %) to decreasing capacitive charge in polyvinyl chloride nanocomposites during variant temperatures (20°C - 80°C).

7 HIGH VOLTAGE METALLIZED FILM CAPACITORS

Nanoparticles enhance industrial, electrical, and thermal properties of composites at moderately low concentrations. The critical properties of composites as a rule will be their dielectric properties which have been contemplated extensively. More as of late, however, research has a tendency to recommend that the dielectric properties of the interphase aid as inhomogeneous, changing with respect to the spiral separation from the focal point of the spherical consideration (Dong et al., 2003; Milton, 2002; Yu & Gu,

2005). Such an inhomogeneous move will be because of the holding instruments happening in the space between the Incorporation and matrix. A regulated plan of the spatial fluctuating property additionally empowers persons to power in the general property of composites (Yu & Gu, 2005). Outcomes to the Dielectric constant have been distributed for two-phase composites for which impeccable holding may be expected existing in the middle between the consideration and the matrix. The outcomes start with Maxwell-Garnett hypothesis (Dong et al., 2003) with the same consequence determined later by utilizing the composite spheres collection model (Wei & Tang, 2004). This same aftereffect can additionally be utilized for the electrical and thermal conductivity, magnetic permeability and diffusivity, provided that round inclusions and matrix are isotropic. There has been a primary standards methodology formed to figure the viable dielectric reaction of composites for a weaken suspension of graded spherical particles. Vo and Shi measured the dielectric properties of composites as a function of inclusion concentration utilizing a suggested simulation model dependent upon successful medium principle (Todd & Shi, 2002; Vo & Shi, 2002; Wei & Tang, 2004).

An expansive high voltage film capacitor commonly comprises various "windings" associated in arrangement parallelly to attain the fancied voltage and capacitance rating. The discharge properties of such capacitor have been powered by the equivalent series resistance (ESR) and equivalent series inductance (ESI). The inductance, ESR, and meanings thereof of pulsed power applications on solitary area metalized film capacitor windings were talked about in a later production (Ho et al., 2008b). High voltage metalized film capacitor windings typically utilize various segments in arrangement dependent upon the utilization of gliding electrodes, which introduce a significant number of issues past the individual issues in the self-evident advantage of a various segment winding for the help of high voltages in a compact structure. The disadvantages are several including, firstly, the point when clearing happens to one section, the voltage of that area is exchanged to different segments. For a little number of sections, this can cause clearing over an additional area. In the extremely worst case, constant clearing might wreck the capacitor and the possibility between areas which happens along the surface of the film and the middle of wound layers. The breakdown strength of the film may be laminar system; the breakdown strength along the layer surface may be feeble and relies on the interfacial pressure which targets the weight between layers after "curing" of the winding, i.e., thermal unwinding of industrial anxieties. In general, the interfacial weight expands for the decreasing span inside the winding. This contemplated metalized capacitor for nanocomposite films was recently currented and completely portrayed. Nanocomposite film enhances charge thickness and execution of the capacitor. As of late, an investigation of the association from equivalent series resistance (ESR) and equivalent series inductance (ESI) to winding outline in view of having current from person conclusion of the winding go of the other, along the outside span alternately through the focal point of the winding (Qin & Boggs, 2009). The previous for the most part brought about easier inductance over the last. As might be expected, ESR might have been lessened by utilizing narrower film to process all the "pancake" such as winding, which likewise had a tendency to lessen the ESI. In this way, the connections around film metallization resistivity, film dielectric constant, winding design, ESR, and ESI are demonstrated. Current rise time recommended less than a couple hundred is doubtful in view of metalized film windings, in spite of the fact that high peak current can be figured out although the limit associations could support them (Qin & Boggs, 2003). The end associations will be an intricate subject past the extension of the exhibit exchange (Michalczyk & Bramoulle, 2003; Slenes et al., 2001; Zhou et al., 2007).

Upgrading the scattering of nano-sized ceramic particles occurs in order to accomplish higher dielectric constant and thereby higher capacitance thickness in nanocomposites. It has been observed that high solids stacking prompts entanglement about porosity in the microstructure, which lowers the successful dielectric constant of the films. Furthermore, it is clear that the transaction at the middle of the rheological properties of the nanoparticles and polymer matrix has compelling elements on the dielectric properties of the film capacitors. Hence, the utilization about various segments inside a metalized film capacitor facilitates high voltage and diminished pro-fractional arrangement resistance inside a compact winding. Likewise, an aftereffect for high voltage and low inductance, vast crest ebbs and flows can, in principle, be allowed. The constraint of the peak current and energy density is the ability of the end connection to withstand the peak current. The energy density for multi-section windings may be also restricted by the edges which are required between sections, the downright width may be set exclusively toward the rated voltage and edge working field. Nanocomposites of organically changed nanoparticles in an epoxy matrix have been synthesized and assessed similarly as dielectrics for the fabrication of integral film capacitor arrays. Natural adjustment of the polymer inorganic interface has been utilized concerning illustration of a configuration device to power the cross-connection thickness of the polymeric matrix and the interfacial cooperation (Kim et al., 2004; Lee et al., 2007; Qi & Boggs, 2004; Qin & Boggs, 2010; Ramesh et al., 2003; Windlass et al., 2003). In this work, another dielectric nanocomposite which comprises of a polymer and different sorts of nanoparticles has been formed for application in slim film metalized capacitors. The nanocomposite polymer is custom figured to high dielectric constant and strength for the least dielectric losses. The nanocomposite could be wound and transformed utilizing customary wound film capacitor manufacturing equipment. Moreover, the current study presents tables on the impact of including nanoparticles to high voltage capacitor for upgrading ESR and ESI, i.e., the study is directed to ESR and ESI whose outcomes start with interfacing various windings in arrangement and parallelly as necessary to meet what is prescribed for voltage and capacitance rating.

7.1 Theory and Effective Parameters

In order to minimize inductance, the associations must be made during a particular case end of the winding, so that the magnetic field outside the winding could be minimized. To quit offering that one alternative for accomplishing this, two coaxial windings divided toward a coaxial insulin have to be utilized, so that the current dives down one winding, transfers the other winding, and comes back of the same wind of the capacitor, as indicated in Fig. (14. a). This lessens the obliged width of the winding by an element of two and lessens inductance marginally (Qin & Boggs, 2010). In turn, choice is one; extremely and totally winding with the return around a coaxial conductor, Fig. (14.b). To a single winding as demonstrated in Fig. (14), the downright film width is:

$$W = N\,(Ws + Wm) \tag{2}$$

Where N is the number of sections, Ws is the electrode overlap width for each section, and Wm is the margin width between sections as illustrated in Fig. (14.b) for a 4-section winding. In the case of two coaxial windings, the width of each winding is half of W.

Figure 14. Construction of high voltage metalized film capacitor (Qin & Boggs, 2010)

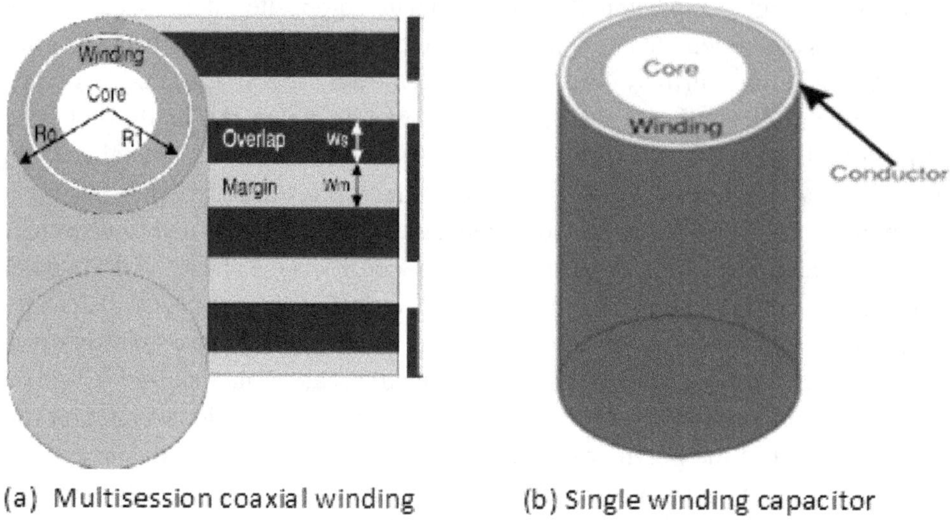

As section electrode overlap increases, the winding length decreases, increasing the ESR in turn. The contribution to the ESR caused by the overlap regions is given by (Dong et al., 2003),

$$ESRo = \frac{4N.\varepsilon.\rho o.\varepsilon o.Ws.}{3.Df.Co} \tag{3}$$

And the resistance caused by the resistivity of the margin regions between winding sections is:

$$ESRm = \frac{N.\rho m.Wm.}{Ls} \tag{4}$$

Where the unit of ρm and ρo is Ω or Ω/sq. The total ESR is the sum of these two contributions.

$$ESRm = ESRm + ESRo \tag{5}$$

Assuming the current through a capacitor is I, the inductance can be obtained from integrating the energy in the magnetic field. In this work, Ampere's law was applied to approximate the magnetic field from the current and to derive the magnetic energy density.

$$L = \frac{1}{I^2.\mu o}\int B^2\, dV = \frac{2Wm}{I^2} \tag{6}$$

The inductance of the winding with similarity to the coaxial winding structure is given by the outer return structure and can be obtained with reasonable accuracy as follows:

$$L2 = \int_{Ri}^{R1} \frac{\mu\pi W\left(R^2 - Ri^2\right)^2}{8.R.Ls^2.Df^2} dR + \int_{Ri}^{R1} \frac{\mu\pi W}{2\pi R} dR \tag{7}$$

Similarly, for the coaxial winding structure discussed above, the inductance is given by

$$L = \int_{Ri}^{R1} \frac{\mu\pi W\left(R^2 - Ri^2\right)^2}{8.R.Ls^2.Df^2} dR + \int_{R1}^{R1+Dg} \frac{\mu\pi W}{2\pi R} dR + \int_{R1+Dg}^{Ro} \frac{\mu\pi W\left(Ro^2 - R^2\right)^2}{8.R.Ls^2.Df^2} \tag{8}$$

In any application, the load resistance should be substantially greater than the capacitor winding ESR in order to transfer most of the energy stored in the capacitor winding to the load. To a significant part of the examination below, it has been traditional that "as indicated in ref (Qin & Boggs, 2010)", the load Resistance may be 9 times more excellent than the ESR, so that the downright resistance in the out will be 10 times the ESR. Since the load impedance might be fixed, the load impedance might direct the most extreme capacitor ESR. We might infer the run through to peak current, Tp, and the peak current, Ip.

$$Tp = \frac{Ln\left(\frac{p1}{p2}\right)}{p2 - p1} \left\{ p1,2 = \frac{-R}{2ESI} \pm \sqrt{\frac{R^2}{4ESI^2} - \frac{1}{ESI.C}} \right\} \tag{9}$$

The peak current:

$$Ip = \frac{-Uo}{ESI\left(P2 - P1\right)}\left(e^{p1.Tp} - e^{p2.Tp}\right) \tag{10}$$

The overall volume of the capacitor:

$$V1 = \pi.Ro^2 \frac{N}{2}\left(Ws + Wm\right) \tag{11}$$

If the core volume is excluded, the overall volume is

$$V2 = \pi\left(Ro^2 - Ri^2\right).\frac{N}{2}.\left(Ws + Wm\right) \tag{12}$$

The multi-section coaxial winding structure has two segments of volume which increment in size to the area overlap, i.e., the volumes of (i) the center and (ii) the coaxial insulin between the two coaxial windings. The energy density in peak current will can be as follows: The place R_{load} will be the load resistance, here 9 times the ESR. A metalized film capacitor (MFC) will ensure precise guaranteeing in power industry that has the point of self-clearing (Boggs et al., 2010), which powers the breakdown that happens in the metallization layers encompassing the breakdown webpage in a certain lifestyle that keeps the capacitor working typically for a little decrease in the aggregate capacitance of the capacitor. The objective of this chapter will be to consider the power capacitors and their electrical and dielectric aspects by utilizing the universal materials and analyzing these properties of utilizing singular nanocomposites materials and the new intended multi-nanocomposites materials. The progressive over universal power capacitors materials properties have been deduced when the recommended new plan is connected by including new multi-nanocomposites polypropylene materials in light of multi-nanoparticles system. Moreover, the impact for the multi-nanoparticles plan in the new capacitor plan is equipped towards providing for exceptional execution with the same cost or less cost. The nanocomposite framework comprises three parts as shown in Fig. (15): (polymer, nanoparticle) and an interphase district (Samir, 2010). A multi-nanoparticle system might have been exhibited for moving forward the electric and dielectric properties of the polymer (Thabet & Salem, 2017). Thus, the successful dielectric constant (ε_{eff}) of the consideration and interphase might be ascertained discussed earlier (Thabet & Salem, 2017).

Figure 15. Interphase region for multi-nanocomposite system

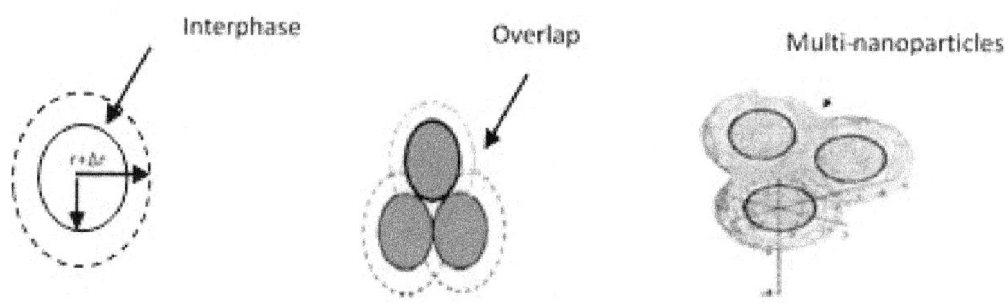

The relative effects identified with the inductance and ESR to pulsed power applications for absolute segment metalized film capacitor windings are recommended dependent upon ref. (Ho et al., 2008a; Kuo et al., 2001). The point of the introduced study will be considered in the conduct about ESI, ESR, volumetric energy density, top discharge current, and so forth. To multi-section, the high voltage windings as an aftereffect of utilizing metallized film inside the capacitor ESR and ESI tend should diminish. In this work, new dielectrics for multi-nanocomposites system have been produced to upgrade resistivity, film dielectric constant, ESR, and ESI in dainty film metalized capacitors. The available fill focuses on impact of including multi-nanoparticles to high voltage capacitor in place to upgrade ESR and ESI, i.e., how to smother the impacts of ESR and ESI which result from interfacing the interior capacitor windings in arrangement and in parallel as required to provide for voltage and capacitance rating for the two model plans, as demonstrated in Fig. (14). The effective conductivity σ_{eff} for the multi-nanocomposite is calculated from (Thabet et al., 2018):

$$\acute{o}_{eff} = \left(1 - V_f\right)\acute{o}_M + V_f \acute{o}_F \tag{13}$$

Where, V_f is the filler volume fraction, $\acute{o}_M$ is the matrix conductivity and $\acute{o}_F$ is the filler conductivity for the multi-nanocomposite where the effective resistivity is used.

With N as the sections number, W_d is the electrode overlap width for each section, and the margin width between sections is W_g as illustrated in Figure (14). Ro is the radius of the outer winding. In the coaxial case it can be expressed as:

$$R_o = \sqrt{\frac{S_c}{\pi} + (R_1 + D_g)^2} \tag{14}$$

$$R_1 = \sqrt{\frac{S_c}{\pi} + R_i^2} \tag{14.i}$$

Where R_1 is the outside radius of the inner winding and D_g is the coaxial winding insulation thickness or between the winding and conductor. As the section electrode overlap increases, the winding length decreases, thus increasing the ESR that can be calculated by ref. (Ho et al., 2008a; Thabet & Salem, 2017). When the current pass through the capacitor can be assumed as I, the inductance can be calculated from the integration of the energy in the magnetic field as follows:

$$L = \frac{1}{I^2_{io}} \int B^2 dV = \frac{2w_g}{I^2} \tag{15}$$

7.2 Nanodielectric for HV Power Capacitors

Clay mineral utilized within the polymer industry is kaolinite. Kaolinite is the fundamental constituent of filler clays; huge amounts of other minerals could frequently be displayed. The molecule shape is the greater part paramount characteristic of kaolinite for polymer applications. The platy nature implies that clay nanoparticles have a more excellent impact on properties, such as viscosity, firmness and strength. Utilizing clay similarly as nanoparticle (10-100nm) provides for large amounts of flame retardancy of all the generated composites, and it's chosen in this section. Clay has an impetus to become the best filler around nanoparticles modern materials. MgO: is the synthetic compound silicon dioxide, otherwise called silica nano-silica with commonplace molecule span of 10~50nm which may be generally utilized as fillers to move forward the industrial and thermal properties of polymers. Magnesium oxide (MgO) film has high auxiliary electron emanation (SEE) yield and it can have applications in level board show and other devices. However, it's worth determinedly relies on the surface conditions and crystalline personal satisfaction for MgO film. It can be discovered in applications at the level board crenation and different devices; it can be utilized within nanowires and nano-capacitors applications. $BaTiO_3$: could be utilized within low and high voltage capacitors applications. $BaTiO_3$ is processing of round $BaTiO_3$ powders toward 700 and 900°C by the spread pyrolysis for ethanol/water results in $BaCl_2$ and $TiCl_4$ that are suspended utilizing two-fluid and ultra-nationalistic atomizers. Polypropylene is one of the most common and versatile thermoplastics in the plastics industry. Polypropylenes can be the best thermoplastic, surpassing others in consolidated electrical properties, heat resistance, toughness, concoction resistance, dimensional stability, and surface gloss in in lesser cost over the greatest part of others. Filling polypropylene with a specific nanoparticle (Clay, MgO, and $BaTiO_3$) extraordinarily expands electrical properties, ductile strength, effect strength, flexural modulus, and redirection temperature under load with a related decrease in prolongation as stated by the sorts and rates of nanoparticles. Furthermore, it will be a standout amongst the majority of vital polymers in case it is utilized as insulation for cutting edge HV capacitors.

Table (1) demonstrates a review of nanoparticles and modern materials that will be currently introduced in this exploration. Magnesium oxide (MgO) has a percentage worthy of upgrades on the nanocomposite which are uncovered with the addition of nano-MgO, such as the change to the crystallization condition,

Table1. Characteristics of Industrial Insulation Materials

Polymers and Nanoparticles	Dielectric Constant
Polypropylene	2.3
Clay	2.8
$BaTiO_3$	3.8
MgO	9.7
Polypropylene + 5wt% Clay	2.34
Polypropylene + 10wt% Clay	2.38
Polypropylene + 15wt% Clay	2.42
Polypropylene + 5wt% $BaTiO_3$	2.41
Polypropylene + 10wt% $BaTiO_3$	2.53
Polypropylene + 15wt% $BaTiO_3$	2.62
Polypropylene + 5wt% MgO	2.71
Polypropylene + 10wt% MgO	3.02
Polypropylene + 15wt% MgO	3.35

Table 2. Electric and Dielectric Specifications of Polypropylene and Nanoparticles (Thabet & Salem, 2017)

Materials	Dielectric Constant	Specific Surface Area m2/g	Conductivity S/Cm	Dielectric Losses
PP	2.3	0.14676	10^{-21}	0.0005
Fumed Silica	4.5	225	10^{-13}	0.0091
Clay	2.8	1.8	$5x10^{-4}$	0.26
ZnO	2.5	27	10^{4}	0.035
MgO	9.7	600	$5x10^{-9}$	0.0009
Fe_2O_3	25	386	$0.358x10^{-12}$	0.004
$BaTiO_3$	3.8	5.4	$20x10^{-9}$	0.0008
TiO_2	11	19.16	10^{-21}	0.0015
Silica	3	24	10^{-16}	0.0015

the upgrade of dc BDs, the diminish of space charge accumulation, the concealment of electric distortion, the diminish of permittivity et cetera. Barium titanate ($BaTiO_3$) has a favorable element of profit identified with their particles suspended in a medium that indicates dielectric unwinding. The dielectric properties of $BaTiO_3$ pottery rely on the molecule size, figure, as well as the kind of the dopants utilized (Kuo et al., 2001). Clays can be acknowledged as a critical modern mineral; and with the late researches of nanotechnology, they have been discovered in diverse applications. "Around these Nano clays, montmorillonite (MMT) indicates a change in the number of properties of polymeric matrices, such as the versatile modulus, flexural and pliable properties and high temperature resistance. Tables (1, 2) depict the characteristics of industrial insulation materials and dielectric specifications of polypropylene and nanoparticles (Thabet & Salem, 2017).

Fumed particles prompt a change in industrial thermal and electrical properties, its composites are broadly utilized in coatings, potting, adhesives, exemplification of semiconductors, laminates and so on. Including fumed silica of the polymer matrix empowers decrease to coefficients of thermal expansion, provides for upgrade in thermal conductivity and other industrial properties. Furthermore, nano-scaled titanium dioxide is the practically researched photo catalyst, since it is a generally sheltered and modest semiconductor that is stable to photo corrosion in an expansive ph size. On the other hand, Polypropylene (PP) may be a standout amongst the majorly utilized business plastics and it has a high-volume resistivity, high breakdown strength and low dielectric loss.

7.3 Characterization of HV Power Capacitors

A 100 kV capacitor winding might be dependent upon 4 layers of 7μm capacitor film, two metalized and two un-metalized. For an absolute winding capacitor, the winding width is the width of the edges in addition to the downright width of the cover of the areas. In the coaxial configuration, the aggregate winding width will be a large fraction of this strength. A metallization resistivity in the cover areas of ρ_o=100 Ω/sq and ρ_m=10 Ω/sq has been expected in the edge districts. High resistivity in the cover locales enhances clearing efficiency and low resistivity in the edge locales lessens the ESR. The winding diameter might rely on few variables, including film thickness and period of film edge. Backing the 100 kV possibility contrasts crosswise over the limit connections, thickness of the insulin layer can't be under ~3 mm. Moreover, with respect to the multi area structure of high voltage capacitor, every last one of the parameters can be communicated as capacities for segment cover width for examining the impacts of different nanoparticles on high voltage metalized film capacitors. Figures ((16. a) - (16. c)) show the Equivalent Series Resistance (ESR), Equivalent Series Inductance (ESI), energy density (excluding core), and energy density (including core) respectively for the metallized film capacitors as a function of section overlap width by using different nanocomposite film materials PP with $BaTiO_3$ nanoparticles. Figure (16.a) indicates the impact of clay nanoparticles on the Equivalent Series Resistance (ESR). It is clear that increasing the rate of $BaTiO_3$ nanoparticles in the polypropylene matrix expands the equivalent series. Figure (16.b) indicates the impact of $BaTiO_3$ nanoparticles on the Equivalent Series Inductance (ESI). It is shown that increasing the rate of $BaTiO_3$ nanoparticles in the polypropylene matrix obviously declines the equal arrangement inductance worth in the high voltage metalized film capacitor. The impact of $BaTiO_3$ nanoparticles may indicate that increasing the rate of $BaTiO_3$ nanoparticles in the polypropylene matrix has low impact on decreasing the current rise time of the long haul in the high voltage metalized film capacitor. The same conduct of crest current has manifested the impact of $BaTiO_3$ nanoparticles on the current rise time in the metallized film capacitors. It can be demonstrated that increasing the rate of $BaTiO_3$ nanoparticles in the polypropylene matrix has low impact to decreasing the crest current in the high voltage metalized film capacitor. The impact of $BaTiO_3$ nanoparticles on the fraction for functional volume (excluding core) in the metallized film capacitors is indicated that increasing the rate of $BaTiO_3$ nanoparticles in the polypropylene matrix has low impact on decreasing the fraction of suitable volume (excluding core) in the high voltage metalized film capacitor. Accordingly, the impact of $BaTiO_3$ nanoparticles on the fraction of suitable volume (counting core) in the metallized film capacitors indicates that increasing the rate of $BaTiO_3$ nanoparticles in the polypropylene matrix has an acceptable impact on decreasing the fraction of useful volume (counting core) in the high voltage metalized film capacitor. Figure (16.c) indicates the impact of $BaTiO_3$ nanoparticles with respect to energy density (excluding core) in the metallized film capacitors. This indicates that increasing the rate of $BaTiO_3$ nanoparticles in the polypropylene matrix has an acceptable impact on increasing energy density (excluding core) in the high voltage metalized film capacitor. Figure (16.d) indicates the impact of $BaTiO_3$ nanoparticles on energy density (counting core) in the metallized film capacitors. It can be indicated that increasing the rate of $BaTiO_3$ nanoparticles in the polypropylene matrix has an acceptable impact on decreasing the energy density (counting core) in the high voltage metalized film capacitor.

Figure 16. Effects of $BaTiO_3$ nanoparticles concentration on characteristics of metalized high voltage capacitors

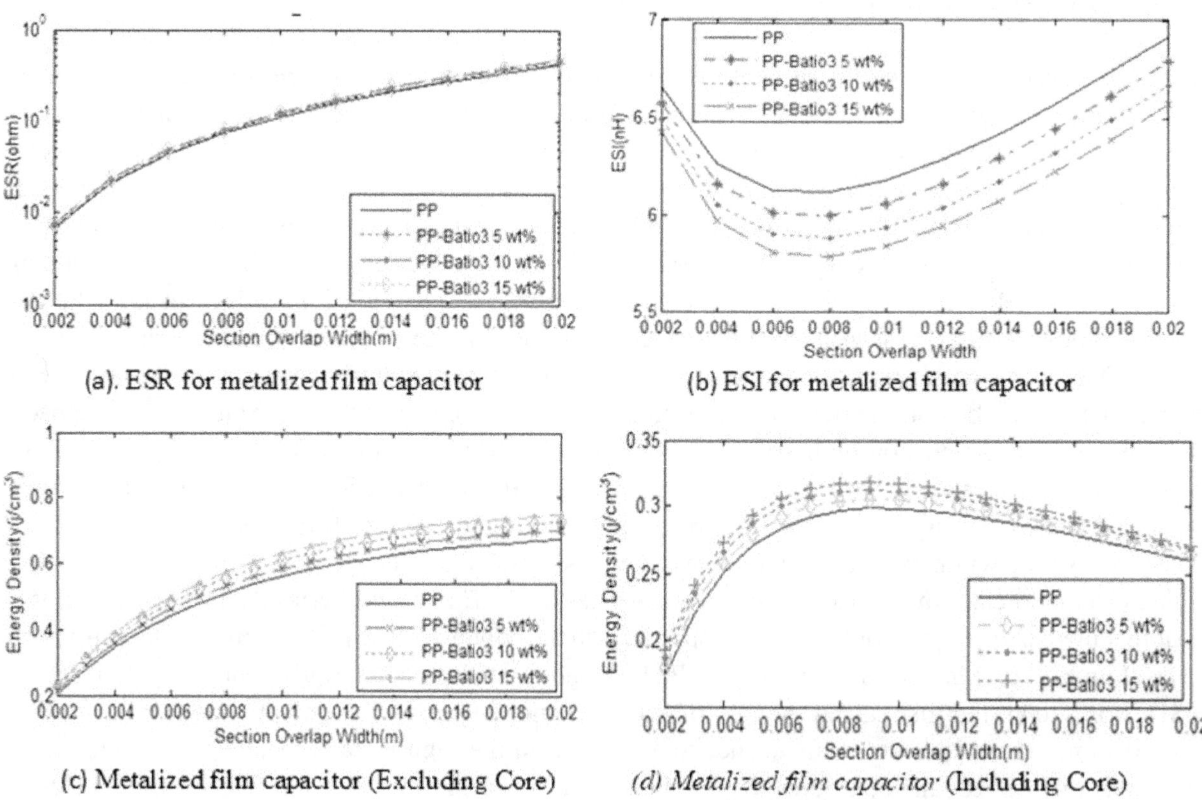

Figure 17. Effects of MgO nanoparticles concentration on the characteristics of metalized high voltage capacitors

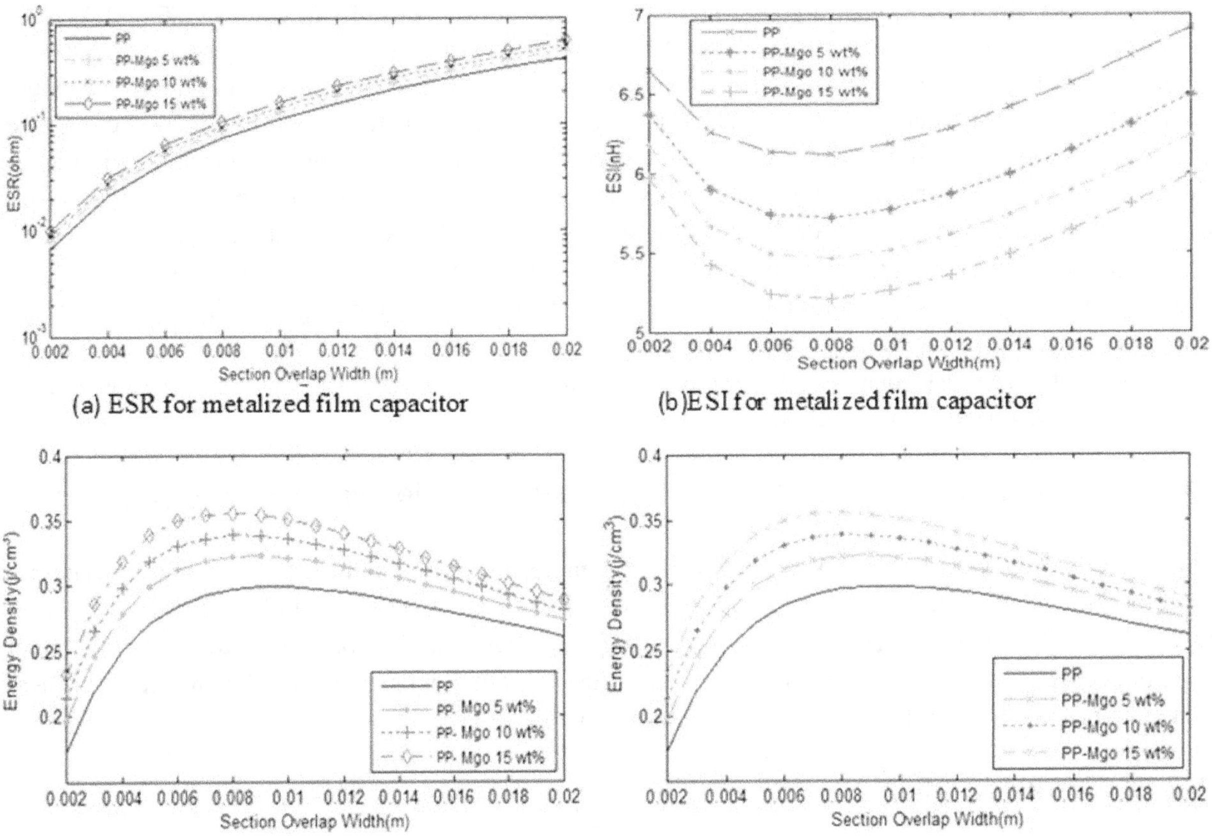

(a) ESR for metalized film capacitor

(b)ESI for metalized film capacitor

(c) Metalized film capacitor (Excluding Core)

(d) Metalized film capacitor (Including Core)

Figures ((17.a) - (17.d)) delineate the Equivalent Series Resistance (ESR), Equivalent Series Inductance (ESI), current rise time, the peak current, fraction of useful volume (excluding core), fraction of useful volume (including core), energy density (excluding core), and energy density (including core) respectively for the metallized film capacitors as a function of section overlap width by using different nanocomposite film materials PP with MgO nanoparticles. Figure (17. a) indicates the impact of clay nanoparticles on the equivalent series resistance (ESR). It is clear that the increasing rate of MgO nanoparticles in the polypropylene matrix obviously expands the equivalent series resistance. Figure (17. b) indicates the impact of MgO nanoparticles on the equivalent series inductance (ESI). It is obvious that increasing the rate of MgO nanoparticles in the polypropylene matrix abates the equivalent series inductance esteem in the high voltage metalized film capacitor really unmistakably. The impact of MgO nanoparticles on the current rise time run in the metallized film capacitors demonstrates that the increasing rate of MgO nanoparticles in the polypropylene matrix has high impact on decreasing the current rise time in the high voltage metalized film capacitor. The same conduct technique for peak current has shown the impact of MgO nanoparticles on the current climb duration of the time in the metallized film capacitors. It can be inferred that increasing the rate of MgO nanoparticles in the polypropylene matrix has a high impact on decreasing the peak current in the high voltage metalized film capacitor. The impact of MgO nanoparticles on the fraction of the functional volume (excluding core) in the metallized film capacitors has evidenced that increasing the rate of MgO nanoparticles in the polypropylene matrix has a high impact on decreasing the fraction of the suitable volume (excluding core) in the high voltage metalized film capacitor. Furthermore, the impact of MgO nanoparticles on the fraction of functional volume (counting core) in the metallized film capacitors has manifested that increasing the rate of MgO nanoparticles in the polypropylene matrix has an acceptable impact on decreasing the fraction of service volume (counting core) in the high voltage metalized film capacitor. Notably, Figure (17.c) indicates the impact of MgO nanoparticles by looking into energy density (excluding core) in the metallized film capacitors. It is indicated that increasing the rate of MgO nanoparticles in the polypropylene matrix has an acceptable impact on increasing the energy density (excluding core) in the high voltage metalized film capacitor. Fig. (17. d) indicates the impact of MgO nanoparticles on energy density (counting core) in the metallized film capacitors. It will be shown that increasing the rate of MgO nanoparticles in the polypropylene matrix has an acceptable impact on decreasing the energy density (counting core) in the high voltage metalized film capacitor.

7.4 Trends of Nanoparticles and Multiple Nanoparticles in HV Power Capacitors

Our research endeavor is a 100-KV capacitor dependent upon 4 layers of 7μm capacitor films, including 2 un-metalized film and 2 metalized films with cover resistivity of 100 Ω/sq and edge resistivity of 10 Ω/sq (Butler & Concordia, 1937). Every one of the parameters can be communicated as works about segment cover width with respect to multi area structure of high voltage capacitor for mulling over ideal nanoparticles for high voltage metalized film capacitor.

A. ESI Characterization

Figure (18.a) makes it self-evident that the addition of separate nanoparticles in the polypropylene loaded for fumed silica matrix declines the Equivalent Series Inductance (ESI) for every multi-nanoparticle, but the best consequence is provided for MgO nanoparticles. Fig. (18.b) clarifies the impact of $BaTiO_3$/PP nanocomposite on distinctive nanoparticles which diminishes the ESI contrasted with the pure PP. Additionally, Fig. (18.c) indicates the impact of expansion of different nanoparticles on the polypropylene matrix. It is recognized that Fe_2O_3 nanoparticles diminish the pro-fractional arrangement inductance to the various nanoparticles, but the best aftereffect will be provided for MgO nanoparticles. Figure (18.d) exhibits the impact of Clay/PP nanocomposite with separate nanoparticles which diminish the ESI contrasted with the pure PP, but for (5%wt. Clay+10% wt. ZnO)/PP that expands the ESI marginally against pure PP, and the Clay/PP nanocomposite give the best ESI regarding shortages for MgO nanoparticles.

B. ESR Characterization

Fig. (19.a) indicates the impact of nanoparticles on the Equivalent Series Resistance (ESR); it is self-evident that expansion of diverse nanoparticles in the polypropylene loaded with fumed silica matrix decreases the equivalent series resistance for all multi-nanoparticle identified with scattered fumed silica nanoparticles in the primary PP matrix. On the other hand, Fig. (19.b) illustrates the impact of utilizing $BaTiO_3$ as opposed to fumed silica concerning illustration of a fundamental nanoparticle scattered inside the matrix and depicts the decrease of ESR. Hence, ZnO nanoparticles provides the best result when it

Figure 18. ESI for Poly propylene nanocomposites and multiple nanocomposites with different nanoparticles, (a) (F. Silica/PP) with different nanoparticles, (b) ($BaTiO_3$/PP) with different nanoparticles (c) (Fe_2O_3/PP) with different nanoparticles, (d) (Clay/PP) with different nanoparticles

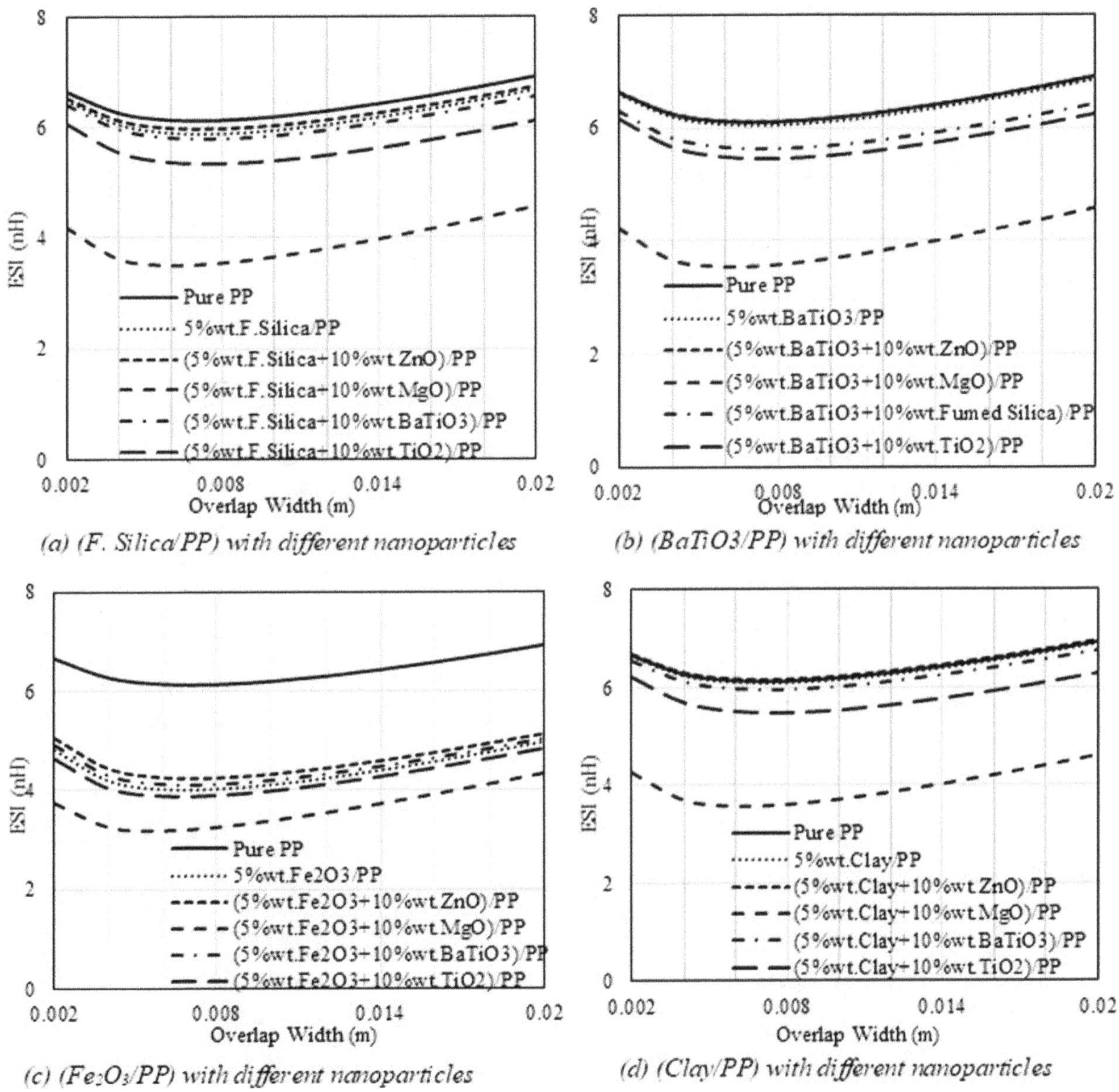

is utilized as a second nanoparticle. Fig. (19.c), indicates that the addition of separate nanoparticles in the polypropylene filled with Fe_2O_3 matrix declines the equivalent series resistance (ESR) to the greater part multi-nanoparticles strategy identified with scatter Fe_2O_3 as an essential nanoparticle in the primary PP matrix. Moreover, Fig. (19.d) illustrates the impact of utilizing clay as a principal nanoparticle scattered inside the matrix which depicts the decrease of ESR. Thus, ZnO nanoparticles provide the best after effect when it is utilized as a second nanoparticle.

Figure 19. ESR for Poly propylene nanocomposites and multiple nanocomposites with different nanoparticles, (a) (F. Silica/PP) with different nanoparticles, (b) ($BaTiO_3$/PP) with different nanoparticles (c) (Fe_2O_3/PP) with different nanoparticles, (d) (Clay/PP) with different nanoparticles

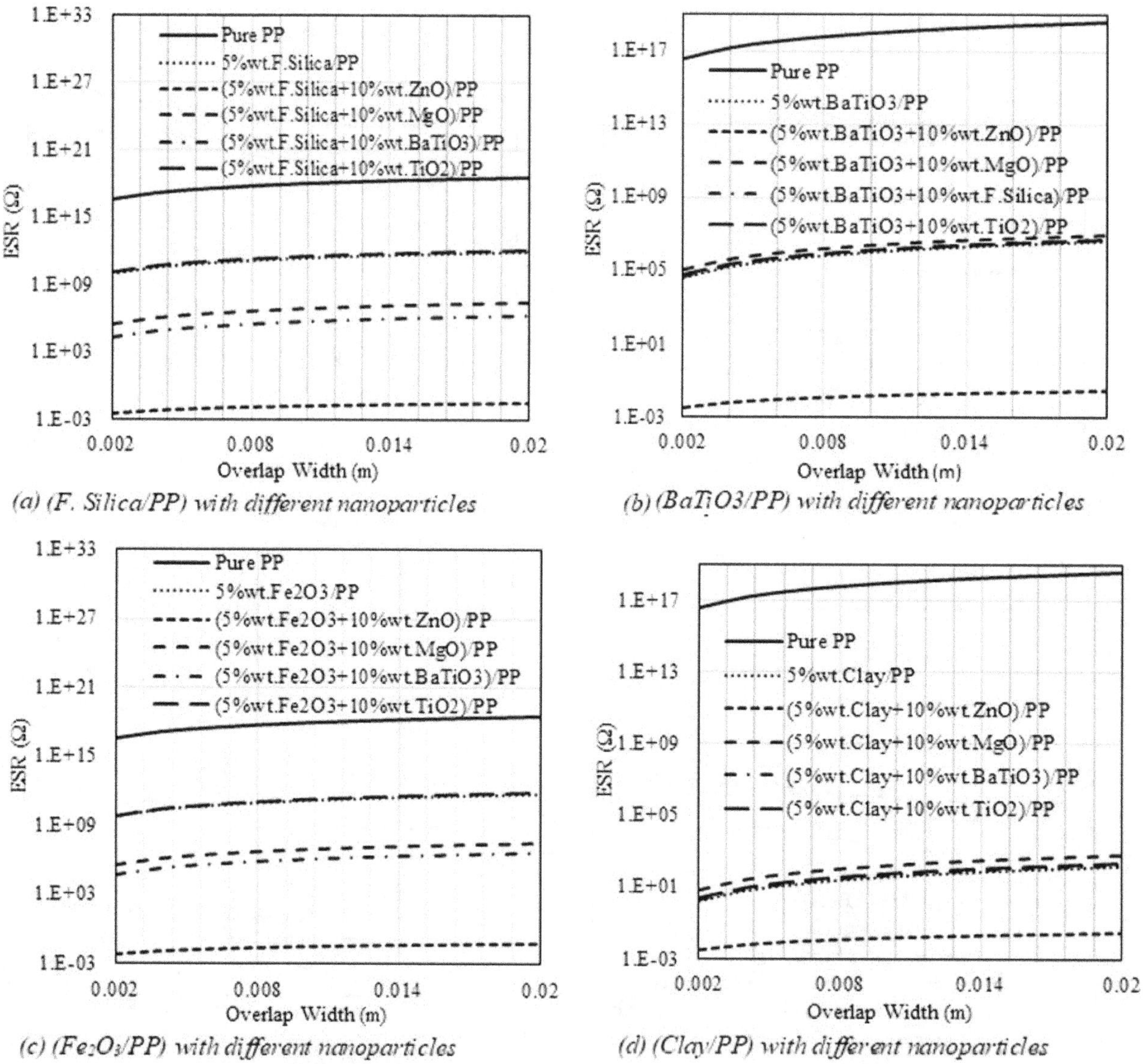

(a) (F. Silica/PP) with different nanoparticles

(b) (BaTiO3/PP) with different nanoparticles

(c) (Fe_2O_3/PP) with different nanoparticles

(d) (Clay/PP) with different nanoparticles

C. Energy Density Characterization

The upgrade for energy density qualities of the metallized film capacitors has been illustrated with reference to various nanoparticles sorts utilized and contrasted with pure and nanocomposite PP for diverse scattered prim nanoparticles, as shown in Fig. (20). Moreover, it is recognized that the nanoparticles (MgO, TiO_2) enhance obviously the energy density to the point when these nanoparticles are utilized as second nanoparticles respectively.

Figure 20. Energy density for Poly propylene nanocomposites and multiple nanocomposites with different nanoparticles, (a) (F. Silica/PP) with different nanoparticles, (b) ($BaTiO_3$/PP) with different nanoparticles (c) (Fe_2O_3/PP) with different nanoparticles, (d) (Clay/PP) with different nanoparticles

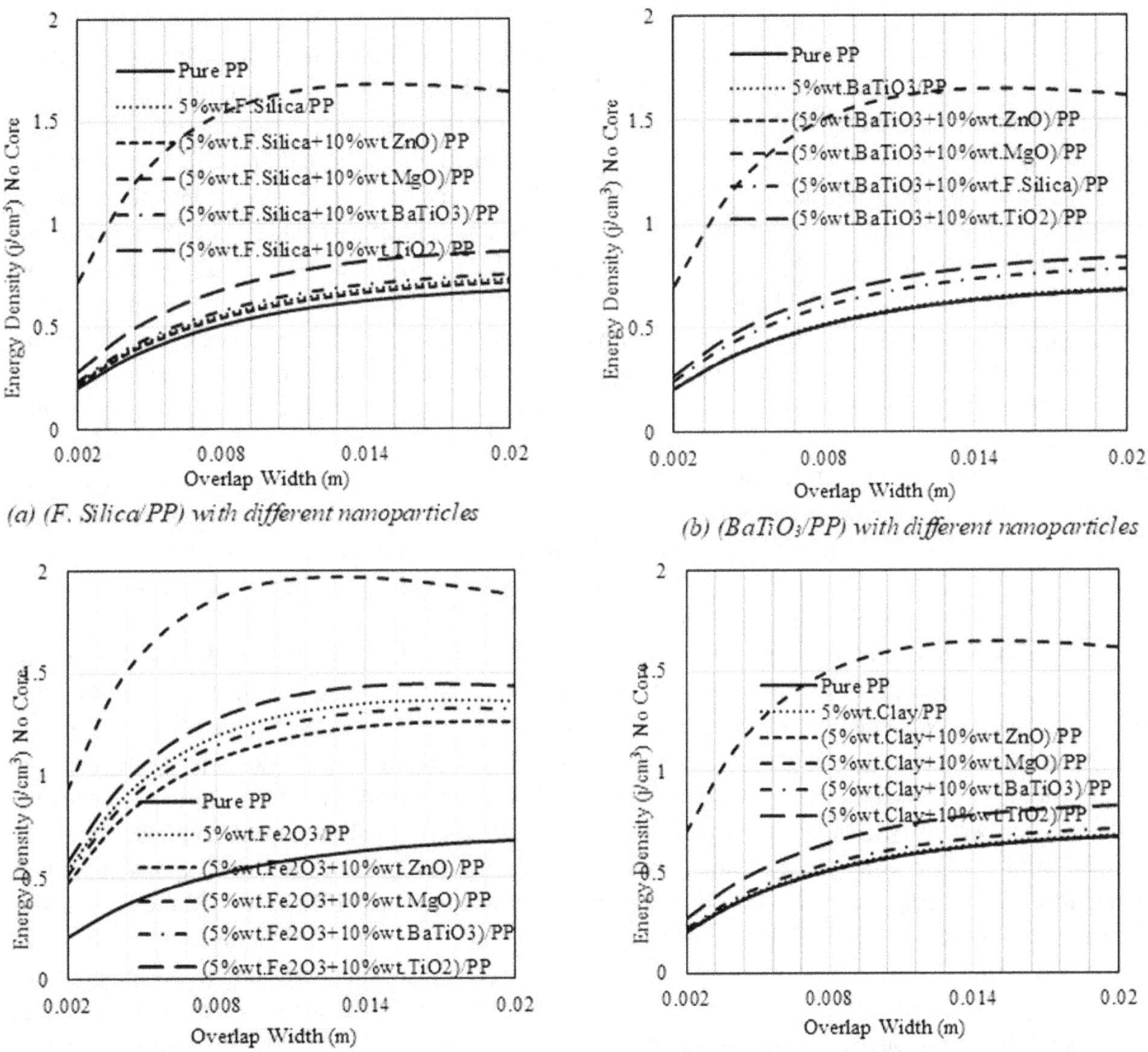

(a) (F. Silica/PP) with different nanoparticles

(b) ($BaTiO_3$/PP) with different nanoparticles

(c) (Fe_2O_3/PP) with different nanoparticles

(d) (Clay/PP) with different nanoparticles

8 FORECAST AND RECOMMENDATIONS

Clay and titanium dioxide nanoparticles decrease the capacitive charge with respect to pure polyvinyl chloride specimens; titanium dioxide may be the best nanoparticles in concentration (1 wt. %) for decreasing capacitive charge in polyvinyl chloride nanocomposites at variant temperatures (20°C - 80°C). A closed capacitive charge happens toward low temperatures, particularly at high power frequencies. Impacts of fumed silica nanoparticles show up in decreasing the capacitive charge with respect to pure polyvinyl chloride specimens. The measured capacitive charge increments in the nanocomposites with increasing concentration of fumed silica nanoparticles more than 5wt. %. Divergence capacitive charge occurs at low temperatures, particularly during high power frequencies. Zinc oxide nanoparticles may cause a greater amount of decrease in capacitive charge in the nanocomposites for increasing nanoparticles concentration (5 wt. %). Divergence capacitive charge outcomes occur in low temperatures, particularly in high energy frequencies. For a multi section, high voltage, low inductance capacitor, peak current and top power thickness are inclined and overwhelmed by limit association current thickness. Such windings have the possibility of current densities and energy densities in a wide margin in abundance that the available limit association of engineering organization might withstand. Including nanoparticles (Clay, MgO and $BaTiO_3$) expand ESR, ESI and energy density and decrease peak current and ascent run providing for high voltage capacitors for useful constraint about limit association current. The Impact of MgO on Nanoparticles is higher than fumed silica and clay with respect to the characteristics of high voltage capacitor, including MgO nanoparticles to polypropylene, providing for better values of ESR, ESI, energy density and peak current within the investigated range. For a multi section, high voltage, low inductance capacitor, peak current and peak power density are likely to be dominated by end connection current density. Such windings have the possibility of current densities and power densities significantly to abundance that the available end association innovation might withstand. Including nanoparticles (Clay, MgO and fumed Silica) expand ESR, ESL and energy density and lessen peak current and climb run through which high voltage capacitors are provided for beneficial constraint about wind connection current. Impact of MgO round nanoparticles may be higher than fumed silica and clay for characteristic high voltage capacitor, including MgO nanoparticles to polypropylene providing for the best qualities of ESR, ESL, energy density, peak current and rise time.

Multi-nanoparticles technology provides an incredible capability and changes the qualities of the metallized film capacitor contrasted with the standalone nanoparticles that are utilized inside the polypropylene matrix. MgO and ZnO spherical nanoparticles provide higher upgrade of ESR, ESI, energy density, peak current and current rise time in polypropylene base matrix. Expansion of specific concentration of Fe_2O_3 as an essential nanoparticle of the PP matrix provides the best shortages enhancing ESI and energy density. At the same time, the change will be marginally on ESR contrasted with the utilization of MgO nanoparticles as an essential nanoparticle for the same wt. %. Expansion of clay nanoparticles concerning the illustration of a prime filler enhances the ESI, ESR and energy density. At the same time, the change will be more effective when clay is utilized within a multi-nanocomposite framework with the same wt. %. Energy density has a tendency to be expanded when multi-nanoparticle system might have been applied. Through a change of ESI, ESR and peak current, enhancing the energy density will be a resultant of utilizing different nanoparticles contrasted with utilizing customary materials or singular nanoparticles method. The new multi-nanoparticles utilized inside polypropylene matrix is receiving more consideration. Applying multi-nanoparticles inside polypropylene matrix grants an ideal property to outline a capacitor for great properties, such as high strength, reliability and low cost. From the above results, it is shown that it has the ideal aspects of a metallized film capacitor utilizing multi-nanoparticles technobabble to decrease ESI and ESR and increment energy density toward utilizing certain concentrations of nanoparticles. As stated by the previously mentioned results, it can be shown that distinct nanocomposite, such as fumed silica, Fe_2O_3, clay and $BaTiO_3$ diminish the electrical arrangement resistance ESR contrasted with the pure polypropylene. Moreover, it is also shown that utilizing fumed silica, Fe_2O_3, Clay and $BaTiO_3$ concerning illustration of a singular nanoparticle lessens electrical arrangement inductance ESI contrasted with the pure PP. Additionally, utilizing fumed silica, Fe_2O_3, clay and $BaTiO_3$ as a singular nanoparticle bases the energy density plainly. In the event of utilizing multi-nanoparticle technique, a colossal change of the properties of the dielectric materials contrasted with the universal single person nanoparticles procedure has been investigated. Accordingly, it is self-evident that utilizing ZnO as a second nanoparticles with Clay, fumed silica, Fe_2O_3 and $BaTiO_3$ clearly diminishes the ESR of the capacitor. Likewise, utilizing TiO_2 nanoparticles as a second nanoparticle has a slight impact during the same %wt. Added to that, applying multi-nanoparticle system diminishes the ESI extraordinarily. Moreover, the energy density is enhanced significantly in case of utilizing multi-nanoparticles method, contrasted with the unique nanoparticle to the same %wt. For example, utilizing MgO concerning illustration as a second nanoparticle for all connected prime nanoparticles obviously enhances energy density and ESI.

REFERENCES

Aman, A., Yaacob, M. M., & Ab, M. (2013, February). Polymeric composite based on waste material for high voltage outdoor application. *International Journal of Electrical Power & Energy Systems*, *45*(1), 346–352. doi:10.1016/j.ijepes.2012.09.004

Andrei, R. G. (1999). A Novel Fuseless Capacitor Bank Design Using Conventional Single Bushing Capacitors. *IEEE Transactions on Power Delivery*, *14*(3), 1124–1133. doi:10.1109/61.772383

Baker, P. P. (2002). Ultra Capacitors for Use in the Power Strength and Distributed Resource Applications. *IEEE Power Engineering Society Summer Meeting, 1*, 316–320.

Bhargava, B., & Haas, R. G. (2002). Thyristor protected series capacitors project. *IEEE Power Engineering Society, Summer Meeting, 1*, 241–246.

Boggs, S. A., Ho, J., & Jow, T. R. (2010). Overview of laminar dielectric capacitors. *IEEE Electrical Insulation Magazine*, *26*(2), 7–13. doi:10.1109/MEI.2010.5482550

Bois, L., Chassagneux, F., Parola, S., Bessueille, F., Battie, Y., Destouches, N., Boukenter, A., Moncoffre, N., & Toulhoat, N. (2009). Growth of ordered silver nanoparticles in silica film mesostructured with a triblock copolymer PEO–PPO–PEO. *Journal of Solid State Chemistry*, *182*(7), 1700–1707. doi:10.1016/j.jssc.2009.01.044

Butler, J. W., & Concordia, C. (1937). Analysis of series capacitor application problems. *Transactions of the American Institute of Electrical Engineers*, *56*(8), 975–988. doi:10.1109/T-AIEE.1937.5057677

Chatterjee, A., Bhattacharjee, P., Roy, N. K., & Kumbhakar, P. (2013, February). Usage of nanotechnology based gas sensor for health assessment and maintenance of transformers by DGA method. *International Journal of Electrical Power & Energy Systems*, *45*(1), 137–141. doi:10.1016/j.ijepes.2012.08.044

Christie, C. V. (1938). *Electrical Engineering*. McGraw-Hill.

Coursey, P. R. (1927). *Electrical Condensers*. Sir Isaac Pitman& Sons.

Dong, L., Gu, G. Q., & Yu, K. W. (2003). *First Principles Approach to Dielectric Response of Graded Spherical Particles* (Vol. 67). Physics Review B.

Ebnalwaled, A. A., & Thabet, A. (2016, July). Power ling the optical constants of PVC nanocomposite films for optoelectronic applications. *Synthetic Metals Journal, 220*, 374–383. doi:10.1016/j.synthmet.2016.07.006

Gouda, O., Thabet, A., Mubarak, Y. A., & Samir, M. (2014, March). Nanotechnology Effects on Space Charge Relaxation Measurements for Polyvinyl Chloride Thin Films. *International Journal of Electrical Engineering and Informatics Engineering and Technology, 6*(1), 1–12.

Hammond, P. W. (1988). *A Harmonic Filter Installation to Reduce Voltage Distortion*. Academic Press.

Ho, J., Jow, T. R., & Boggs, S. A. (2008a). Implications of Advanced Capacitor Dielectrics for Performance of Metallized Film Capacitor Windings. *IEEE International Power Modulators and High-Voltage Conference*, 556-559. 10.1109/IPMC.2008.4743718

Ho, J. S., Jow, R., & Boggs, S. A. (2008b). Implications of Advanced Capacitor Dielectrics for Performance of Metalized Film Windings. *IEEE Transactions on Dielectrics and Electrical Insulation, 15*(6), 1754–1760. doi:10.1109/TDEI.2008.4712681

Kamalapur, G. D., & Udaykumar, R. Y. (2011, March). Rural electrification in India and feasibility of Photovoltaic Solar Home Systems. *International Journal of Electrical Power & Energy Systems, 33*(3), 594–599. doi:10.1016/j.ijepes.2010.12.014

Kim, S. J., Li, M., Ding, Sh., Yu, M. B., & Kwong, D. (2004, August). Improvement of Voltage Linearity in High MIM Capacitors Using HfO2–SiO2 Stacked Dielectric. *IEEE Electron Device Letters, 25*(8), 538–540. doi:10.1109/LED.2004.832785

Kimbark, E. W. (1966). Improvement of system stability by switched series capacitors. *IEEE Transactions on Power Apparatus and Systems, PAS-85*(2), 180–188. doi:10.1109/TPAS.1966.291555

Kuo, D.-H., Chang, C.-C., Su, T.-Y., Wang, W.-K., & Lin, B.-Y. (2001). Dielectric behaviours of multi-doped BaTiO3/epoxy composites. *Journal of the European Ceramic Society, 21*(9), 1171–1177. doi:10.1016/S0955-2219(00)00327-7

Lee, S., Hyun, J., Kim, H., & Paik, K. (2007, August). A Study on Dielectric Constants of Epoxy=SrTiO3 Composite for Embedded Capacitor Films (ECFs). *IEEE Transactions on Advanced Packaging, VOL., 30*(3), 428–433. doi:10.1109/TADVP.2007.898614

Longland. (1984). *Power capacitor handbook*. Butterworths Publisher.

Longland, T., Hunt, T. W., & Brecknell, W. A. (1984). *Power Capacitor Handbook*. Butterworth-Heinemann. doi:10.1016/B978-0-408-01427-4.50023-3

Ma, P., Nie, W., Yang, Z., Zhang, P., Li, G., Lei, Q., Gao, L., Ji, X., & Ding, M. (2008). Preparation and Characterization of Polyimide/Al2O3 Hybride films by Sol-Gel Process. *Journal of Applied Polymer Science, 108*(2), 705–712. doi:10.1002/app.27540

Maini. (1998). *Electronic Projects for Beginners* (2nd ed.). Pustak Mahal.

Masuda, Tanaka, Watanabe, Nanao, & Mochizuki. (n.d.). *Nippon Kinzoku Gakki-si/J. Jpn*. Academic Press.

Masuda, M., Nanao, S., Masuko, N., & Mochizuki, T. (1986). Keikinzoku/J. *Jpn. Inst. Light Metals*, *36*(10), 633. doi:10.2464/jilm.36.633

Michalczyk, P., & Bramoulle, M. (2003, January). Ultimate Properties of the Polypropylene Film for Energy Storage Capacitors. *IEEE Transactions on Magnetics*, *39*(1), 362–365. doi:10.1109/TMAG.2002.806413

Milton, G. W. (2002). *The Theory of Composites*. Cambridge University Press. doi:10.1017/CBO9780511613357

Natarajan. (2019). *Power system Capacitors* (1st ed.). CRC Press.

Phillips, K. J. (1994). Conducting a power factor study. *Consulting-Specifying Engineer*, *10*, 54–58.

Porter, G. A., & McCall, J. C. (1990). *Application and Protection Considerations in Applying Distribution Capacitors*. Pennsylvania Electrical Association.

Qi, X., & Boggs, S. (2004, December). Analysis of the Effects of End Connection Strength on the Dielectric Loss of Metallized Film Capacitors. *IEEE Transactions on Dielectrics and Electrical Insulation*, *11*(6), 990–994. doi:10.1109/TDEI.2004.1387822

Qin, S., & Boggs, S. A. (2009). Application of a Quasi-Static EM Solver to Optimization of Low Inductance Film Capacitors. *17th IEEE Pulsed Power Conference*, 790-794.

Qin, S., & Boggs, S. A. (2010, August). Limits to the Performance and Design of High Voltage Metalized Film Capacitors. *IEEE Transactions on Dielectrics and Electrical Insulation*, *17*(4), 1298–1306. doi:10.1109/TDEI.2010.5539702

Qin, X., & Boggs, S. A. (2003, October). Electrothermal failure of metallized film capacitor end connections–computation of temperature rise at connection spots. *Journal of Applied Physics*, *94*(7), 4449–4456. doi:10.1063/1.1602947

Ramesh, S., Shutzberg, B. A., Huang, Ch., Gao, J., & Giannelis, E. P. (2003, February). Dielectric Nanocomposites for Integral Thin Film Capacitors: Materials Design, Fabrication and Integration Issues. *IEEE Transactions on Advanced Packaging, VOL.*, *26*(1), 17–24. doi:10.1109/TADVP.2003.811365

Ramo, S., Whinnery, J. R., & Van Duzer, T. (1994). *Fields and Waves in Communication Electronics* (3rd ed.). John Wiley and Sons.

Samir, O. (2010). *A Simulation Model for Calculating the Dielectric properties of Nano-Composite Materials and Comprehensive Interphase Approach*. 14th International Middle East Power Systems Conference, Egypt.

Sarathi, R., & Umamaheswari, R. (2013). Understanding the partial discharge activity generated due to particle movement in a composite insulation under AC voltages. *International Journal of Electrical Power & Energy Systems*, *48*(June), 1–9. doi:10.1016/j.ijepes.2012.11.017

Se'vigny, R., Me'nard, S., Rajotte, C., & McVey, M. (2000). Capacitor Measurement in the Substation Environment: A New Approach. *Proceedings of the IEEE 9th International Conference on Transmission and Distribution*, 299–305. 10.1109/TDCLLM.2000.882836

Shi, T. M., Xie, C. G., Huang, S. M., Williams, R. A., & Beck, M. S. (1991, October). Capacitance-based instrumentation for multi-interface level measurement. *Measurement Science & Technology*, *2*(10), 923–933. doi:10.1088/0957-0233/2/10/006

Slenes, K. M., Winsor, P., Scholz, T., & Hudis, M. (2001, January). Pulse Power Capability of High Energy Density Capacitors Based on a New Dielectric Material. *IEEE Transactions on Magnetics, 37*(1), 324–327. doi:10.1109/20.911847

Thabet. (2013a). Influence of Cost-Less Nanoparticles on Electric and Dielectric Characteristics of Polyethylene Industrial Materials. *International Journal of Electrical Engineering and Technology, 4*(1), 58-67.

Thabet. (2013b). Experimental Investigation on Thermal Electric and Dielectric Characterization for Polypropylene Nanocomposites Using Cost-fewer Nanoparticles. *International Journal of Electrical Engineering and Technology, 4*(2), 1-12.

Thabet, A. (2015a, January). Experimental enhancement for dielectric strength of polyethylene insulation materials using cost-fewer nanoparticles. *International Journal of Electrical Power & Energy Systems, 64*, 469–475. doi:10.1016/j.ijepes.2014.06.075

Thabet, A. (2015b, March). Experimental Study of Space Charge Characteristics in Thin Films of Polyvinyl Chloride Nanocomposites. *International Journal on Electrical Engineering and Informatics*, *7*(1), 1–11. doi:10.15676/ijeei.2015.7.1.1

Thabet, A. (2015c, June). Experimental Verification for Improving Dielectric Strength of Polymers by Using Clay Nanoparticles. *Advances in Electrical and Electronic Engineering Journal*, *13*(2), 182–190.

Thabet. (2016a). Theoretical Analysis for effects of nanoparticles on dielectric characterization of electrical industrial materials. *Electrical Engineering (ELEN) Journal*. doi:10.100700202-016-0375-4

Thabet, A. (2016b). Thermal experimental verification on effects of nanoparticles for enhancing electric and dielectric performance of polyvinyl chloride. *Journal of the International Measurement Confederation*, *89*, 28–33. doi:10.1016/j.measurement.2016.04.002

Thabet, A., Abdelhady, S., & Abdel-Rahim, A.-M. M. (2018). Design of Multi-Nanoparticles Technique for Enhancing Magnetic Characterization of Power Transformers Cores. *Advances in Electrical and Electronic Engineering*, *16*(2), 167–177. doi:10.15598/aeee.v16i2.2354

Thabet, A., & Mubarak, Y. A. (2011). *Novel Nanocomposite Insulation Materials for the Enhancing Performance of Power Cables*. In 21st International Conference and Exhibition on Electricity Distribution, "CIRED2011", Frankfurt, Germany.

Thabet, A., & Mubarak, Y. A. (2015, February). Experimental Dielectric Measurements for Cost-fewer Polyvinyl Chloride Nanocomposites. *Iranian Journal of Electrical and Computer Engineering*, *5*(1), 13–22. doi:10.11591/ijece.v5i1.pp13-22

Thabet & Mubarak. (2016). The Effect of Cost-Fewer Nanoparticles on the Electrical Properties of Polyvinyl Chloride. *Electrical Engineering (ELEN) Journal*. doi:10.100700202-016-0392-3

Thabet, A., Mubarak, Y. A., & Ismail, S. (2010). *Exponential Power Law Model for Predicting Dielectric Constant of New Nanocomposite Industrial Materials.* In International Conference On Materials Imperatives in the New Millennium, (MINM 2010), Cairo, Egypt.

Thabet, A., & Salem, N. (2017). Optimizing dielectric characteristics of electrical materials using multi-nanoparticles technique. *2017 Nineteenth International Middle East Power Systems Conference*, 716-721.

Todd, M. G., & Shi, F. G. (2002). Validation of a Novel Dielectric Constant Simulation Model and the Determination of its Physical Parameters. *Microelectronics Journal, 33*(8), 627–632. doi:10.1016/S0026-2692(02)00038-1

Ue, M., Takeda, M., Suzuki, Y., & Mori, S. (1996). Chemical stability of γ-butyrolactone-based electrolytes for aluminum electrolytic capacitors. *Journal of Power Sources, 60*(2), 185–190. doi:10.1016/S0378-7753(96)80009-7

Vo, H. T., & Shi, F. G. (2002). Towards Model-Based Engineering of Optoelectronic Packaging Materials, Dielectric Constant Modeling. *Microelectronics Journal, 33*(5-6), 409–415. doi:10.1016/S0026-2692(02)00010-1

Watanabe, K., Sakairi, M., Takahashi, H., Hirai, S., & Yamaguchi, S. (1999). Article. *Journal of Electroanalytical Chemistry (Lausanne, Switzerland)*, 473.

Wei, E. B., & Tang, S. P. (2004). Dielectric Response of Graded Cylindrical Composites. *Physics Letters. [Part A], 328*(4-5), 395–399. doi:10.1016/j.physleta.2004.06.033

Windlass, H., Markondeya Raj, P., Balaraman, D., Bhattacharya, S. K., & Tummala, R. R. (2003, February). Polymer Ceramic Nanocomposite Capacitors for System-On-Package (SOP) Applications. *IEEE Transactions on Advanced Packaging, 26*(1), 10–16. doi:10.1109/TADVP.2003.811369

Xiu, Y., Zhu, L., Moon, J., Hess, D. W., & Wong, C. P. (2007). Eutectic liquid in Sol-Gel process for superhydrophobic silica thin films — antistiction of MEMS devices. *IEEE 12th International Symposium on Advanced Packaging Materials: Processes, Properties, and Interfaces.*

Yu, K. W., & Gu, G. Q. (2005). Effective Conductivity of Composites of Graded Spherical Particles. *Physics Letters. [Part A], 345*(4-6), 448–452. doi:10.1016/j.physleta.2005.07.037

Zhou, X., Chu, B., Neese, B., Lin, M., & Zhang, Q. M. (2007, October). Electrical energy density and discharge characteristics of a poly(vinylidenefluoridechlorotrifluoroethylene) copolymer. *IEEE Transactions on Dielectrics and Electrical Insulation, 14*(5), 1133–1138. doi:10.1109/TDEI.2007.4339472

Chapter 5
Nano-Tech OHTL Insulators

ABSTRACT

This chapter detects types and structure of ceramic and polymeric insulators. It also sheds light on improving efficiency and high voltage distribution through insulators string toward utilizing nanotechnology techniques. Furthermore, this chapter takes interest in displaying the effects of nanoparticles on overhead transmission lines (OHTL) insulators and high voltage distribution on suspension insulators. The importance and features of using polymeric and non-ceramic suspension insulators have been recently detailed. Thus, this chapter tackles and designs recent models for using individual and multiple nanoparticles in polymeric OHTL suspension insulators. It draws forecasts and recommendations for developing suspension insulators by using nanotechnology techniques.

1 NANO-TECH VISION

The type and concentrations of nanoparticles are the primary components for regulating electric and dielectric properties; therefore, the efficiency and high voltage distribution are regulated through the insulators string. Moreover, the ideal sorts and concentrations of nanoparticles for upgrading efficiency and high voltage distribution of insulators string are also displayed. COMSOL Multi-physics reproduction as 2D geometry has been indicated as the most extreme high voltage concentrations on new suspension insulators (66kV, 132kV, 220kV, and 500kV) nanocomposites against customary insulators. A similar consideration can be to detail the simulation plan and determine new suspension insulators nanocomposites. The irregular distribution of voltage and electric field stress along the composite insulation strings cause an awful impact on their electrical and mechanical properties undermining the security of the power system. In this chapter, another outline is presented for utilizing blending and covering nanoparticles strategies for expanding self-capacitance of silicon rubber (SiR) insulators, thereby the voltage and electric field distribution end up additionally uniform along the string of these lines moving forward SiR material properties. The three-dimensional simulation model of SiR composite insulators was set up to figure out the electric field conveyance toward the Finite Element Method (FEM) which is utilized by ANSYS product. This chapter also displays the suitableness results for SiR insulators string issues in manufacturing and declining the connected electrical field stress at the live end about com-

DOI: 10.4018/978-1-7998-8536-8.ch005

posite insulation up to 49.2. An electrical insulation can be a material whose interior electric charges don't stream freely, making it almost incomprehensible to direct electric current under the impact of an electric field. This contrasts with materials, semiconductors and conductors, which direct a greater amount of electric current effectively. The property by which insulation is recognized is its resistivity; insulators have higher resistivity over semiconductors or conductors. An impeccable insulation doesn't exist because a considerable number of insulators holds numerous little numbers about portable charges (charge carriers) which might convey current.

In addition, all insulators turn into electrically conductive when a sufficiently large voltage is applied, so that the electric field tears electrons away from the atoms. This is known as the breakdown voltage of an insulation. Some materials, such as glass and Teflon, which have high resistivity, extremely aid great electrical insulators. A considerably bigger class of materials, despite the fact that they might have high resistivity will still be sufficient to keep high current from streaming at typically utilized voltages and accordingly will be utilized as insulation for electrical wiring and cables. Cases incorporate rubber-like polymers and most plastics. Insulators will be utilized within electrical gear with backing and separate electrical conductors without permitting current through themselves. A protecting tape utilized within high wrap electrical cables or other gear is known as insulation. The expression insulation may be utilized more particularly to allude to insulating backs used to join electric power distribution or transmission lines on utility poles and transmission towers. They help the weight of the suspended wires without permitting the current to stream through the tower with ground (Taylor, 2003).

2 CERAMIC INSULATORS

Overhead conductors to high-voltage electric power transmission would bare and be insulated toward the encompassing air. Conductors of low voltages in distribution might have exact insulation regardless of what they regularly uncover. Insulating supports, called insulators, are during the concentrations underpinned via utility poles or transmission towers. Insulators are also required when wire enters structures or electrical devices, for example, so that transformers or circuit breakers will have material for protecting the wire. These empty insulators with a conductor inside them are called bushings. Insulators utilized for high-voltage power transmission are made from glass, porcelain or composite polymer materials. Porcelain insulators are produced from clay, quartz or alumina and feldspar, and are secured with a smooth birch glaze to shed water. Insulators made from porcelain rich over alumina are utilized when high mechanical strength is the model. Porcelain has a dielectric strength approximately of 4-10 kV/mm (Nzenwa & Adebayo, 2019). Glass has a high dielectric strength; it attracts buildup and the thick irregular shapes necessary for insulators are challenging to cast without internal strains (Cotton, 1958). Insulation producers quit settling on glass insulators in the late 1960s and replaced them with ceramic materials. Recently, a few electric utilities have been converted to polymer composite materials for a percentage of insulators. These are normally created of a national pole made of fiber fortified plastic and an external climate shed produced for silicone rubber or ethylene propylene diene monomer rubber (EPDM). Composite insulators are less costly, lighter in weight and have a phenomenal hydrophobic capability. This blending makes them perfect gas for administration to dirtied territories. However, these materials don't have the long-haul turned-out administration existence of glass and porcelain. The electrical breakdown of an insulation because of over-the-top voltage can occur remarkably in two ways:

1. A puncture arc is a breakdown and conduction of the material of the insulator, creating an electric bend through the inner part of the insulator. The heat resulting from the arc usually damages the insulator irreparably. Puncture voltage is the voltage across the insulator (when installed in its normal manner) that causes a puncture arc.
2. A flashover arc may be a breakdown and conduction of the air around or along the surface of the insulator and has an arc along the outside of the insulation. They are typically planned to withstand this without harm. Flashover voltage may be the voltage that makes a flash-over arc.

Practically high voltage insulators can be outlined in view of an easier flashover voltage over puncture voltage, with the goal that they streak through in front of the puncture to evade harm. Clay, pollution, salt, and especially water on the surface of a high voltage insulation can make a conductive path crosswise over it, causing leakage currents and flashovers. The flashover voltage can be decreased by more than half the point, when the insulation is wet. High voltage insulators for open air use formed to expand the length of the leakage way along the surface from one limit of the other, known as the Creepage length, to minimize these leakage currents (Holtzhausen, 2008). To finish this, the surface may be molded under an arrangement of corrugations or concentric disc shapes. These generally require that you quit offering on that one or all the more sheds; descending confronting container molded surfaces that go about as umbrellas to guarantee that the only surface leakage way under the 'cup' endures tight dry wet climate. Least Creepage distances are 20–25 mm/kV, but all must be expanded for high contamination or airborne sea-salt regions. Regarding the sorts of insulators, the pin sort insulation is mounted around a pin on the cross-arm of the pole. There is a groove on the upper conclusion of the insulation. The conductor passes through this groove and is tied to the insulation for a tempered wire of the same material as the conductor. Pin sort insulators are utilized for transmission and distribution of communications, and the electric power during voltages is up to 33kV. Past the operating voltage for 33kV, the pin kind insulators are excessively cumbersome and consequently uneconomical.

For suspension insulation to voltages more stupendous than 33 kV, it may be a common act to utilize suspension sort insulators, comprising of an amount of glass or porcelain discs joined in arrangement by metal joins in the structure of a string. The conductor will be suspended at the bottom limit of this string, at the same time when the main conclusion is secured in the cross-arm of the tower. The amount of disc units utilized relies on the voltage. Strain insulation - a deadlock or family pole or tower- may utilize a straight area of line ends or angles off in an additional bearing. These poles must withstand the parallel (horizontal) strain of the long straight area of wire. For backing this parallel load, strain insulators are utilized. For low voltage lines (less over 11 kV), shackle insulators are utilized as strain insulators. However, to the high voltage transmission lines, strings about cap-and-pin (suspension) insulators are used, appended at the cross-arm level course. When the strain load in lines is exceedingly high, for example, at in length waterway spans, two or additional strings are utilized parallelly. In shackle insulation- right on time days- the shackle insulators are utilized concerning illustration strain insulators. In early days, they were frequently utilized for low voltage distribution lines. Such insulators could be possibly utilized on a level position or in a vertical position. They can be straightforwardly altered of the pole for a jolt or of the cross arm. Bushing empowers you to quit offering on that one; otherwise, few conductors will pass recipient segment, such as a divider or a tank, and insulate the conductors (IEC 60137:2003, 2003).

2.1 Line insulators

The overhead offering conductors can be upheld on the poles or towers to the extent that currents and flows from conductors don't stream to earth through supports; i.e. transport conductors must be appropriately insulated from supports. This will be attained toward securing line conductors on backs with the help of insulators. The insulators furnish fundamental insulation between transport conductors and backs, and along these lines forestall whatever leakage current from conductors with earth (Richardson, 2011). In general, the insulators have the accompanying alluring properties:

i. High mechanical strength in order to withstand conductor load and wind load.
ii. High electrical resistance of insulation material in order to avoid leakage of current flows to earth.
iii. High relative permittivity of insulation material in case dielectric capacitance is high.
iv. The insulation material ought to be non-porous, allowed from impurities and cracks; generally, the permittivity can be brought down.
v. High ratio of puncture strength to flashover. The large fraction regularly utilized material for insulators of overhead transport is porcelain, glass, steatite and exceptional piece materials that can be utilized to a set degree. Porcelain is processed by terminating toward a high engineering mixture of kaolin, feldspar and quartz. It may be stronger mechanically than glass, causing less issues from leakage and less influenced towards progressions of temperature.

2.2 Pin Type Insulator

The fruitful operation for an overhead line relies to a great degree upon the fitting determination of insulators. There are a few sorts of insulators; but the most regularly utilized are the pin type, suspension type, strain insulation and shackle insulation. Pin type insulators are utilized for transmission and distribution of electric power during voltages up to 33 kV. Past the working voltage of 33kV and beyond the operating voltage of 33 kV, the pin type insulators become too bulky and hence uneconomical (ANSI C29.1-1988, 1988; ANSI C29.11-1989, 1989; ANSI/IEEE Std 4-1978, 1978). The pin kind insulation is secured of the cross-arm on the pole. There is a groove on the upper limit of the insulation to lodge the conductor. The conductor passing through this groove is also certain toward the tempered wire of the same material as the conductor. Insulators are required to withstand both mechanical and electrical stresses. The later sort is fundamental because of line voltage and might cause the breakdown of the insulation.

The electrical breakdown of the insulation can happen in any case by flash-over or puncture. To flashover, an arc happens in the middle of the line conductor, insulation pin and the release hops crosswise over the air gaps, accompanying shortest separation. In the event of flash-over, the insulation will proceed to enactment over its best possible ability unless the amazing high temperature handled by the arc wrecks the insulation reason for insulation failure. In case of puncture, the release happens from conductor with pin through the figure of the insulation. When such breakdown is involved, the insulation is lasting press fabric wrecked because of unreasonable heat. In practice, addition thickness for porcelain will be given in the insulation to evade puncture toward the line voltage. The ratio of puncture strength to flashover voltage is known as security variable. This can allude to the fact that the strength of security variable is high, so the flash-over takes spot preceding the insulation gets punctured. To pin kind insulators, the worth of wellbeing component is around 10.

2.3 Suspension Type Insulator

The expense of pin sort insulation increments quickly as the attempting voltage is expanded. Therefore, this kind for insulation is not prudent Past 33 kV. For high voltages (>33 kV), it may be an ordinary act to utilize suspension kind insulators. They comprise many porcelain discs associated in arrangement with metal joins in the manifestation of a string. The conductor is suspended toward the base wind of this string while the opposite end of the string is secured of the cross-arm of the tower. Every unit or disc is intended to low voltage; for instance, if the working voltage is 66 kV, at that point six discs in arrangement will be given on the string (Gross, n.d.; IEC 815, ; Looms, n.d.; LWIWG-01, ; LWIWG-02, ; LWIWG-03,).

Advantages

1. Suspension kind insulators are less expensive than pin sort insulators for voltages Past 33 kV.
2. Each unit or disc of suspension kind insulation is planned to low voltage, usually 11 kV. Depending upon the working voltage, the wanted number of discs can be joined in arrangement.
3. In case the disc is damaged, the entirety string is not pointless because the harmed disc might be reinstated by the callous one.
4. The suspension plan gives more stupendous adaptability of the line. The association during the cross arm is that the insulation string is allowed to swing in whatever course and could consume the position in case mechanical stresses are base.
5. In the event of expanded interest in the transmission line, it is found more palatable with supplying the more excellent request by raising the offering voltage over to furnish other situated conductors. The extra insulation required to the raised voltage could be effectively acquired in the suspension course of action by including the required number of discs.
6. The suspension kind insulators are in the most part utilized with steel towers. The conductors run underneath the earthed cross-arm of the tower; therefore, this course of action gives halfway security from lightning.

Disadvantages

1. Suspension insulation is string costlier than pin and post sort insulation.
2. Suspension string obliges more tallness of the supporting structure than that to pin or post insulation to look for the same ground freedom of the present conductor.
3. The amplitude of allowed swing of conductors is bigger for the suspension insulation system; hence, more divisions between conductors and bolstering ought to be furnished.

2.4 Strain Insulators

When there is a dead end of the line or there is corner or sharp curve, the line is subjected to greater tension. In order to relieve the line of excessive tension, strain insulators are used. For low voltage lines (< 11kV), shackle insulators are utilized as strain insulators. However, to high voltage transmission lines, strain insulation comprises a gathering for suspension insulators. The discs of strain insulators are utilized within the level plane. When the strain for lines is exceedingly high, as in long waterway spans, two or more strings will be utilized parallely (Cherney, n.d.; Flink & Beaty, n.d.; Weedy & Cory, n.d.).

2.5 Shackle Insulators

In the past, the shackle insulators were utilized as strain insulators. At present, every now and then, they are utilized for low voltage distribution lines. Such insulators can be utilized possibly over a level position or for a vertical position. They could be straightforwardly altered of the pole with a jolt or of the cross arm. A shackle insulation altered of the pole and the conductor in the groove may be altered for a delicate tying wire (Ahmed et al., n.d.).

2.6 Potential Distribution Over Suspension Insulator String

A string of suspension insulators comprises many porcelain discs joined in arrangement through metallic joins. For a 3-disc string of suspension insulators, the porcelain part of every disc is amid two metal joins. Therefore, each disc types a capacitor "C". This may be known as shared capacitance or self-capacitance. Assuming that there is shared capacitance alone; afterwards, charging current might have been the same through every last one of the discs. Hence, the voltage over each unit might be the same V/3. However, in genuine practice, capacitance also exists between metal fitting of each disc and tower or earth. This is known as shunt capacitance C_1. Because of shunt capacitance, the charging current may not be the same through every last disc of the string. Consequently, voltage crosswise over every disc will be separate. Obviously, the disc closest to the line conductor will have the most extreme voltage. Subsequently, it can be inferred that V_3 will be substantially more than V_2 or V_1. The accompanying concentrations might be noted regarding the possibility conveyance over a string of suspension insulators:

1. The voltage inspired upon a string of suspension insulators doesn't disseminate itself uniformly crosswise over the single disc because of the vicinity of shunt capacitance.
2. The disc closest to the conductor has the greatest voltage crosswise over it. As we move towards the cross-arm, the voltage crosswise over every disc is driven towards decreasing.
3. The unit closest to the conductor is under most extreme electrical stress and will be prone to puncture. Therefore, methods must be furnished to rule out the possibility crosswise over each unit.
4. Though the voltage inspired crosswise over the string is dc, that point voltage crosswise over every unit might be the same. It may be on account that insulation capacitances are insufficient to dc.

2.7 String Efficiency

The voltage connected over the string of suspension insulators is not uniformly disseminated over different units or discs. The disc closest to the conductor has a higher possibility over the opposite discs. This unequal possibility distribution may be undesirable and is generally communicated as far as string efficiency. The ratio of voltage crosswise over the entire string of the item for the number of discs and the voltage crosswise over the disc closest to conductor is known as string efficiency (92 WM 265-9 PWRD,). String efficiency =Voltage crosswise over the string/ (n ×Voltage over disc closest on conductor). The place n = number of discs in the string. String efficiency is an important consideration, since it decides the potential distribution along the string. High string efficiency is just one of the more uniform voltage conveyances. Accordingly, 100% string efficiency is a perfect instance to which the voltage crosswise over every disc will make precisely the same. In spite of the fact that it is difficult to accomplish 100% string efficiency, deliberations ought to be made to enhance it concerning illustration near this strength as could be allowed. Possibility distribution on a string for suspension insulators is not uniform. The most extreme voltage is seen in the insulation closest to the line conductor and abates progressively as the cross arm is approached. If the insulation of the most elevated pushed insulation breaks down or streak-over takes place, the breakdown of other units will occur progressively. This necessitates equalizing the possibility over the different units of the string to enhance the string efficiency as follows:

1. By utilizing more cross-arms, the worth of string efficiency relies upon the strength of k, ratio for shunt capacitance will be common capacitance. The lowest worth of k is the string efficiency, and the voltage distribution may be one of the most uniform. The worth of k can be diminished by decreasing the shunt capacitance. In order to decrease the shunt capacitance, the separation of conductor from tower must be expanded i.e., more drawn out cross-arms ought to be utilized. However, limits for expense and strength of tower don't permit the utilization of Verwoerd long cross-arms. Practically, k = 0-1 which is the limit that can be attained by this technique.
2. Toward evaluating the insulators. In this method, insulators of distinctive measurements are used with the goal that each has a different capacitance. The insulators are capacitance graded; i.e., they are collected in the string in such an approach that the main unit has the base capacitance, expanding progressively as the bottom unit (nearest to conductor) is arrived at. Since voltage is conversely logical for capacitance, this system has a tendency to rule out the possibility distribution crosswise over the units in the string. This technique has the disservice represented in the extensive number of the different-sized insulators it requires. However, great results can be acquired by utilizing standard insulators for the greater part of the string and bigger units close to that of the offering conductor.
3. By utilizing a watch ring. The possibility crosswise over each unit in a string can be adjusted by utilizing a watch ring that is a metal ring electrically associated with the conductor and encompassing the base insulation, and the watch ring introduces capacitance between the middle of metal fittings and the line conductor. The watch ring is contoured over such an approach that shunts capacitance currents and flows, and so on. The equivalent to metal fitting is offering capacitance currents and flows and so on. The line shortages are the charging current i that streams through every unit of string. Consequently, there is a chance for uniform possibility distribution crosswise over the units.

3 POLYMERIC INSULATORS

3.1 Construction of polymeric Insulators

Polymeric insulators have a number of points of interest in the ceramic and glass insulators, as shown in Fig. (1), such as beneficial execution in defiled environment, light weight, no difficult handling, support free, and considerably low expense.

Core: the center may be the inside insulating and only a composite insulation. It is exceptional for conveying the mechanical load. It is comprised mostly of glass fibers positioned over a resin matrix in such a way as to attain most extreme rubbery.

Housing: the lodging is outside the center and protects it from the climate. It might be prepared for climate sheds. A percentage of plans about composite insulators utilize a sheath made of protecting tape in the middle between the climate sheds and the center. This sheath may be only the lodging.

Climate sheds: - climate sheds are insulating parts, projecting from the lodging or sheath, proposed with expansion for the the leakage separation to give a hindered way for water drainage.

Wind Fitting: wind fitting transmits the mechanical load of the center. They are typically constructed out of metal.

Coupling Zone: The coupling zone is only the limit fitting that transmits the load of the line, of the tower, or of any other insulation. It doesn't incorporate the interface between the middle of the center and the fitting.

Interface: - An interface is the surface between different materials. Examples of interfaces in composite insulators are as follows: a. Glass fiber/impregnating resin, b. Filler/polymer, c. Core/housing, d. Housing/weather sheds, i.e., Housing/end fitting.

The ceramic insulators have been utilized within the energy industry for more than a decade. This material endures harsh environmental. These are stable structures due to their atoms' seal pack configuration and strong ionic bonding between them which helps to resist environmental stresses. Therefore, ceramic housing and bolstering ought not be influenced by concentrations similar to ultimo violet rays, surface electrical action and the firm nature of material. The housing utilized to bushing around high voltage gears give acceptable self- materials for strength. The fundamental crude materials of these insulators have clays, feldspar and quartz which are modest and promptly open (Amin, n.d.). Since 1970, the utilization of polymeric insulators for open air high voltage insulation has expanded so dramatically that ceramic insulators are becoming outdated in the energy industry. Electrical organizations have begun utilizing polymeric insulators in distribution voltages et cetera bit by bit toward low voltage transmission but towards 765 kV ac and 500 kV dc (Manias et al., 2001). The polymeric insulators have exhibited best sullying streak in abilities and consequently would be routinely deployed in different situations. The purpose behind the notoriety of polymeric insulation materials are their various advantages, so that the thickness of polymeric materials may be lesser than porcelain materials, which makes constructions and erection for insulators simpler and snappier. The lighter weight permits the utilization of less expensive structures and settling game plans, and additionally has, lesser transportation cost. Creepage separation for every unit length is generally higher in polymeric insulators than porcelain insulators. AC wet flashover will be avoided by giving work to exchanging breadth climate sheds to polymeric insulators which produce their structure, as one of the most intricate issues. The composite insulators uncover an

Figure 1. Insulator Construction (Manias et al., 2001)

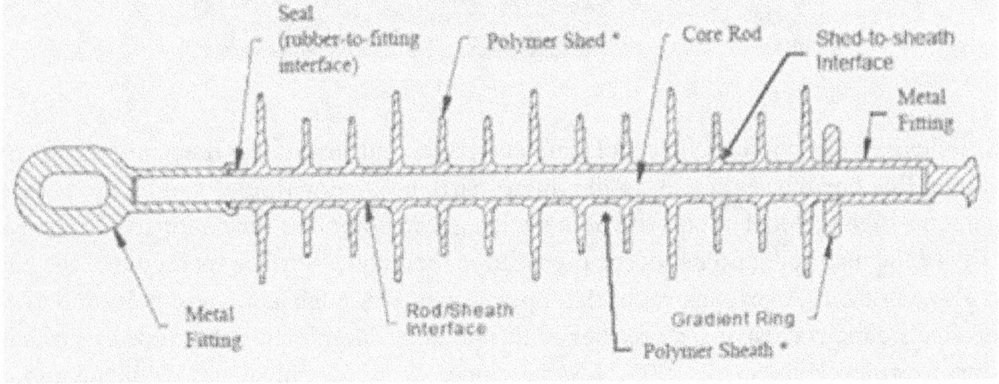

exceptional electrical execution dirtied nature's domain because of their hydrophobic properties. The conductive tainting suspended inside the water globules may be spasmodic because of framing of water globules on the polymeric surface. This causes easier leakage current and the likelihood of dry band arcing. The physical properties of the polymer material direct that these materials won't crush like porcelain materials. For an instigative of an internal fault, the required failure system may be rupturing, prompting an outside streak through and dispersal of the vitality outside of the lodging. The acceleration stresses dependent upon the quit offering on one gram might be withstood without harm by polymeric insulators because of their lighter weight, high damping element and high strength plan uniqueness.

3.2 Nano-Dielectric Composite Materials

For outdoor insulation, two materials are regularly utilized for the lodging ethylene propylene rubber and silicone rubber. SiR is favored because of its characteristic hydrophobicity and its useful execution in dirtied situations. A small number of a little measure (2wt.% up to 5wt%.) of inorganic nanoparticles to polymers ought to be added for mechanical and thermal strength and execution change (Dengke et al., 2004). A polymer nanocomposite can be defined as a system in which at least one dimension of filler is on a nano metric scale. In the polymer matrix, a uniform filler distribution is always desired. The creation of a positive position connection between the middle of the polymer and the nanoparticle is obligatory to avoid phase separation and agglomerated, as shown in Fig. (2).

Figure 2. SEM characterization for structure of polymer nanocomposite

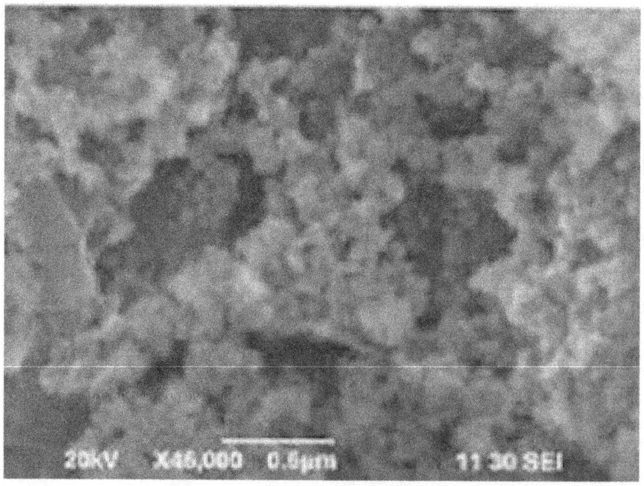

Nanoparticles exist in spherical, tube and platelets shapes and are not less than nanometers unless you quit offering on that one measurement. In the micro-particles, nanoparticles can be made by softening dependent upon a high molecule. It is critical to avoid agglomeration of the nanoparticles and guarantee great security of the matrix regardless of the preparation technique. Surface medicine of the nano extent particles is also workable. An expansive reach for polar operators, such as silanes, poly alcohols and non-polar agents like stearic corrosive and greasier acids will be economically accessible as surface medication operators for fillers (Fuse et al., 2005). LS/clay minerals are the only class of silicate minerals, such as phyllosilicates and they incorporate characteristic and engineered clays, such as asmica, bentonite, magadiite, laponite and fluorohectorite. LS are the practically and generally utilized 2-D nanoparticles in different fields. The structure of LS comprises a 2-D layer of two combined silicate tetrahedral sheet with an edge-shared octahedral sheet of metal iotas, such as AL or Mg. Polymer/silicate nanocomposites provide a magnetic technique to enhance the polymer properties, such as stiffness, strength and obstruction properties without any progress in the preparation strategy. Around distinctive sorts of nanotubes, carbon nanotubes are practically the broadly used, carbon nanotubes made of allotropes of carbon also have a place in the fullerene structural gang. For nanotubes, the diameters are in the range of a couple of nanometers; however, they are millimeters or indeed going centimeters in length. Therefore, these nanotubes have a high part ratio, thereby conferring high strength of the polymer with a little weight percent. The fantastic properties of carbon nanotubes can have an outcome regarding its holding nature. Carbon nanotube-rein powered composites enjoy specific enthusiasm in the field of material science to create fundamentally lightweight solid materials. A major issue in this field may be the failure to accomplish a homogeneous scattering for nanotubes in the polymer matrix because of these amassed tubes. Nanoparticles with three measurements in the nanometer administration are the circular nanoparticles for the most part obtained by a sol-gel procedure (Mark, 1996). In sol-gel processes, the organic /inorganic hybrid materials can be formed by the condensation reaction between the functionalized prepolymers and the metal alkoxides leading to the formation of a chemical bond between the polymer and the inorganic filler. Consequently, the consolidation of filler particles in polymers through the sol-gel methodology avoids the amass of filler. Silica, TiO_2, $CaSO_4$, $CaCO_3$, ZnO, $ZnFe_2O_4$, are generally utilized in inorganic nanoparticle in the polymer field.

4 NANOPARTICLES FOR OVER HEAD TRANSMISSION LINE INSULATORS

4.1 Theory and Effective Parameters

Outdoor insulators and Open-air insulators structure essential analytics, and only these transmission systems have noteworthy sway with respect to system dependability (Looms, 1987). The most recent couple A long time has seen the expanded utilization of composite (Also known as non-ceramic or polymeric) insulators for the greater part voltages because of preferences for light weight, simplicity of establishment and easier expense. Composite insulators have phenomenal contamination execution because of the hydrophobicity of the lodging material (Gorur et al., 1999). The hydrophobicity of lodging materials, such as silicone rubber might be a foundation crown in the amazing binds of the water droplets and in the prompt restricted debasement of the lodging (Phillips et al., 1999). Numerous sorts of insulation strings are utilized within overhead transmission line for protecting material overhead line conductors presenting with the supports like: Pin type, suspension kind and strain kind strings (Doshi et al., 2011). Nanotechnology engineering will be the later inquiry about the advances to upgrading execution for material results and modern requisitions. In this field, distinctive hypotheses and techniques are concentrated on model, design, manufacture and portray nano-engineered structures and gadgets of sensors, electronics, bio-techniques, and propelled materials.

As of now, a consistent advancement of insulators will be exceptional by utilizing nanotechnology techniques. The principal resin gets for this contemplate may be to upgrade the physical and electrical properties of insulator materials by utilizing of nanoparticles of metals which exacerbate settled inside an insulation dielectric matrix. Produces of the new suggested insulators can be altered and synthesized by utilizing low devices and costs. Size, structure, and centralization of nanoparticles have exhibited the electric and optic properties of insulator materials that make them proficient (Ebnalwaled & Thabet, 2016; Kumagai & Yoshimura, 2004; Mehta, 2005; Patel & Singh, 2015; Thabet, 2015b; Thabet, 2016; Thabet, 2017; Thabet & Ebnalwaled, 2017; Thabet & Mubarak, 2017a; Thabet & Mubarak, 2017b). Therefore, the successful nanoparticles and their concentrations for upgrading voltage distribution for universal insulators string have been predicted. The fundamental issue for suspension insulators hosting a string of indistinguishable twin insulation discs will be non-uniform distribution voltage through the string. The closest insulation unit of the offering conductor has the most extreme voltage gradient; thus, they are pushed to their most extreme suitable in contrast to the others. The string efficiency will be a measure of the usage of materials in the string (Doshi et al., 2011; Kumagai & Yoshimura, 2004; Mehta, 2005; Patel & Singh, 2015). The line unit may be constantly under the most extreme stress. A percentage deliberation can be aggravated to decrease the ratio of shunt capacitance to self-capacitance, capacitance grading, and static protecting (Mehta, 2005). In this research, we have succeeded in upgrading the efficiency and high voltage distribution over insulators string by utilizing the nanotechnology techniques. Furthermore, the ideal sorts and concentrations of nanoparticles for upgrading efficiency and high voltage distribution of insulators string have also been specified. Nanotechnology techniques have been attained by decreasing the ratio of shunt capacitance to self-capacitance toward utilizing specified nanoparticles. As demonstrated in Fig. (3.a), a string for suspension insulators comprises a number of porcelain or glass discs joined in arrangement through metallic joins. Because of shunt capacitance (C1) of the middle metal fitting of each disc and tower or earth, the charging current is not the same through every last disc of the string, as demonstrated in Fig. (3.b). Hence, the cross arm or tower and the voltage over every unit is V_1, V_2 and V_3 respectively (Mehta, 2005; Patel & Singh, 2015).

Figure 3. Voltage distribution over three discs insulating string

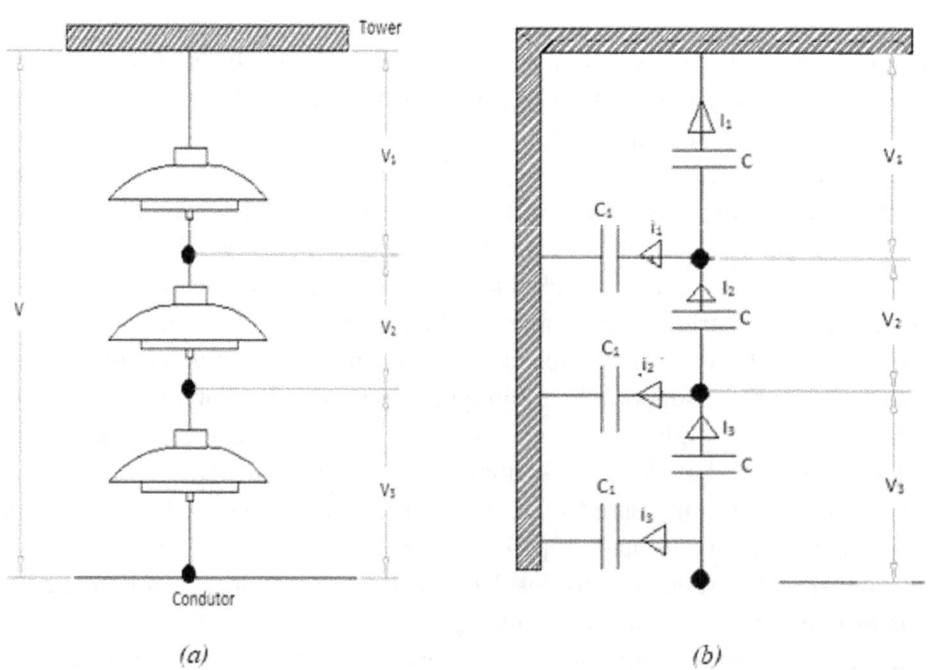

In case the self-capacitance of each disc is C and the shunt capacitance C1 is some fraction K of self-capacitance i.e., C1 = KC. The string voltage is defined as follows:

$$V_{ph} = V_1 + V_2 + \dots + V_n \tag{1}$$

$$V_2 = V_1(1 + K) \tag{2}$$

$$V_3 = V_1(1 + 3K + K^2) \tag{3}$$

The following points may be noted from the above mathematical analysis: The voltage across discs increases from top to down progressively as the cross-arm is approached. The value of (K=C1/C) is the guiding parameter for the potential across the discs and the string efficiency. The short string has more efficiency than the larger one.

$$\zeta_{string} = \frac{V_{ph}}{n \times V_n} \times 100 \tag{4}$$

4.2 Disc insulator Nanocomposites

Power law model relationships are utilized within dielectric demonstrating of composite systems (Brosseau et al., 2001; Gershon et al., 2001; Karkkainen et al., 2000; Nelson, 2001; Ozmusul & Picu, 2002). The general two segments of the power-law model for complex permittivity have been successful in an extensive variety of materials systems, including air-particulate composites (Nelson, 2001), ceramic-ceramic composites (Gershon et al., 2001) and polymer-ceramic composites (Brosseau et al., 2001). The composite system holding filler, interphase and matrix districts is illustrated in Fig. 4. The element resin parts for composite system are matrix, filler and the interphase locale that is subordinate upon the aspects of the filler part and its thickness encompassing every nanoparticle (Ozmusul & Picu, 2002).

Figure 4. The interphase region surrounding the filler particles in a composite system

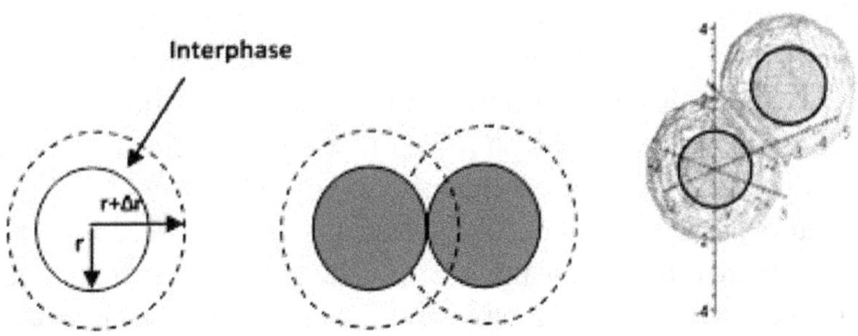

The successful permittivity model of the two-component systems utilizes the volume fraction part (Karkkainen et al., 2000). In addition, dielectric models for mixtures of composite constituents have been developed, including the Maxwell-Garnett model and the Bruggeman model (Tagami et al., 2010). The two-component power-law association is effortlessly broadened to a three-part composite system, as depicted by:

$$\varepsilon_c^{\beta} = \varnothing_f \varepsilon_f^{\beta} + \varnothing_i \varepsilon_i^{\beta} + \varnothing_m \varepsilon_m^{\beta} \tag{5}$$

Note that a composite model that holds an interphase region, ϕf, ϕi, and ϕm representable in the volume fraction of the filler component, the interphase locale and the matrix part from the composite, respectively, εc, εf, εi and εm has aided the complex permittivities of the composite, the filler, as well as the interphase and matrix districts of the composite, respectively. β is a dimensionless parameter referring to the state and introduction of the filler particles inside the mass composite. A large number of researches have focused on the powerful permittivity of the composite in view of interphase volume fraction, ϕi, and may be dictated to monodisperse, circular particles (Thabet & Mobarak, 2010; Todd & Shi, 2003).

4.3 Specification of Disc Insulator Nanocomposites

COMSOL Multi-physics is a finite element analysis, solver and simulation software / FEA software package for different material science and engineering applications, particularly coupled phenomena, or multi-physics. For traditional physics-based client interfaces, COMSOL Multi-Physics additionally permits entering coupled systems of incomplete differential equations (PDEs). Fig. (5.a) indicates measurement of porcelain and glass disc of insulation string size. Fig. (5.b) reveals the polymeric disc of insulation string extent at finally 1604mm. Thus, the impacts of nanoparticles in high voltage distribution of porcelain or glass insulators string with the Emulating specifications are taken into consideration as follows:

- 66 KV insulating string has 7 discs, the diameter of each is 260mm, and its height is 170mm.
- 132 KV insulating string has10 discs, the diameter of each is 270mm, and its height is 137mm.
- 220 KV insulating string has 18 discs, the diameter of each is 325mm, and its height is 180mm.
- 500 KV insulating string has 30 discs, the diameter of each is is 350mm, and its height is 194mm.

Porcelain insulators are produced from clay, quartz or alumina and feldspar, and are secured with a smooth birch glaze with shed water. Porcelain has a dielectric strength of around 4–10kV/mm. Porcelain is a fantastic insulation to use in high voltage applications, particularly over open-air provisions like terminals to high-voltage cables, bushings of power transformers, insulation of high back antennas and a significant number different segments. Silicon rubber may be utilized as high voltage insulation (10-500kV) and is known as polymer insulator or silicone composite insulator. On the other hand, silicone rubber composite insulators have many points of interest such as fantastic anti-pollution performance, suitableness to high dirtied area, free of cleaning…etc. The recommended nanoparticles are the best nanoparticles around feathery powders for a greatly low thickness. Thus, the proposed nanoparticles have round molecule shape and have the most imperative aspects to upgrading insulators materials. Consequently, electric and dielectric properties of insulators and nanoparticles materials have been measured and nitty gritty, as demonstrated in Table 1.

Figure 5. Disc of insulator string dimension

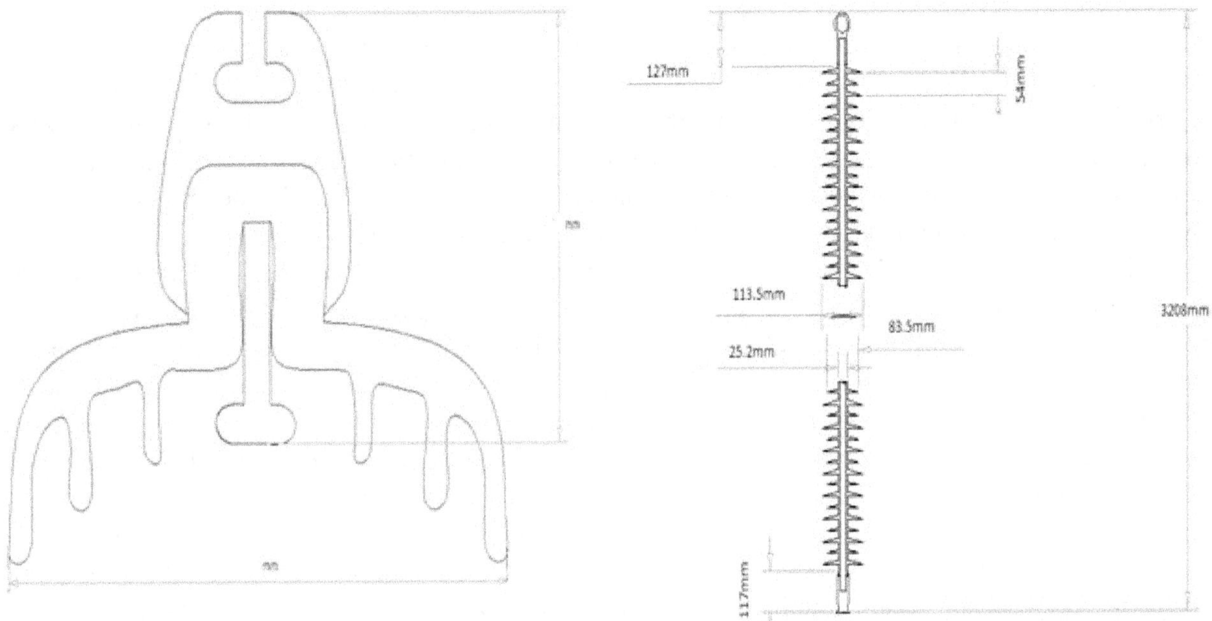

(a) Porcelain design

(b) Silicon rubber design

Table 1. Electric and dielectric properties of polymeric and nanoparticles materials

Materials & Nanoparticles	Dielectric Constant
Porcelain	5.56
SiR	4.3-6.5
Clay	1.8
ZnO	1.7-2
MgO	9-9.7
SiO_2	4.5
TiO_2	10-48
Al_2O_3	9-9.5
Silica	2.5-3.2
Ba_2TiO_3	3.8-4.5
Fe_2O_3	25
Fiber Glass	7

4.4 Effects of Nanoparticles on OHTL Insulators

One of the main challenges in manufacturing high execution for glass insulators nanocomposites is accomplishing planning and kind of the nanoparticles for base matrix. This model is indispensable on account of beneficial suitableness nanoparticles, and their concentrations in the base matrix is the groundwork for acquiring guaranteeing material properties. Currently, in this section, the numerical and simulation results connected to the new nanocomposites insulators string to aggravating efficiency and high voltage distribution. On the other hand, the polymeric insulators have easier thickness over porcelain materials and lighter weight that permits the utilization of less expensive structures and altering courses of action for easier transportation expense. Creepage separation for every unit period may be normally higher for polymeric insulators than porcelain and glass insulators. Polymeric nanocomposites materials accomplish outlining and determination of the nanoparticles in a polymeric matrix. The emulating outcomes indicate the numerical and simulation results that, connected on the new nanocomposites insulators, string to aggravating efficiency and high voltage distribution. As shown in Fig. (6), the controlled capacitance of disc insulation porcelain in view of the nanoparticles sorts and their concentrations are indicated. It is self-evident that titanium di-oxide and alumina nanoparticles have the capacity of expanding capacitance for porcelain insulators. However, different nanoparticles help to diminish the capacitance of string insulators. High permittivity of nanoparticles is a powerful property for expanding disc capacitance and upgrading the string efficiency. The impact of nanoparticle sorts and their concentration effects on efficiency of porcelain insulators string under 66kV & 132kV individually.

The efficiency of porcelain insulating strings nanocomposites (66 kV and 132kV) increments for expanding nanoparticles concentrations of titanium di-oxide (TiO2) and magnesium oxide (MgO), but

Figure 6. Effect of nanoparticles on disc capacitance of Porcelain suspension insulators

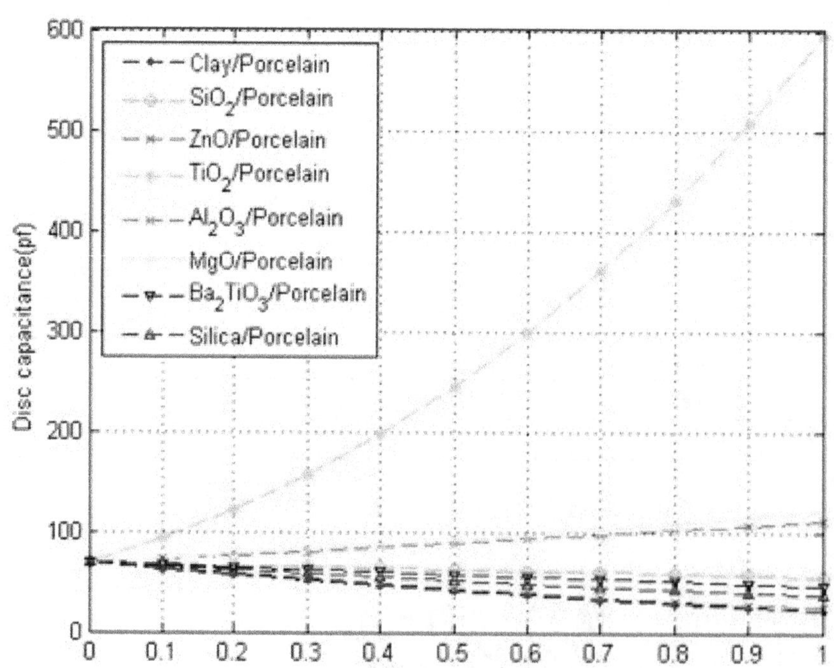

Figure 7. Efficiency of Porcelain suspension insulators nanocomposites string at 500 kV

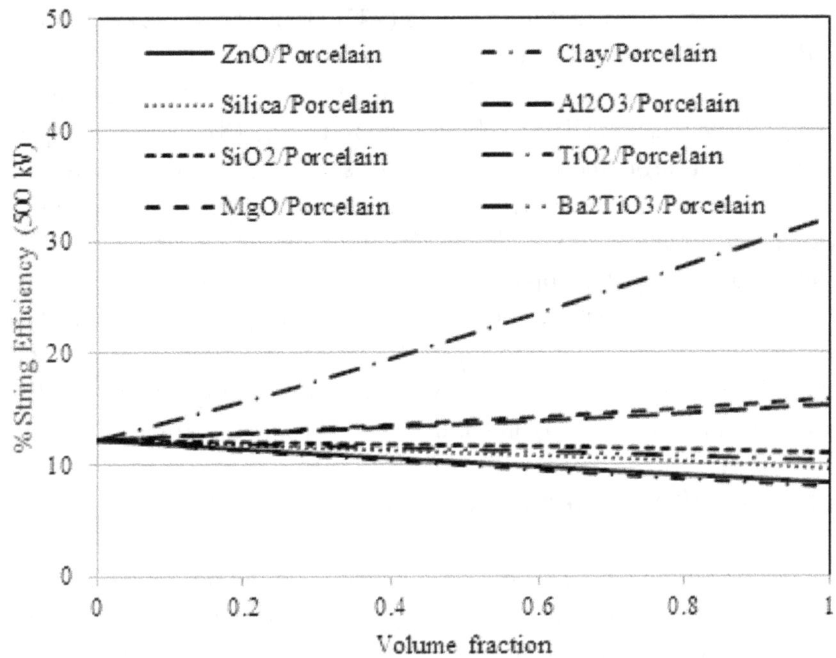

Figure 8. Efficiency of silicon rubber suspension insulators nanocomposites string at 220kV

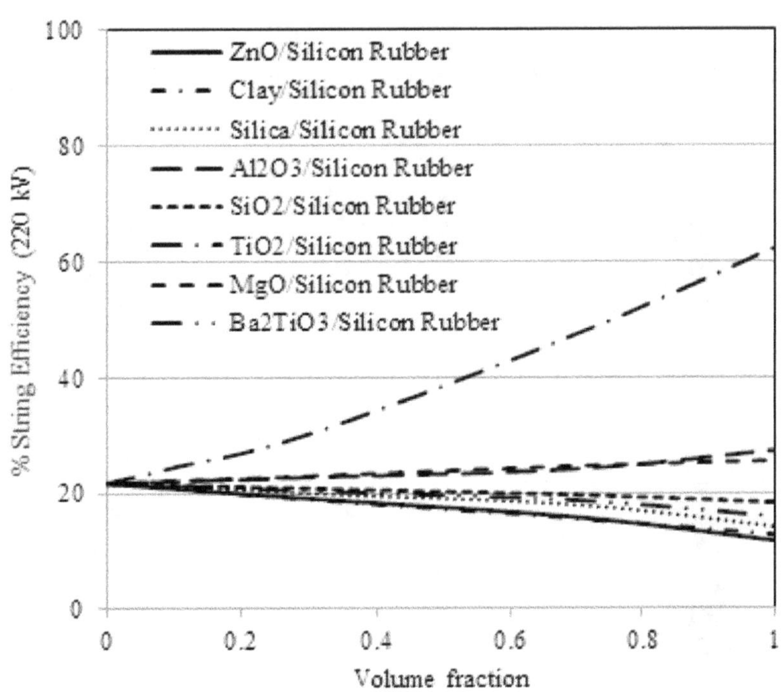

the different nanoparticles have a negative impact on the same porcelain suspension string insulators appraisals. On the other hand, the impact for alumina (Al2O3) nanoparticles is shown for porcelain insulating strings under 500kV as demonstrated in Fig. (7) respectively. The efficiency upgrade rate that occurs for porcelain suspension string under 66kV is higher than the same suspension string nanocomposites under 132kV, 220kV, and 500kV. For rubber insulators, Fig. (8) shows the impact of nanoparticle sorts and their concentrations around efficiency insulators string (220kV) individually. The efficiency of rubber suspension insulators nanocomposites increments for expanding nanoparticles concentrations of titanium di-oxide (TiO_2), magnesium oxide (MgO) and alumina (Al_2O_3) nanoparticles. Constantly, different nanoparticles have a negative impact on the same glass suspension string insulators appraisals and the efficiency upgrade rate, so that the rubber suspension string under 66kV will be higher than the same suspension strings nanocomposites under 132kV, 220kV and 500kV. Finally, it is delineated that the titanium di-oxide (TiO_2) nanoparticles are the best inclusions to upgrading the efficiency for insulators string, then, magnesium oxide (MgO) and alumina (Al_2O_3) nanoparticles.

4.5 High Voltage Distribution over Suspension Insulators

Figures (9) shows the high voltage distribution for pure and nanocomposites porcelain suspension insulators string at 132kV. It is noteworthy that high voltage distribution to pure porcelain insulation string may be non-uniform; high voltage on the closest disc from the existing conductor and low voltage on the top disc are indicated looking into shades. In case of utilizing titanium di-oxide nanoparticles, it can be discovered that titanium di-oxide nanoparticles upgrade high voltage distribution to decrease the high voltage on the closest disc from live conductor and expand voltage on the top disc, as indicated in shades for 30wt. % TiO_2/Porcelain nanocomposites. Titanium di-oxide nanoparticles are additionally elucidating

Figure 9. High voltage distribution for Porcelain suspension insulators string at 132kV

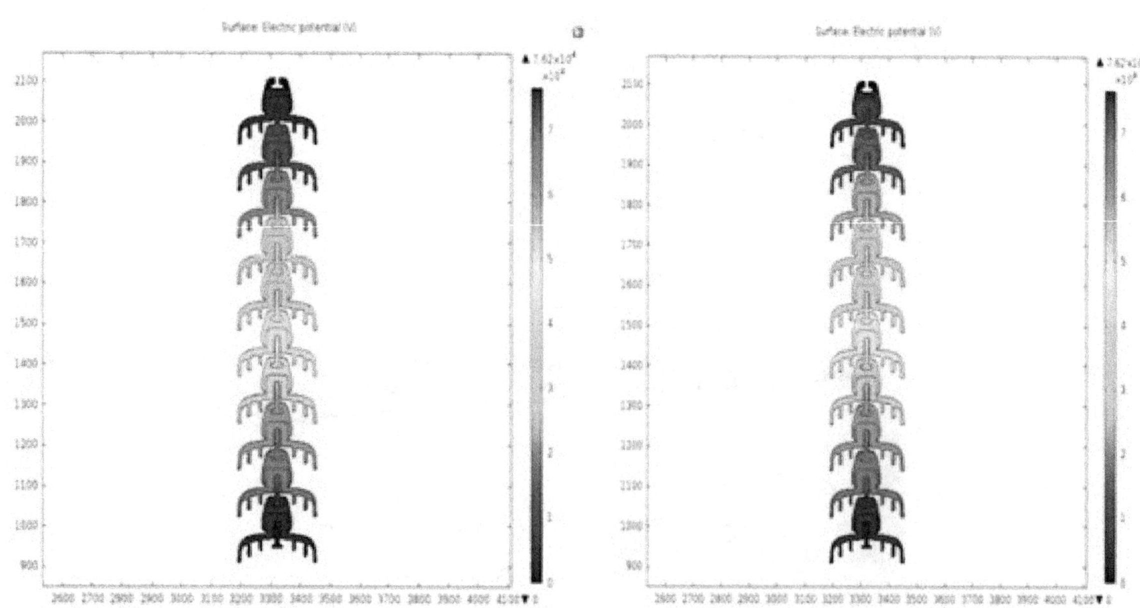

(a) Pure Porcelain_132kV (b) 30wt.% TiO2/ Porcelain_132kV

Figure 10. High voltage distribution for Porcelain suspension insulators string at 220kV

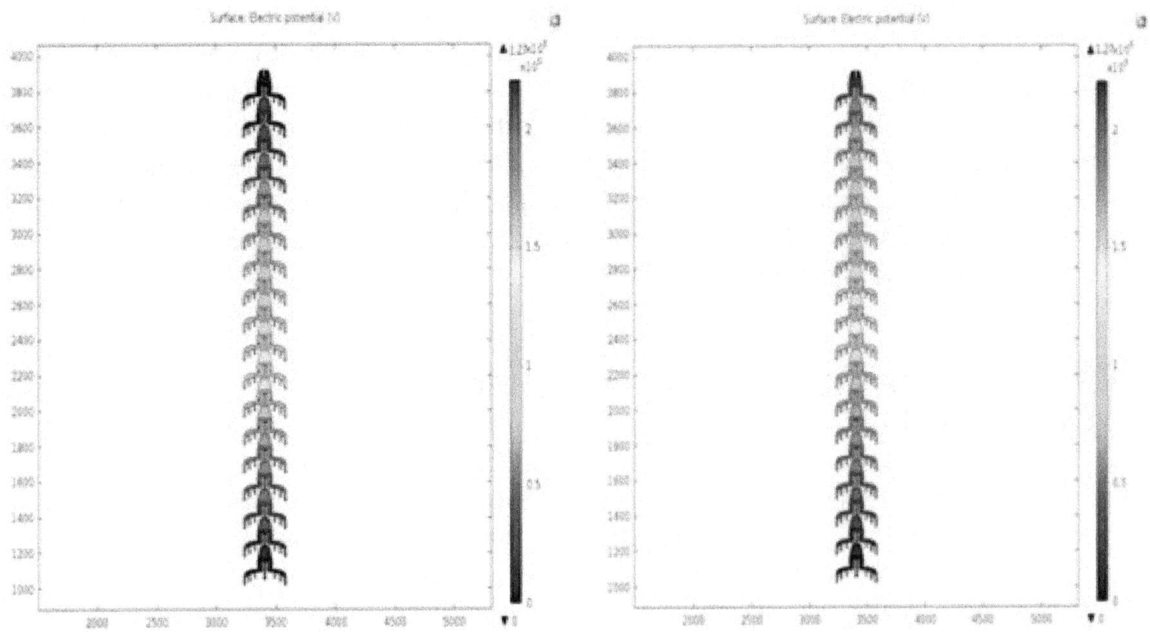

(a) Pure Porcelain_ 220kV

(b) 30wt.% TiO2/ Porcelain_220kV

Figure 11. High voltage distribution for Silicon Rubber suspension insulators nanocomposites string at 220Kv

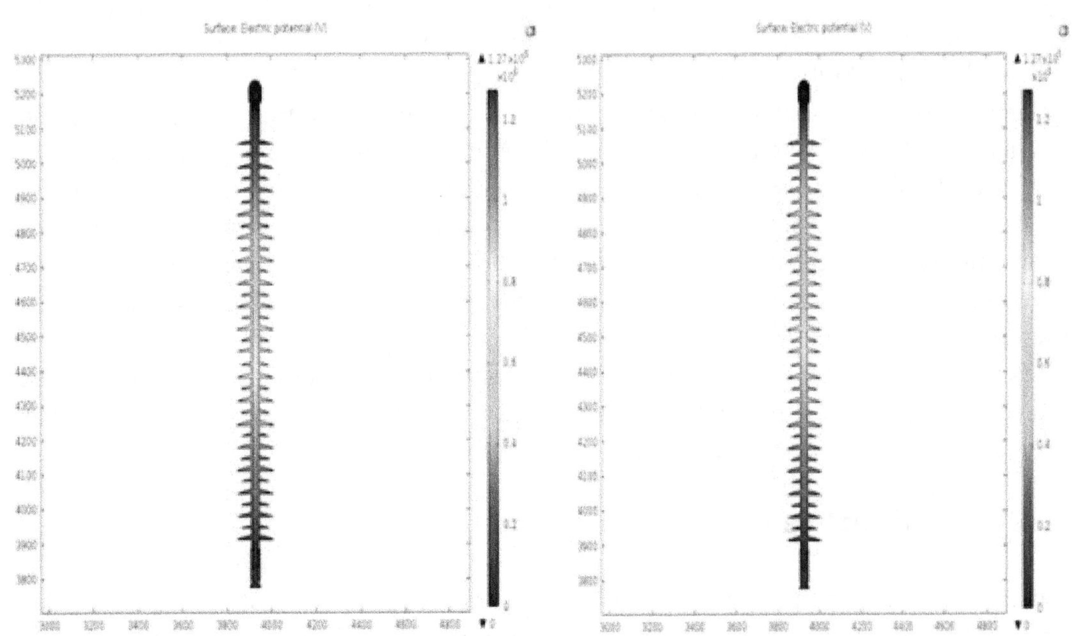

(a) Pure Si, Rubber _220kV

(b) 30%wt. TiO_2/Si. Rubber_220kV

and proficient for upgrading insulators string over upgrade high voltage distribution in insulators string. On the other hand, it is recognized that high voltage distribution to pure polymeric insulation string is non-uniform; the high voltage on the closest disc from live conductor and the low voltage on the top disc concerning illustration is indicated on shades. In the event of utilizing titanium di-oxide and magnesium oxide, as shown in Fig's (10, 11), the high voltage distribution is more uniform decreasing high voltage on the closest disc from carrying the conductor and expanding the voltage on the top disc, as indicated in shades. In any case, the opposite recommended nanoparticles expand the non-uniform high voltage distribution to a high voltage on the closest disc from live conductor and decrease voltage on the main disc of silicon rubber nanocomposites.

5 MULTIPLE NANOPARTICLES FOR OVER HEAD TRANSMISSION LINE INSULATORS

Composite insulation causes few issues throughout operation as stated by operation reports, such as following and disintegrating of the shed material, chalking and crazing of insulation surface, decreasing of tainting flashover stress and crumbling of mechanical stress (Schneider et al., 1989). The primary reason behind the vast majority of issues in composite insulation are the irregular distribution of voltage and electric field along the insulation string. It is known that 70% of the connected voltage concentration from the beginning 22% of string length close to live end are contrasted with the center territory of an insulation string, for a 34.5 kV composite insulation (Que et al., 2007). Encounters show the voltage distribution along the entirety insulation string seems with a "U" style distribution (Dominguez et al., 2013; Mei et al., 2011; Thabet & Mobarak, 2016). On the other hand, later researches have indicated the capacity of nanotechnology to enhance the electrical properties of polymer materials. Recent nanotechnology techniques have helped in the new creation of insulators materials (Karkkainen et al., 2000; Thabet, 2016; Thabet & Mubarak, 2017b; Todd & Shi, 2003).

This chapter displays new strategies to move forward the irregular voltage conveyance of composite insulation dependent upon two techniques. The main technique is inserting nanoparticles and different nanoparticles in the silicone rubber climate shed. The second strategy is covering composite insulation for nanoparticles materials. These two strategies have expanded the self-capacitance of composite insulator for moving forward electric field stress connected on insulators. Moreover, three sorts of 500 kV composite insulators models required to be planned based on silicon rubber polymer have been investigated; the first type is silicon rubber with nanoparticles, the second type is silicon rubber with multiple nanoparticles, and the third type is silicon rubber coated with nanoparticles. The insulation model for tower and conductor has been secured, and the voltage conveyance and electric field stress have been ascertained toward limited component strategy (FEM) programming ANSYS.

5.1 Theory and Effective Parameters

Power law model relationships are utilized within dielectric demonstrating the composite systems, and a multi-nanoparticles procedure has been suggested for creating the electric and dielectric properties of polymer. Additionally, the interphase thickness has been evaluated (Karkkainen et al., 2000; Thabet & Salem, 2017; Thabet et al., 2018; Todd & Shi, 2003).

The effective dielectric constant of the inclusion and interphase can be expressed as:

$$\varepsilon_E = \frac{\varepsilon_j(b)+T(b)}{\varepsilon_j U(b)+V(b)+V(b)} \tag{6}$$

Where εj is the dielectric constant of the second Nanoparticle, the dielectric constant of the interphase sphji can be determined as follows:

$$\varepsilon_E(b) = \varepsilon_{phji} + \frac{\left(\frac{a^3}{b^3}\right)}{\frac{1}{\varepsilon_j+\varepsilon_{phji}} + \frac{\left(1-\frac{a^3}{b^3}\right)}{3*\varepsilon_{phji}}} \tag{7}$$

Eq. (2) is used to find the component composite system, as described by

$$\varepsilon_{effj}^{\beta} = \varphi_j \varepsilon_j^{\beta} + \varphi_{effi} + \varphi_{effi}\varepsilon_{effi}^{\beta} \tag{8}$$

Where, ϕj is the volume fraction of the second Nanoparticle component of the multi-composite system. The interphase volume fraction of multi-composite system is calculated by:

$$\varphi_{phij}(1+F)(S_j \bullet \nabla r)\bullet \rho_j \bullet \varphi_j \tag{9}$$

Where Sj is the specific surface area of second Nanoparticle measured in (m^2/g).

Composite insulators typically produce a fiberglass pole to mechanical load and a polymeric material lodging which comprises climate sheds and sheath. The insulators have been prepared with metal fittings at both lines and ground finishes. The streamlined geometry and extents of the insulators to be displayed to this investigation are demonstrated in Fig. 12. There are 44 climate sheds on the lodging. An evaluating gadget (ring) may be put in both the carry on with and the ground conclusion (Zhao & Comber, 2000).

Figure 12. Simplified geometry and dimensions of the non-ceramic insulators to be modeled: (a) fiberglass rod; (b) polymeric material weather sheds; (c) polymeric material sheath; (d) metal end fitting.

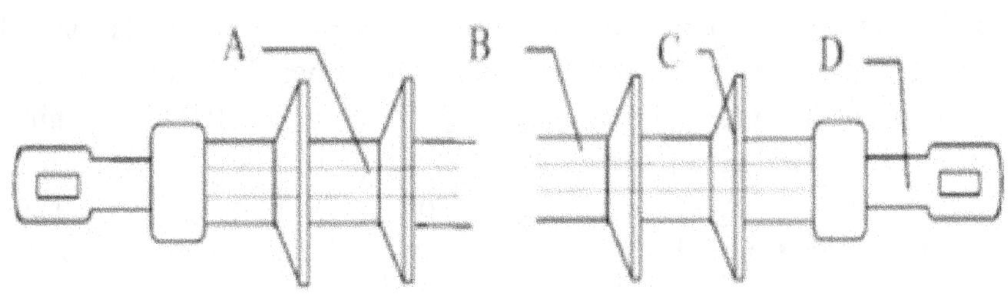

A 500 kV model setup according to actual dimensions of 3-D model designed by TCI Company catalog for type (FXBW1-500/100) composite insulator, as shown in Table 2.

Table 2: Dimension of Non-Ceramic Insulators String 500 kV

Insulator Components	Dimensions
Rated Voltage	500 kV
Section height	4030 mm
Min. arcing distance	3600 mm
Min arcing distance	171/85 mm
Shed Spacing	79 mm
Dia. corona ring	400 mm

5.2 Polymeric and Nano-Metric Specifications

As shown in Table 1, it specifies the dielectric constant of recommended nanoparticles utilized for upgrading the execution of silicon rubber nanocomposites and multi-nanocomposites insulators string for admiration to the altered power law model (Thabet & Salem, 2017; Thabet et al., 2018).

5.3 Effects of Multiple Nanoparticles on OHTL insulators

The electric field conveyance along insulation has been evaluated, and the numeric product ANSYS has been utilized in the count (McCalley & Krishnan, 2014; Zhao & Comber, 2000; Zhao et al., 2015). A limited component system may be utilized to comprehend these useful issues about electrical field distribution on non-ceramic insulators string. It has been investigated by looking into simulation for non-ceramic three models of insulators nanocomposite, multi-nanocomposites and covering system about composite insulators that were set up. The cross-arm and conductor were taken into account, and the genuine operation circumstances in insulation model materials have been recreated. Calculations of electric field stress on SiR insulation string through longitudinal and width areas have been illustrated, as indicated in Fig's (13-15). Advanced models show the impact of nanocomposite insulators materials upon the electrical field distribution for insulation string. It is recognized that the quarter of string has high electric field stress because of introducing carry on with end at the same time of including a concentration "3wt.% Fe_2O_3 or 5wt.% Fe_2O_3".

Nanoparticles to pure silicon rubber diminish the electric field stress during the quarter of string by 5.8% or 8.4% individually. The kind and concentration of nanoparticles evolving dielectric constant of the insulation piece material disc that evolves the self- capacitance of insulation will transform the electric field distribution over every last one of the insulation strings. In this way, it is clear that including "5wt.% Clay" nanoparticles expand the electric field stress up to 1.4%. Concerning the illustration of recent researches, the impact of nanoparticles by looking into the electric field stress with respect to the sir insulation string varies in view of their sorts and concentration qualities. As shown in simulated models, the calculations of electric field stress around sir insulation string again longitudinal and width areas for sir insulation string. Concerning recent researches (Thabet, 2015a; Thabet, 2020; Thabet et al., 2019), the impact of utilizing multi-nanocomposite insulators materials on electrical field distribution of insulation string has been investigated. It is recognized that the begin with quarter for string has high electric field stress because of introducing carry on with wind at the same time of including a concentration "(3wt.% Clay+5wt.% Fe_2O_3) or (3wt.%Clay+10wt.%Fe_2O_3) for multi-nanoparticles to pure silicon rubber, diminishing the electric field stress during the to begin with quarter of string to 8.2% or 12.4% respectively. Hence, it is clear that including (3wt.%Clay+10wt.%ZnO) multi- nanoparticles expand the electric field stress up to 12.2%.

The impact of nanoparticles for electric field stress on sir insulation string varies in view of their sorts, concentration values and others in the game plans for nanoparticles inside build matrix of different nanocomposites materials. This is in terms of what is stated by the calculations of electric field stress looking into sir insulation string over longitudinal and width areas for SiR. In addition, it is recognized that the to begin with quarter for string has high electric field stress because of available carry on with limit at the same time of including a concentration "(3wt.% Al_2O_3+10wt.% Fe_2O_3)" of multi-nanoparticles to pure silicon rubber, declining the electric field stress during the begin with quarter of string to 12.4%. Moreover, is shown that including (3wt.% Al_2O_3+10wt.% ZnO) multi-nanoparticles expand the electric field stress dependent upon 12.6. Additionally, the electric field distribution around sir insulation string has been transformed in light of type, concentration and course of action of multi-nanoparticles that will change the dielectric constant of build matrix SiR polymer material. Moreover, the impact of multi-nanocomposite insulators materials with respect to electrical field distribution of sir insulation string on longitudinal and width areas is investigated. Thus, the to begin with quarter of string has high electric field stress because of the display of live end while including a concentration "(3wt.% ZnO+10wt.% Clay)" of multi- nanoparticles that decline the electric field stress during the to begin with quarter of

Figure 13. Electric field distribution across composite insulator with multi- nanocomposites materials

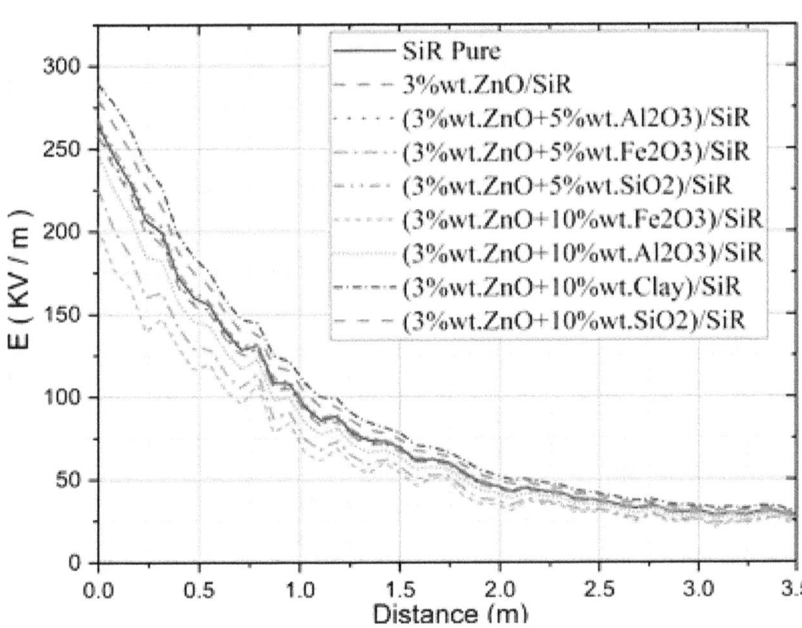

Figure 14. Electric field distribution at wide section with multi-nanocomposites materials

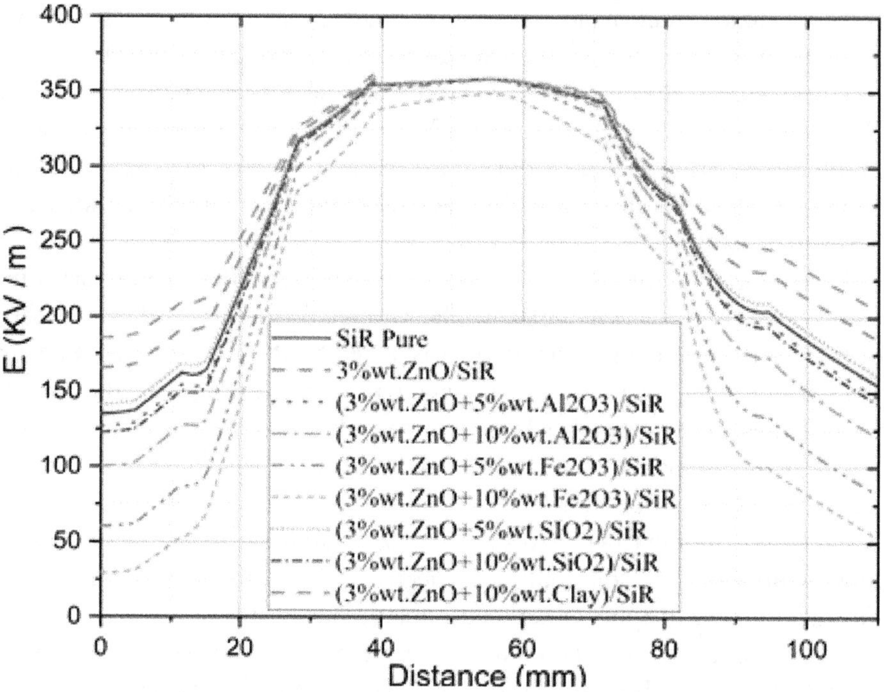

Figure 15. Effect of nanoparticles coating layer percentage on electric field distribution across SiR insulator

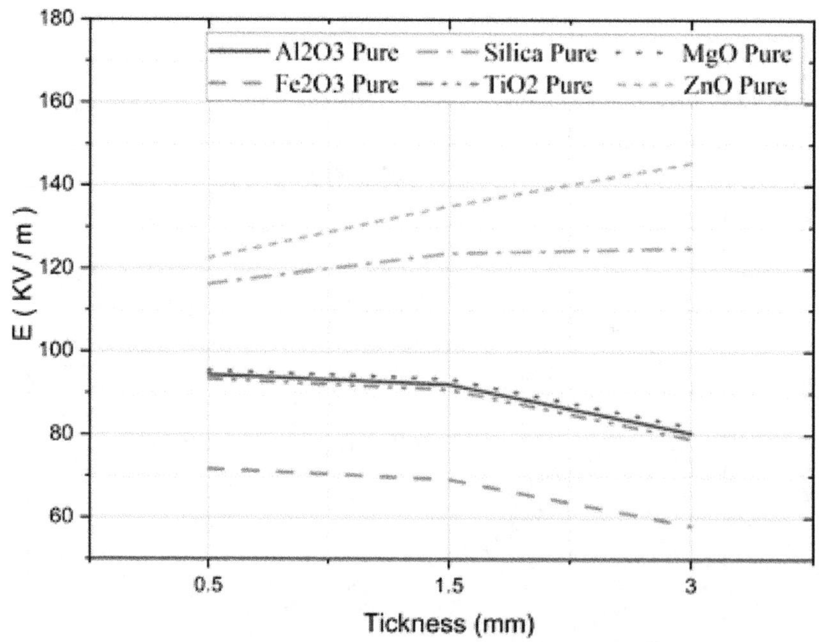

string to 12. 4%. Additionally, it is evident that including (3wt.% ZnO+10wt.% Clay) multi-nanoparticles expand the electric field stress dependent upon 5%, as demonstrated in Figs. (13, 14).

The impact of covering nanoparticles layer "thickness 0.5mm" with SiR insulation string around the electrical field stress through the insulation string is shown. For example, the layers of nanoparticles "ZnO and Fe_2O_3" to pure silicon rubber insulation decrease the electric field stress toward the to begin with quarter of string to 38. 6% or 48. 8% individually. The sort of nanoparticles changes the rational dielectric constant for insulation material climate shed that is transformed by insulation self-capacitance; henceforth, the electric field distribution occurs again for every last one of the insulation strings. Fig. (15) indicates the conduct of electric field stress because of covering pure SiR insulation string for variant nanoparticles. It has been recognized that expanding covering layer thickness impacts self-capacitance for non-ceramic sir insulation that transforms the electric field stress crosswise over string. The sort and thickness of nanoparticles covering layer may be the primary variable to characterize the conduct technique of electric field stress in non-ceramic sir insulation string, as indicated in Fig. (15).

6 TRENDS OF NANOPARTICLES INSIDE SUSPENSION INSULATORS

The above outcomes indicate regulating in capacitance of disc insulation (porcelain, glass, and silicon rubber) in the light of the nanoparticle's sorts and their concentrations; it has been indicated that titanium di-oxide and alumina nanoparticles have the capacity of expanding capacitance of porcelain insulators; however, the opposite nanoparticles support diminishing the capacitance for string insulators. For string insulators, the impact of nanoparticles sorts and their concentrations for efficiency for porcelain and silicon rubber insulators strings has been demonstrated; the efficiency of high voltage insulating strings nanocomposites builds for expanding nanoparticles concentrations of titanium di-oxide (TiO_2), magnesium oxide (MgO) and alumina (Al_2O_3) nanoparticles.

The efficiency upgrade rate occurring to the suspension string under 66kV may be higher than the same high voltage suspension strings nanocomposites. In any case, the high voltage conveyance for pure and nanocomposites of high voltage insulators strings has been conveyed; utilizing titanium di-oxide nanoparticles upgrade high voltage conveyance to decrease high voltage on the closest disc from carry on with conductor and expand the voltage on the top disc. Titanium di-oxide and alumina nanoparticles are to a great extent elucidating and effective for upgrading the 66kV insulators string over upgraded high voltage distribution 500kV insulators string. On the other hand, clay and silica nanoparticles expand high voltage on the closest disc from live conductor and decrease voltage on the top disc as indicated in shades, and the high voltage distribution may be more non-uniform. It is noted that the high voltage distribution for pure polymeric insulation string is non-uniform. However, utilizing titanium di-oxide (TiO_2), magnesium oxide (MgO) and alumina (Al_2O_3) nanoparticles improve the high voltage distribution and diminish high voltage on the closest disc from carry on with conductor and expand the voltage on the highest point disc as indicated in shades. Moreover, the opposite suggested nanoparticles expand the non-uniform high voltage distribution on the closest disc from carry on with conductor and decrease voltage on the highest point disc, as indicated in colors. This ought to be thought about carefully to upgrade the string efficiency and the high voltage distribution for suspension insulators under variant high voltage appraisals dependent upon nanotechnology techniques; i.e., dielectric consistent characterization increments for expanding concentration of titanium di-oxide (TiO_2), magnesium oxide (MgO) and alumina (Al_2O_3) nanoparticles.

This chapter concentrates on upgrading the string efficiency and the high voltage distribution of suspension insulators under variant high voltage appraisals dependent upon nanotechnology techniques; i.e., dielectric constant characterization builds for expanding concentration of titanium di-oxide (TiO_2), magnesium oxide (MgO) and alumina (Al_2O_3) nanoparticles.

7 FORECAST AND RECOMMENDATIONS

High permittivity nanoparticle can be the viable property for expanding disc capacitance and upgrading the string efficiency; therefore, it can be controlled in the capacitance of disc insulation in view of the nanoparticles' sorts and their concentrations. A certain sum of titanium di-oxide, magnesium oxide or alumina expands string efficiency in diverse values about volume fraction. The upgrade ratio for string efficiency can be to a greater extent elucidating in insulating strings nanocomposites under medium voltage compared by high voltage nanocomposites insulating strings. High voltage distribution through insulators string will be improved by utilizing the best proficient nanoparticles; i.e., Titanium di-oxide (TiO_2), magnesium oxide (MgO) or alumina (Al_2O_3). In any case, the high voltage distribution is all the more elucidating and effective under medium voltage insulators string than high voltage conveyance of insulators string.

The kind and concentration of multi-nanoparticles will revolve around the dielectric constant of insulation arrangement material climate shed that evolve the self-capacitance of insulation, after those impacts the electric field distribution on every last one of the insulation strings. Utilizing nanocomposites and multi-nanocomposites as opposed to pure sir insulation diminish the electric field stress distribution over (8.4wt.% -12.4wt.%) respectively. Nanocomposite, multi-nanocomposite, and covering layers of nanoparticles have been standard electric field stress distribution in SiR insulation string, as stated by the model of investigation and have succeeded in achieving useful effects on decreasing the values of electric field stress distribution. Nanoparticles covering layer has a capability to decline the electric field stress distribution up to (48. 8%) regarding the utilization of pure sir insulation string. The kind and thickness of nanoparticles covering layer will be the primary variable to characterize the conduct technique of electric field stress through non-ceramic SiR insulation string.

REFERENCES

92 WM 265-9 PWRD. – "A report of the IEEE Non-Ceramic History".

Ahmed, Brown, & Srinivas. (n.d.). *Testing of Polymer Insulators used in Electric Power Distribution Circuits*. Academic Press.

Amin. (n.d.). *Methods For Preparation Of Nano – Composites For Outdoor Insulation Application*. COMSATS Institute of Information Technology Wah, Pakistan.

ANSI C29.1-1988. For Electrical Power Insulators-Tests Methods, 1988.

ANSI C29.11-1989. For Composite Suspension Insulators for Overhead Transmission Line-Tests, 1989.

ANSI/IEEE Std 4-1978. Techniques for High Voltage Testing, 1978.

Brosseau, C., Queffelec, P., & Talbot, P. (2001). Microwave Characterization of Filled Polymers. *Journal of Applied Physics*, *89*(8), 4532–4540. doi:10.1063/1.1343521

Cherney. (n.d.). *Polymer Insulators for the Distribution Electricity System*. Academic Press.

Cotton, H. (1958). *The Transmission and Distribution of Electrical Energy*. English Univ. Press. copied on Insulator Usage, A.C. Walker's Insulator Information page.

Dengke, C., Hui, Y. J., Xishan, W., & Lei, L. (2004). *International conference on solid dielectrics*. Academic Press.

Dominguez, D. C., Espino-Cortes, F. P., & Gomez, P. (2013). Optimized design of electric field grading systems in 115 kV non-ceramic insulators. *IEEE Transactions on Dielectrics and Electrical Insulation*, *20*(1), 63–70. doi:10.1109/TDEI.2013.6451342

Doshi, T., Gorur, R. S., & Hunt, J. (2011). Electric field Computation of Composite Line Insulators up to 1200 kV AC. *IEEE Transactions on Dielectrics and Electrical Insulation*, *18*(3), 861–867. doi:10.1109/TDEI.2011.5931075

Ebnalwaled, A. A., & Thabet, A. (2016). Controlling the optical constants of PVC nanocomposite films for optoelectronic applications. *Synthetic Metals Journal*, *220*, 374–383. doi:10.1016/j.synthmet.2016.07.006

Flink & Beaty. (n.d.). *Standard Handbook For Electrical Engineers* (13th ed.). McGraw Hill.

Fuse, N., Kozako, M., Tanaka, T., & Ohki, Y. (2005). *IEEE Conference on Electrical insulation and dielectric phenomena*. IEEE.

Gershon, D., Calame, J. P., & Birnboim, A. (2001). Complex Permittivity Measurements and Mixing Laws of Alumina Composites. *Journal of Applied Physics*, *89*(12), 8110–8116. doi:10.1063/1.1369400

Gorur, R. S., Cherney, E. A., & Burnham, J. T. (1999). *Outdoor Insulators*. Ravi S. Gorur Inc.

Gross. (n.d.). *Power System Analysis* (2nd ed.). John Wiley & Sons.

Holtzhausen, J.P. (2008). *High Voltage Insulators*. IDC Technologies.

IEC 60137:2003. 'Insulated bushings for alternating voltages above 1,000 V.' IEC, 2003.

IEC 815. Guide for Selection of Insulators in respect of Polluted Conditions.

Karkkainen, K. K., Sihvola, A. H., & Nikoskinen, K. I. (2000). Effective Permittivity of Mixtures: Numerical Validation by the FDTD Method. *IEEE Transactions on Geoscience and Remote Sensing*, *38*(3), 1303–1308. doi:10.1109/36.843023

Kumagai, S., & Yoshimura, N. (2004). Leakage current characterization of estimating the condition of ceramic and polymeric insulating surfaces. *IEEE Transactions on Dielectrics and Electrical Insulation*, *11*(4), 681–690. doi:10.1109/TDEI.2004.1324357

Looms. (n.d.). *Insulators For High Voltages*. Peter Peregrinus.

Looms, J. S. T. (1987). *Insulators for High Voltages*. Peter Peregrinus Ltd.

LWIWG-01 (96) – CEA Purchasing Specification "Dead-end / Suspension Composite Insulator for Overhead Distribution Lines".

LWIWG-02 (96) – CEA Purchasing Specification "Line Post Composite Insulator for Overhead Distribution Lines".

LWIWG-03 (96) – CEA Purchasing Specification "Guy Strain Composite Insulator for Guy Wires".

Manias, E., Touny, A., Wu, L., Strawhecker, K., Lu, B., & Chung, T. C. (2001). Polypropylene/Montmorillonite Nanocomposites. Review of the Synthetic Routes and Materials Properties. *Chemistry of Materials, 13*(10), 3516–3523. doi:10.1021/cm0110627

Mark, J. E. (1996). Ceramic-reinforced polymers and polymer-modified ceramics. *Polymer Engineering and Science, 36*(24), 2905–2920. doi:10.1002/pen.10692

McCalley, J. D., & Krishnan, V. (2014). A survey of transmission technologies for planning long distance bulk transmission overlay in US. *International Journal of Electrical Power & Energy Systems, 54*, 559–568. doi:10.1016/j.ijepes.2013.08.008

Mehta. (2005). *Principles of Power Systems*. Academic Press.

Mei, H. W., Peng, G. M., Dai, H. Q., Wang, L. M., Guan, Z. C., & Cao, L. F. (2011). Installing insulation jacket to improve outdoor insulation performance of composite insulator. *IEEE Transactions on Dielectrics and Electrical Insulation, 18*(6), 2004–2012. doi:10.1109/TDEI.2011.6118638

Nelson, S. O. (2001). Measurement and Calculation of Powdered Mixture Permittivities. *IEEE Transactions on Instrumentation and Measurement, 50*(5), 1066–1070. doi:10.1109/19.963159

Nzenwa, E. C., & Adebayo, A. D. (2019). Analysis of Insulators for Distribution and Transmission Networks. *American Journal of Engineering Research, 8*(12), 138-145.

Ozmusul, M. S., & Picu, R. C. (2002). Elastic Moduli of Particulate Composites with Graded Filler-Matrix Interfaces. *Polymer Composites, 23*(1), 110–119. doi:10.1002/pc.10417

Patel & Singh. (2015). Investigation of Ceramic and Non-Ceramic Suspension Insulator under Different Polluted Condition. *Global Journal of Engineering Science and Researches, 2*(12).

Phillips, A. J., Childs, D. J., & Schneider, H. M. (1999). Aging of Non- Ceramic insulators due to corona from water drops. *IEEE Transactions on Power Delivery, 14*(3), 1081–1089. doi:10.1109/61.772357

Que, W. G., Sebo, S. A., & Hill, R. J. (2007). Practical Cases of Electric Field Distribution Along Dry and Clean Nonceramic Insulators of High- Voltage Power Lines. *IEEE Transactions on Power Delivery, 22*(2), 1070–1078. doi:10.1109/TPWRD.2007.893190

Richardson. (2011). *Insulators: Types, Properties & Uses*. Nova Science Publishers Inc.

Schneider, H. M., Hall, J. F., Karady, G., & Renowden, J. (1989). Nonceramic insulators for transmission lines. *IEEE Transactions on Power Delivery, 4*(4), 2214–2221. doi:10.1109/61.35649

Tagami, N., Hyuga, M., Ohki, Y., Tanaka, T., Imai, T., Harada, M., & Ochi, M. (2010). Comparison of Dielectric Properties between Epoxy Composites with Nanosized Clay Fillers Modified by Primary Amine and Tertiary Amine. *IEEE Dielectrics and Electrical Insulation Transactions, 17*(1), 214–220. doi:10.1109/TDEI.2010.5412020

Taylor. (2003). *Function of Grading rings to Composite Insulator.* Academic Press.

Thabet. (2015a, January). *Electrical Composite Materials for Industrial Applications: Design and Investment.* LAP LAMBERT Academic Publishing.

Thabet, A. (2015b, June). Experimental Verification for Improving Dielectric Strength of Polymers by Using Clay Nanoparticles. *Advances in Electrical and Electronic Engineering Journal, 13*(2), 182–190.

Thabet, A. (2015c, January). Experimental enhancement for dielectric Strength of polyethylene insulation materials using cost fewer nanoparticles. *International Journal of Electrical Power & Energy Systems, 64*, 469–475. doi:10.1016/j.ijepes.2014.06.075

Thabet, A. (2016). Thermal experimental verification on effects of nanoparticles for enhancing electric and dielectric performance of polyvinyl chloride. *Journal of the International Measurement Confederation, 89*, 28–33. doi:10.1016/j.measurement.2016.04.002

Thabet, A. (2017, June). Theoretical Analysis for effects of nanoparticles on dielectric characterization of electrical industrial materials. *Electrical Engineering (ELEN) Journal, 99*(2), 487–493.

Thabet. (2020). *Design and Investment of High Voltage NanoDielectrics.* IGI Global. doi:10.4018/978-1-7998-3829-6

Thabet, A., Al-Sharif, M. R., Abdel-Moamen, M. A., & El-Nobi, A. (2019). Improvement of Electrical Field Distribution on Non-Ceramic Insulators String Using Nanoparticles. *IEEE International Middle East Power System Conference MEPCON.*

Thabet, A., & Ebnalwaled, A. A. (2017, November). Improvement of surface energy properties of PVC nanocomposites for enhancing electrical applications. *Journal of the International Measurement Confederation, 110*, 78–83. doi:10.1016/j.measurement.2017.06.023

Thabet, A., & Mobarak, Y. (2016, February). Predictable Models and Experimental Measurements for Electric Properties of Polypropylene Nanocomposite Films. *Iranian Journal of Electrical and Computer Engineering*, *6*(1), 120.

Thabet, A., & Mobarak, Y. A. (2010). Dielectric Characteristics of New Nano-Composite Industrial Materials. *International Conference on High Voltage Engineering and Application "ICHVE".* 10.1109/ICHVE.2010.5640767

Thabet, A., & Mubarak, Y. A. (2017a). Experimental Enhancement for Electric Properties of Polyethylene Nanocomposites under Thermal Conditions. *Advances in Electrical and Electronic Engineering Journal, 15*(1), 55–62. doi:10.15598/aeee.v15i1.1727

Thabet, A., & Mubarak, Y. A. (2017b, June). The Effect of Cost-Fewer Nanoparticles on the Electrical Properties of Polyvinyl Chloride. *Electrical Engineering Journal, Springer, 99*(2), 625–631. doi:10.100700202-016-0392-3

Thabet, A., & Salem, N. (2017). Optimizing Dielectric Characteristics of Electrical Materials Using Multi-Nanoparticles Technique. *2017 Nineteenth International Middle East Power Systems Conference (MEPCON)*. 10.1109/MEPCON.2017.8301260

Thabet, A., Salem, N., & Essam, E. M. (2018, June). Modern Insulations for Power Cables Using Multi-Nanoparticles Technique. *International Journal on Electrical Engineering and Informatics*, *10*(2), 271–279. doi:10.15676/ijeei.2018.10.2.6

Todd, M., & Shi, F. (2003). Molecular Basis of the Interphase Dielectric Properties of Microelectronic and Optoelectronic Packaging Materials. *IEEE Transactions on Components and Packaging Technologies*, *26*(3), 667–672. doi:10.1109/TCAPT.2003.817862

Weedy & Cory. (n.d.). *Electric Power Systems* (4th ed.). John Wiley & Sons.

Zhao & Comber. (2000). Calculation of Electric Field and Potential Distribution Along Nonceramic Insulators Considering the Effects of Conductors and Transmission Towers. *IEEE Trans. Power Delivery*, *15*(1).

Zhao, C., Mei, H., Zhu, M., Dai, H., Wang, L., & Guan, Z. (2015, June). Influence on Voltage Distribution of Composite Insulator by Inserting Metal Electrodes in Weather Sheds. *IEEE Transactions on Dielectrics and Electrical Insulation*, *22*(3), 1623–1629. doi:10.1109/TDEI.2015.7116358

Chapter 6
Transformers Nanofluids

ABSTRACT

This chapter interprets the impacts of powder nanoparticles on upgrading the electrical and physical properties of pure transformer oils and contains the new technologies for preparation transformers nanofluids. This chapter draws attention to the theories of dynamic charging and effective parameters for nanofluid polarization. Furthermore, this chapter presents recent procedures for estimation and control of nanofluids conductivity and the effects of nanoparticles on nanofluid conductivity. Moreover, this chapter demonstrates the favored methodology of nanofluid preparation and the tested high voltage breakdown. The precision of recommendation and forecast is another addition to qualitative investigations.

1 DESIGN AND INVESTMENT

Transformer oil-based nanofluids with nanoparticles suspensions significantly has certain or negative voltage levels regarding incorporation sorts and their concentrations in pure transformer oil. This chapter presents scattering of low concentrations for nano-sized Zinc oxide circular particles in Diala d insulating oil to upgrade their physical and dielectric properties. Therefore, this chapter displays execution transformer oil nanofluids as elective pure oil insulation for upgrading power transformer characterization; it checks the lab-test comes about dielectric strength and electrical properties of oil nanofluids medium in view of nanotechnology strategies. Foreseeing the compelling dielectric parameters of equal transformer oils nanofluids relies on the procedure utilized that is regularly allowed to infer a formula. Then, it can utilize the prediction recipe dependent upon the situated parametric nanoparticles of known detailing over variant volume fraction. In this chapter, the successful variables of relative permittivity for resultant transformer oils nanofluids are examined, such as (Diala B, and Diala D) dependent upon polarization nanoparticles like (SiO_2, ZnO, TiO_2 and Graphite). The effect of this contribution has reported foreseeing the effects that are exceedingly impacted under variant thermal conditions. A similar examination has been carried on the impacts of utilizing distinct nanoparticles and multi-nanoparticles inside pure transformer oils with diverse radii and volumetric concentrations.

This chapter formulates new multi-nanoparticles procedure to decrease the positive streamer propagation over silicon and vegetables oils nanofluids; transformer oil nanofluids for multi-nanoparticles

DOI: 10.4018/978-1-7998-8536-8.ch006

suspensions is mixed to moderate the propagation for certain streamer more than utilizing distinct nanoparticles. Henceforth, the insulation characteristics are progressed and breakdown voltage nanofluid can be stretched. This chapter manifests the impact of nanoparticle radius, nanoparticles' polarization, relative permittivity of oil, applied field and the charge thickness of nanoparticles interface. It may be deduced that transformer oil-based nanofluids for conductive individual and multi-nanoparticles suspensions have extensively higher effects over ahead moderating positive streamer than non-conductive nanoparticles. A similar study has shown the extent of stored accuse and charging time as outcomes for utilizing individual nanoparticles and multi-nanoparticles inside transformer oil. Furthermore, this chapter investigates the effects of distinct and multi-nanoparticles on compelling conductivity (static and electrophoretic) of transformer oils. Furthermore, it has been investigated on the primary effective variables of electrical conductivity for transformer oil nanofluids, such as nanoparticles, applied electric field and thermal conditions. A reproduction contemplate has been performed to utilize distinct and various nanoparticles (SiO_2, ZnO, TiO_2 and Graphite). In view of static and electrophoretic conductivities, a lab-test fill has been carried out to check the viability of utilizing various nanoparticles for controlling conductivity power transformers nanofluids. Finally, a similar study has been accomplished for specifying the ideal nanoparticles for controlling the conductivity of transformer oil nanofluids under thermal conditions to increment the age life-time transformer oil. This single section displays a specific and inventive system to upgrade the charging progress for the transformer's oil by utilizing multi-nanoparticles. The charging progress procedure for energy transformer oils may be the primary approach for abating down the streamer propagation and upping dielectric strength. In addition, transformer oil's-built suspension multi-nanoparticles (dielectric, conductive, and semi-conductive) have significant levels of saved electrons that are higher than suspension accepted nanofluids. A lab-test worth of effort has been conducted to confirm the viability of utilizing multi-nanoparticles for abating down positive streamer propagation to transformers oils against utilizing individual nanoparticles system or accepted transformers' oil.

2 TRADITION TRANSFORMERS FLUIDS

Oil utilized for insulation in transformers can be mineral oil, and it is acquired by refining rough petroleum. Creature oils and vegetable oils are not utilized for this reason for having these structure greasy acids with respect to heating which will be destructive for the cellulosic utilized within insulation. Mineral oils are being used as fluid dielectrics in electrical supplies in excess of hundred along time currently. Regardless of the accessibility of a mixture for synthetic oils, with distant and more predominant properties, mineral oils held its way because of their abundant accessibility and economy. Properties that are essential to utilize mineral oil as dielectric are: (1) High insulating property, (2) Useful oxidative and aging stability and good heat transferability. The reliable performance of mineral insulating oil in an insulation system depends upon certain oil characteristics, which can affect the overall performance of the electrical equipment. (3) High electric strength will withstand the concentrations in administration. (4) Sufficiently low viscosity so that its capacity with circularize and exchange heat is not impeded. (5) Sufficiently low temperature properties down of the amazing climatic states that it has in its establishment rules. (6) Best possible oxidation imperviousness to guarantee long life in administration. (7) Beneficial imperviousness to emulsion on keeps, considering water suspension in it. (8) Free from inorganic acid, soluble base and destructive figure which cause erosion for metal parts and insulation and quicken the processing for slime. (9) No sledging under ordinary dividing states. (10) High flash/fire concentrations. (11) Low pour perspective - low water substance.

The primary prerequisites of transformer oil as recorded different national and universal norms main state the least prerequisites for transformer oil. Many transformer producers and electricity companies and boards have their own stringent specifications based on the required conditions. With propelled refining techniques, it will be permitted by time to process the transformer oil tailor-made suit of shield as the accurate prerequisites for such determination and with dissipating high temperature. It is to be noted that the misfortunes in a transformer insulating oil gives a medium for high dielectric strength over which the coils also center are submerged. A circular segment might hop crosswise over the same division of inner energized parts at a substantially down voltage, whether the tank has air or not. In addition, oil conducts high temperature far starting with energized parts much superior to air. In time, oil degrades from typical operations because of heat and contaminants. Oil can't hold high dielectric strength if presented to air or dampness. Dielectric strength decreases for absorption of dampness and oxygen. These contaminants weaken the part insulation. For this reason, endeavors are made on keep insulating oil starting with contacting air, particularly looking into bigger energy transformers. Utilizing a firmly fixed transformer tank may be impractical because of weight varieties coming about because of thermal development and withdrawal insulating oil. Common systems of sealing oil-filled transformers are the conservator with a flexible diaphragm, bladder, or a positive pressure inert-gas (nitrogen) system. Recovery transformers are bought with conservators. At the same time, more diminutive station administration transformers have a pressurized nitrogen cover with respect to the highest priority on oil. Some station service transformers are dry-type, self-cooled or forced air cooled.

3 NANOFLUIDS POLARIZATION

This chapter displays regulation dielectric consistent pro-fractional transformer oils nanofluids dependent upon specialized foul determination of nanoparticles. Moreover, it also concentrates on the compelling variables with respect to successful relative permittivity of resultant transformer oils nanofluids (Diala B and Karamay25# oils). New multi-nanofluids have been recommended and heated under thermal conditions and variant volumetric concentrations. A similar endeavor has been done in this Section to show the contrasts in the middle of impact for including distinct and multi-nanoparticles with power transformer oils.

3.1 Dielectric Constant of Nanofluids

Streamer propagation speed transformer oil-based nanofluids can be short of pure oil; however, the breakdown voltage will be higher than that for pure oil. Investigations are performed to think about the sway of nanoparticle volumetric fraction and temperature on the nanofluids (NF) relative permittivity, discovering that transformer oil-based ZnO nanofluid has a to a certain degree higher relative permittivity than that of traditional transformer oil in a low nanoparticle volumetric fraction, the relative permittivity increases straightforwardly for nanoparticle volumetric fraction and abates straightly with temperature (Miao et al., 2013). It has been investigated when utilizing transformer oil–based multi-nanoparticles empower to utilize one sort of nanoparticles for high relative permittivity and the different sort for small relative permittivity to abating streamer propagation and keeping of the effective relative permittivity nanofluids during sensible esteem (Jin, 2015).

The prerequisite of nanofluids is to upgrade the electrical insulation and additionally the thermal conductivity transformer oil (Wong & de Leon, 2010). However, the customary insulating liquids are environmentally poor heat exchange fluids; nanofluids show improved properties. The case in point is similar to thermal exchange liquids and insulation (Fleming et al., 2005). It can be discerned that the interfaces between the dielectric and the nano-metric particles camwood vehemently affects the dielectric properties of the composite material as a rule (Thabet, 2020; Thabet et al., 2012). On the great holders, all faults and degradations initiated by thermal aging are kept and developed for huge amounts because of the low cooling efficiency of insulating systems. Recently, a creative sort of dielectric, transformer oil-based nanofluids (NFs), has attracted great attention to its uncommon transform for both heat exchange efficiency and dielectric properties (). Concerning illustration, it is shown toward the nanoparticle charging hypothesis of nanoparticles to nanofluids to get free electrons and change over these fast-moving electrons on slower adversely charged particles, which slows the streamer and increases the breakdown voltage for transformer oil (Hwang et al., 2010).

In the meantime, it may be trusted that nanoparticles can manufacture the shallow trap thickness in transformer oil and decline the pace streamer propagation through electron getting and de-catching system in shallow traps. The effects of nanoparticles included substances on the electrical properties of fundamental dielectrics are at present of broad premium (Jianlin & Decai, 2010; Kasap, 2009; Lewis, 2004; Pradhan, 2008; Pradhan et al., 2008; Saad et al., 2004; Tagami et al., 2010). Insulating oils which will be utilized as transformer's fluids are a mix of exceptionally refined mineral oils and a fragrant extricate from a distillate. Nanoparticles would choose an extensive variety of permittivities (Thabet et al., 2012). Diala B oil has been chosen as insultingly oil since it offers intrinsic regular resistance with oil corruption and will be suitable for long-term applications as transformers, grid and mechanical transformers dependent upon most extreme load (Product Data Sheet from Shell Lubricants, 2011). Karamay 25# oil may be a profoundly refined mineral oil and a chosen cell reinforcement. The outline of dielectric constant of insulating and cooling transformers, oil-immersed switch gear is shown in Table 1.

Table 1. The relative permittivity and Conductivity of selected insulting oils and nanoparticles at 25 °C (Thabet et al., 2012)

Nanoparticles and Materials	Relative Permittivity	Conductivity (*s/m*)
Diala B oil	2.2	$2*10^{-12}$
Karamay 25# oil	2.135	$1.52*10^{-12}$
Clay	1.8	10^{-2}
ZnO	7.4	10^{-6}
MgO	9.7	$5.011*10^{-1}1$
TiO_2	90	10^{-11}
SiO_2	4.5	10^{-11}
Graphite	15	$5.011*10^{-4}$
$LiTaO_3$	40	$2.511*10^{-9}$

3.2 Theory and Effective Parameters

The relative permittivity nanofluids are basically contingent upon relative permittivity of base oils and relative permittivity for utilized nanoparticles that determines polarizability of nanoparticles under applied field transformer oils. Transformer oil-based nanofluids for electric applied field, both atoms about transformer oils and nanoparticles can be polarized. There are three sorts of polarization in nanofluids, transformer oil molecules' polarization, the nanoparticles' inward polarization of, and the orientation polarization about accused nanoparticles as polar particles. The relative permittivity of transformer oil-based NFs, ε_(r, n) can be communicated By Clausius-Mossotti comparison as as taken in ref.(Miao et al., 2013):

$$\frac{\varepsilon_{rn}-1}{\varepsilon_{rn}+2}=\frac{1}{3\varepsilon_0}\left(N_1\alpha_1+N_2\alpha_2+N_2\alpha_3\right) \tag{1}$$

Where α_1 is the polarizability of a transformer oil molecule, ε_0 is the permittivity value in vacuum, α_2 is the polarizability of a nanoparticle caused by inner polarization, α_3 is the orientational polarizability of a charged nanoparticle treated as a polar molecule, N_1 and N_2 are the numbers of transformer oil molecules and nanoparticles in per unit volume.

The effective relative permittivity model of nanofluid for multi-nanoparticles relies with respect to polarization for transformer oil particles and with respect to polarization nanoparticles on two sorts of polarization by recognizing mono-dispersing nanoparticles for base liquid concerning illustration, as indicated in Fig. (2), the compelling relative permittivity of transformer oil-based multi nanoparticles, ε_(r,mn) could make communicated comparison as takes after. With applying outside electric field looking into oil, both sorts of nanoparticles cause polarized, positive and negative surface polarization charges to assemble toward two sides of a nanoparticle: the surface polarization charge thickness of both sorts camwood can be expressed:

$$\frac{\varepsilon_{r,mn}-1}{\varepsilon_{r,mn}+2}=\frac{1}{3\varepsilon_0}\left(N_1\alpha_1+N_i\alpha_i+N_i\alpha_{i2}+N_j\alpha_j+N_j\alpha_{j2}\right) \tag{2}$$

Where $\alpha 1$ is the polarizability of a transformer oil molecule, α_i is the polarizability of the first type of nanoparticles caused by inner polarization, α_{i2} is the orientational polarizability of a first type of charged nanoparticles as polar molecule, α_j is the polarizability of the second type of nanoparticles caused by inner polarization, α_{j2} is the orientational polarizability of the second type of charged nanoparticles as polar molecule, N_1, N_i and N_j are numbers of transformer oil molecules and first and second nanoparticles in per unit volume respectively. By applying the external electric field on oil, both types of nanoparticles will be polarized, positive and negative surface polarization charges will gather at two sides of a nanoparticle, as shown in Fig. (1), Where, ε_i is the relative permittivity of the first type of nanoparticles, ε_j is the relative permittivity of the second type of nanoparticles, ε_1 is the relative permittivity of transformer oil, a_i is the radius of the first nanoparticle, a_j is the radius of the second nanoparticle, φ_{pi} is the angle between a_i and X a_x is in the first nanoparticle, φ_{pj} is the angle between a_j and X axis in the second nanoparticle,

Figure 1. Surface polarization charges of multi-charged nanoparticles in nanofluid

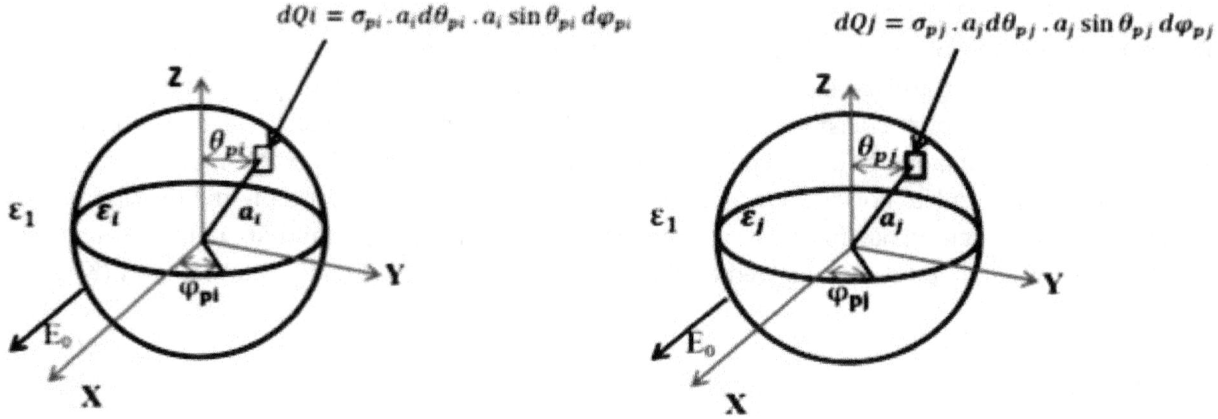

θ_{pi} is the angle between a_i and Z axis in the first nanoparticle, θ_{pj} is the angle between a_j and Z axis in the second nanoparticle.

Then after polarization of nanoparticles, both positive and negative charges will be two polar for nanoparticle with separation equals in breadth of utilized nanoparticle, and they will turn with one another. μ_{ci} and μ_{cj} (C.m), are electric dipole moments of first and second charged nanoparticle when they are polar molecules, as shown in figure (2):

$$\mu_{ci} = 2a_i Q_{i+} \tag{3}$$

$$\mu_{cj} = 2a_j Q_{j+} \tag{4}$$

Figure 2. Charged modeled nanoparticle as a polar molecule

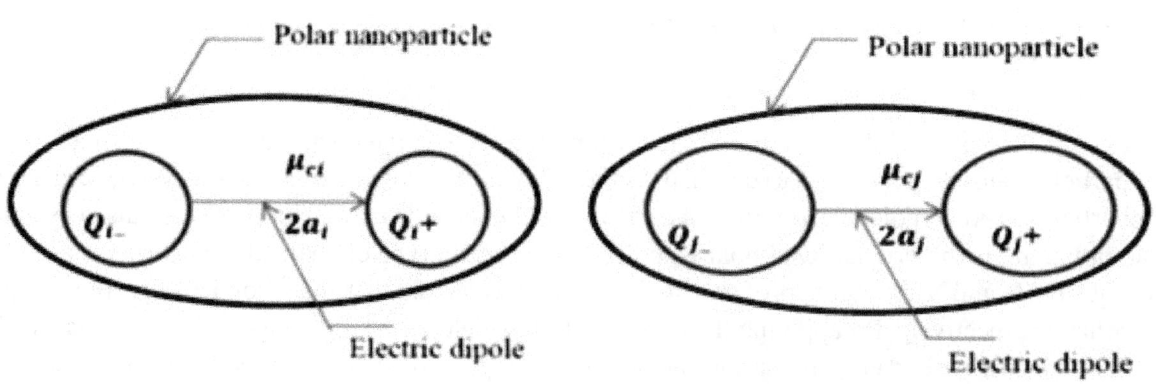

As shown in Fig. (3), in pure transformer oil, only the polarizability α_1 and molecule number N_1 are considered to describe the relative permittivity of pure oil dependent, given in the following equation:

Figure 3. Oil molecule and nanoparticle inner polarization in transformer oil-based nanofluids

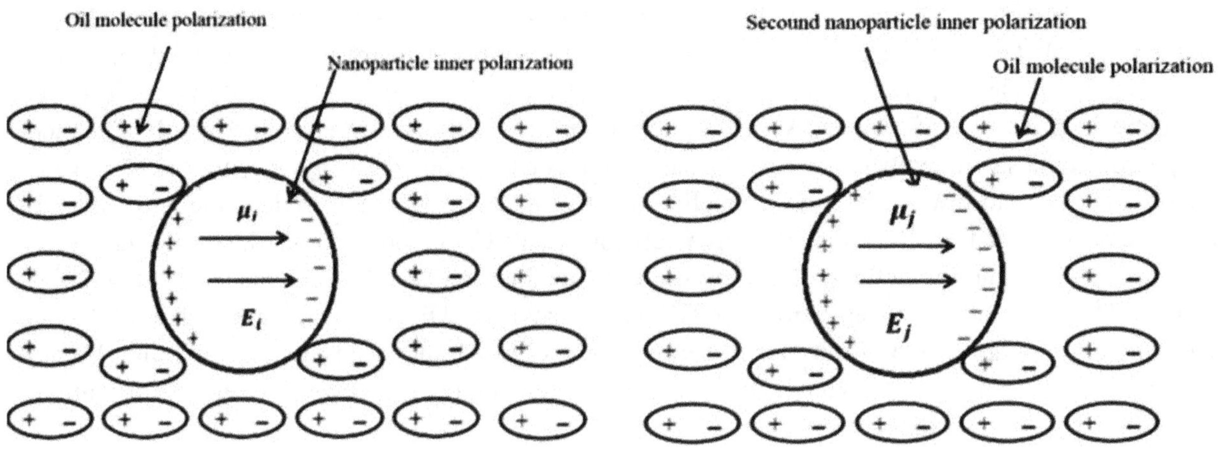

$$\frac{\varepsilon_1-1}{\varepsilon_1+2}=\frac{1}{3\varepsilon_0}(N_1\alpha_1) \tag{5}$$

Inner Polarizability of Nanoparticles

We can get the two types of nanoparticle polarization model of NF effective relative permittivity from:

$$\frac{\varepsilon_{r,mn}-1}{\varepsilon_{r,mn}+2}=\frac{\varepsilon_1-1}{\varepsilon_1+2}+\frac{\varphi_i}{3}(\varepsilon_i-1)+\frac{\varphi_j}{3}(\varepsilon_j-1)+\frac{\varphi_i}{3KT}\pi\varepsilon_0 a_i^3 E_0^2\left(\frac{\varepsilon_i-\varepsilon_1}{2\varepsilon_1-\varepsilon_i}\right)+\frac{\varphi_j}{3KT}\pi\varepsilon_0 a_j^3 E_0^2\left(\frac{\varepsilon_j-\varepsilon_1}{2\varepsilon_1-\varepsilon_j}\right) \tag{6}$$

Small concentrations of nanoparticles inside the transformer oil meet the expectations as electrons scavengers nanofluids that decrease the streamer propagation and increase the breakdown voltage comparing for pure transformer oil. Multi-nanoparticles make the effective relative permittivity of nanofluids minimal higher than the effective relative permittivity for nanofluids that utilize individual nanoparticles, but they absorb tremendous numbers of free electrons.

3.3 Nanofluids Dielectric Characterization

The successful relative permittivity of nanofluid may be the summation polarizability of oil molecules' amount, in addition to the polarizability of nanoparticles' amount and orientation polarizability for accused nanoparticles' number. As the strength nanoparticles' amount to work ahead volumetric concentration, the compelling relative permittivity nanofluid will build with increasing the nanoparticles in transformer oil, and it will be contrasted with the nanoparticle with in-turn, subject to their relative permittivity to the same oil. However, the successful permittivity increments with volume concentration for nanoparticles and declines with temperature. Augmenting the temperature abates the orientation polarizability nanoparticle. The successful permittivity contrasts from oil to another and from nanoparticle to another reliant on their relative permittivity. Figures (4, 5) show the impact of volumetric concentrations for successful relative permittivity of transformer oil-based ZnO for separate temperature Diala b and Karamay25# oils. However, ZnO nanoparticles build toward the same percent to the three sorts of oil, compelling relative permittivities of resultant nanofluids to vary starting with oil to oil, on account of contrasts in their permittivities. According to the same reason, the decrease in strength of compelling permittivities with temperature isn't the same. In this case, when the volume centralization of ZnO nanoparticles expands with 20%, the compelling relative permittivity expands dependent upon 2. 81% and 2. 72% on Diala b and Karamay25# separately toward 20°C. On the other hand, increasing the temperature by 40°C, the compelling relative permittivity declines with 4. 31% in Diala b and for 4. 24% to Karamay25#. Expansion concentrations for TiO_2 nanoparticles to Diala B and Karamay25# oils expand the compelling relative permittivity for nanofluid pointedly with increasing its concentration because of high relative permittivity of TiO_2. Toward increasing temperature, the effective relative permittivity nanofluids have exact small decrement.

Nanoparticles with small relative permittivities can be utilized with volumetric concentration higher than the individuals with high permittivities. High concentrations can be a lab-test to look after the nano-meter measure of nanoparticles because of the alluring energy between nanoparticles, which can lead to the formation of agglomerations that settle out of suspension, so the used concentration of these nanoparticles should also be limited. Including SiO_2 nanoparticles can be done in Diala B or Karamay25# oils and this provides for small increment in the effective permittivity of their nanofluids with the increasing concentration of nanoparticles and high decrement in the increasing temperature. This movement is an after-effect decrement polarizability of nanoparticles because of small relative permittivity of nanoparticles and the opposite proportion between the middle of temperatures and orientation polarization. In the event that Diala B effective permittivity of transformer oil-based SiO_2 increments dependent upon 1.18% and 1.20% over Karamay25#, the increasing centralization of SiO_2 nanoparticles increments up to 20% and declines at the same time by 4.2% for Diala b and by 4.1% for Karamay25#when the temperature expands by 40°C. In case of including two sorts of nanoparticles to transformer oil, the compelling relative permittivity nanofluid is capacity around volumetric concentrations of utilized nanoparticles. Including ZnO and TiO_2 just as rates starting with aggregate volumetric centralization on transformer oils with applied field 10MV/m and the size of nanoparticles is 15nm, the comes about will be as indicated in Figs. (6,7). The effective relative permittivity of Diala b -based ($ZnO+TiO_2$) nanofluid expands dependent upon 14. 43% and toward 12. 68% in karamay25#, the point when concentration increments up to 20%. At the same time, it abates towards 4.67% in Diala b and towards 4.67% in karamay25#, when temperature expands by 40°C.

The radius of nanoparticle has a small impact on increasing or decreasing the effective permittivity nanofluids concerning illustration work with respect to both orientation polarizability and internal polarizability. In this way, the successful relative permittivity amongst resultant nanofluids has been expanded with the increasing size of nanoparticle, but it has different values reliant on the relative permittivity of utilized oil and relative permittivity of nanoparticle. The impact of size on the compelling relative permittivity of Diala B and karamay25# oil-based TiO_2, $LiTaO_3$ Graphite, ZnO, MgO, and SiO_2 during 106 V/m, temperature is 40°C and the centralization is 1wt%. Concerning the increasing size of nanoparticles, the successful relative permittivity resultant nanofluids expand separately, as stated by the permittivity of utilized nanoparticle and chosen oil. The impact of nanoparticle's size on compelling relative permittivity of oil is blended with sorts or a great number of nanoparticles that are varying its effective relative permittivity for the same oil by utilizing each standout amongst them independently. The

Figure 4. Effective relative permittivity with nanoparticle's volume fraction

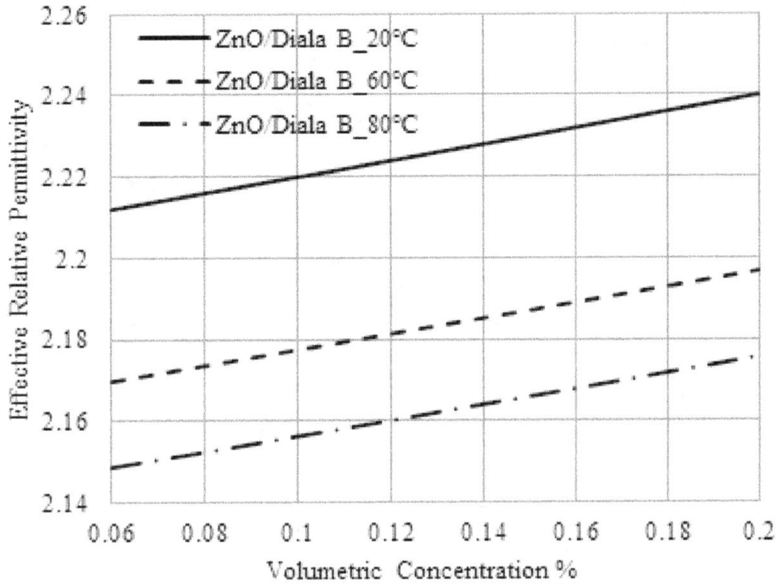

Figure 5. Effective relative permittivity with nanoparticle's volume fraction

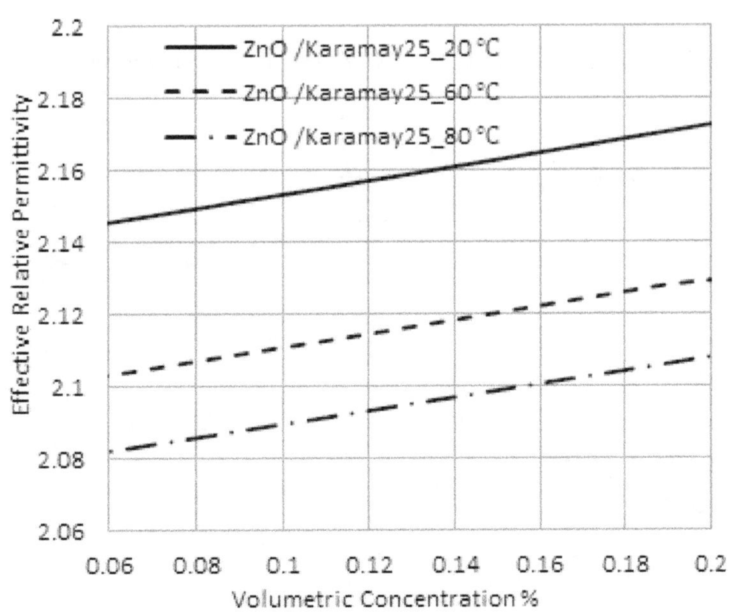

Figure 6. Effective relative permittivity with nanoparticle's volume fraction

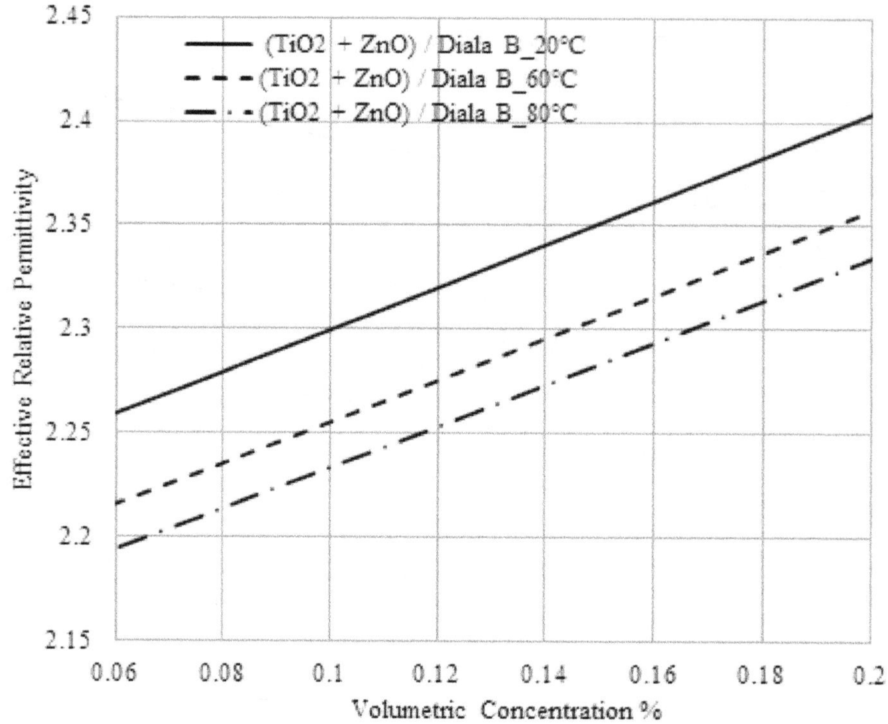

Figure 7. Effective relative permittivity with nanoparticle's volume fraction

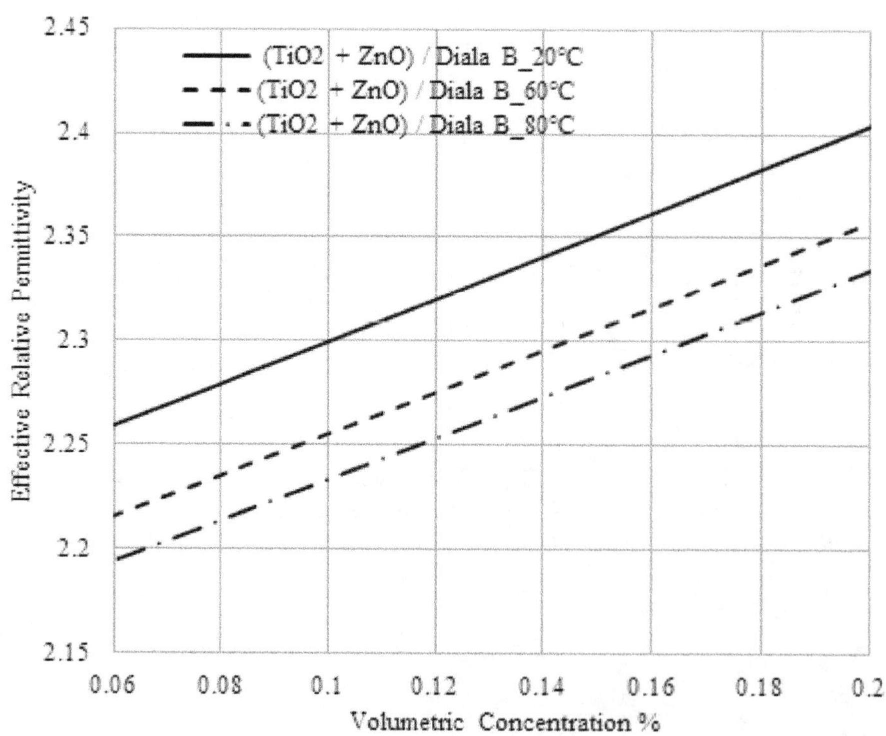

effective permittivity of multi-nanoparticles nanofluid expands with the increasing size of nanoparticles. Utilizing multi-nanoparticles (TiO_2 +ZnO) provides for the most noteworthy increases in the effective relative permittivity of the chosen oils, ($LiTaO_3$ + MgO) and then (Graphite + SiO_2).

Table. 2. Effective Relative Permittivities of Transformer Oils Nanofluids

Nanoparticles	Volumetric Concentrations of Nanoparticles in Transformer oils					
	Diala B			Karamay25#		
	0.08%	0.14%	0.2%	0.08%	0.14%	0.2%
Graphite	2.1797	2.1961	2.2167	2.113	2.1285	2.145
SiO_2	2.1634	2.1675	2.1715	2.0972	2.1012	2.1051
Graphite + SiO_2	2.1715	2.1822	2.192	2.1051	2.11	2.1249
$LiTaO_3$	2.2189	2.2658	2.3138	2.151	2.1964	2.2428
MgO	2.1715	2.1816	2.1918	2.105	2.1149	2.1248
$LiTaO_3$ + MgO	2.1951	2.232	2.2519	2.1279	2.1544	2.182
TiO_2	2.2996	2.4123	2.531	2.2291	2.338	2.452
ZnO	2.1735	2.1852	2.1969	2.107	2.1183	2.129
TiO_2 + ZnO	2.2356	2.282	2.3576	2.1671	2.23	2.2851

Table 2 demonstrates successful relative permittivity of nanofluids after starting to include the chosen interesting and multi-nanoparticles in Diala B and karamay25# in individual volumetric concentrations; the nano-size is 15 nm; the applied field is 10 MV/m and the temperature is 60 °C. In the event of including a rise to the volumetric concentrations of the sorts of nanoparticles chosen for the transformer oil, the resultant effective relative permittivity varies from the relative permittivity oil with each standout amongst them utilized independently with the same concentration. The compelling relative permittivity of nanofluid will increment with increasing the nanoparticles in transformer oil. Additionally, it is in contrast to the nanofluid that is alternately reliant on relative permittivity of utilized oil and different kinds of nanoparticles.

4 DYNAMIC CHARGING IN NANOFLUIDS

Positive streamer propagation in transformer oils nanofluids principally relies on charging elements under nanofluids. Adequacy of charging dynamic methodology under nanofluids relies on looking into the kind of the utilized nanoparticles. The charging element inside nanofluids is recognized by the fundamental venture in controlling for speed of certain streamer propagation under nanofluids; in any case, abating streamer propagation provides for the preference of raising the dielectric strength for nanofluids. This section displays a new theoretical model for mulling over the charging change of diverse sorts of multi-nanofluids. Moreover, this section illustrates the successful components of charging flow over individual and multi-nanofluids based on- Diala B, silicon and ester oils.

A similar examination has demonstrated clearly the impact of different distinct and multi-nanoparticles inside transformer oils on charging dynamic. Looking into nanotechnology may be intrigued by the growth business. Specialists have investigated how electric and dielectric properties have been controlled toward including a small sum of nanoparticles. Low versatility of these nanoparticles can thwart the improvement of a net space charge zone in the streamer tip and smother the propagating of electric field wave that is necessary to drive electric field subordinate atomic ionization and streamer propagation further under the fluid (Lewis, 2004). To acknowledge the electrodynamics procedures that occurs in electrically pushed transformer oil, the nanoparticles absorb electrons with change occurring over them under moderate charged particles (Lesaint & Massala, 1998; Segal et al., 1998). Starting with the customary perspective, conductive substances move forward conductivity and fortify the sullying of fluid dielectric by impurities. The breakdown strength of the transformer oil built nanofluids is altered by the conductive Fe_3O_4 nanoparticles that might have been exceptionally progressed.

Conductive nanoparticles trap electrons, as stated by the time constant of accuse unwinding. The time constant determinedly impacts the degree to which the electrodynamics is formed within the insulation fluid transform. Semi-conductive and dielectric NPs additionally progress insulation strength transformer oils, as it is demonstrated that electrons stick to polarized charges with respect to semi conductive or dielectric nanoparticles (Hwang et al., 2013; Sima et al., 2015; Takada et al., 2008). Fe_3O_4 nanoparticles provide for useful superfluity to transformer oil, which could incredibly enhance the certain drive breakdown strength for transformer oil and AC breakdown voltage, and nanofluids might give acceptable critical upgrades in the dielectric similar to silica which enhances the AC breakdown voltage of mineral oil. These contrasts in streamer propagation characteristics and breakdown strength to pure oil and nanofluid are nearly identified with the transform of space charge distribution initiated by shallow trap in nanofluid. Including a small sum of Al_2O_3 nanoparticles under transformer oils move forward the intend lightening drive breakdown voltages of nanofluids (Rafiq & Li, 2015). It may be uncovered that the propagation characteristics of positive and negative streamers in nanofluid are markedly influenced toward the addition of TiO_2 nanoparticles. Ester and Silicon oils nanofluids for multi-nanoparticles suspensions are mixed to moderate propagation of certain streamer rather than utilizing distinct nanoparticles; henceforth the breakdown voltage of nanofluid is extended (Wedin, 2014).

The absence of mineral oil assets turns into our current reality issue and pushes individuals to find separate vitality sourball. Nevertheless, the provision of mineral oil for transformers also causes issues since there will be a spillage throughout operation and the oil turns into a pollutant of the nature's domain because of its less biodegradable trademark (Yuliastuti, 2010). Vegetable oil has been recognized as useful elective material for transformer oil because of its beneficial biodegradability characteristic, low pour point, high streak point, and high dissolvability (Rudnick & Randles, 2006). STO-50 Silicone transformer oil is portrayed by its high dielectric strength, total administration temperature range, high streak point, low vapor pressure, low viscosity progress at greatly high and low temperatures and dormancy to essentially known substrates. It has a break down voltage worth over 50KV (Lew, 2005).

4.1 Nanoparticles in Transformer Fluids

The chosen oils will be the majority to be utilized within the transformer applications, and they have separate electrical permittivities and conductivities. However, mineral oils are to the greater part pervasive offline. Ester and Silicon oils are represented as the future of the transformer oils due to the persistent deficiency of petroleum stocks. Chosen nanoparticles are picked as they have total ranges for permittivities and conductivities. Diala B has been chosen concerning illustration insultingly oil, as it offers intrinsic common resistance to oil corruption and suitableness for in length existence applications, such as transformers, grid and mechanical transformers up to the greatest load (Product Data Sheet from Shell Lubricants, 2011).

Ester oil (MIDEL7131) has been chosen similarly because it is an exceptionally biodegradable insulating oil for use in environmentally delicate applications. It has fantastic electrical aspects and thermal dependability. It is committed from engineered hydrocarbon liquids that are particularly decided for biodegradation and environment compatibility, they have low vapor weight and have high flash point (Technical Data Sheet from MIDEL, 2010). STO-50 Silicon oil is a 100% straight Polydimethylsiloxane liquid (CAS# 63148- 62-9) for a viscosity of about 50cSt (centistoke) in 25°C. STO-50 Silicon is figured particularly for use concerning illustration of a dielectric liquid and dielectric coolant of electrical transformers. Table 2 shows the relative permittivity of MIDEL7131 and Silicon Oils (Technical Data Sheet from Clearco Products Co, Inc, 2015; Technical Data Sheet from MIDEL, 2010). It is environmentally sheltered, and its high thermal soundness makes it great decision transformers that are found inside (Technical Data Sheet from Clearco Products Co, Inc, 2015). Relative permittivity of MIDEL7131oil is 3.2 and STO-50 Silicon oil is 2.7 (Technical Data Sheet from Clearco Products Co, Inc, 2015; Technical Data Sheet from MIDEL, 2010).

4.2 Theory and Effective Parameters

To model the electrodynamics inside an electrical transformer oil-based nanofluid, it is essential to start with modelling the charging of the nanoparticles for oil. The nanoparticles in transformer oil are prompted or polarized under the applied electric field in a short timescale. Then positive and negative charges are produced toward the more level and upper hemispheres nanoparticle, in as much as no space charge is watched in the heft as indicated in Fig. (8). The unwinding time for conductive nanoparticles may be dismissed, so the prompted surface thickness of superbly nanoparticles (Sima et al., 2014) is:

$$\sigma_s = 3\varepsilon_1 E_0 \cos\theta \tag{7}$$

Where θ is the angle of negative charges on positive hemi-sphere, E0 is the applied field, ε_1 is relative permittivity of transformer dielectric liquid.

On other hand, with dielectric and most of the semi-conductors, the relaxations times are long relative to the timescale of interest for streamer growth. Therefore, the density of these polarized charges on the surface of the dielectric particle is written as follows:

$$\sigma_s = 3\varepsilon_1 E_0 \cos\theta \frac{\varepsilon_2 - \varepsilon_1}{2\varepsilon_1 + \varepsilon_2} \tag{8}$$

Once the nanoparticles are saturated with negative charges, the electron saturation charge for the nanoparticle is in coulombs (Sima et al., 2014).

In case of multi-nanofluids, the applied electric field lines in transformer oil nanofluids are represented as (Thabet et al., 2017a; Thabet et al., 2017b). Firstly, the multi-nanoparticles over transformer dielectric fluids are prompted or polarized under the applied electric field in a short time-scale; as

Figure 8. Charging dynamic of nanoparticle after applying uniform electric field

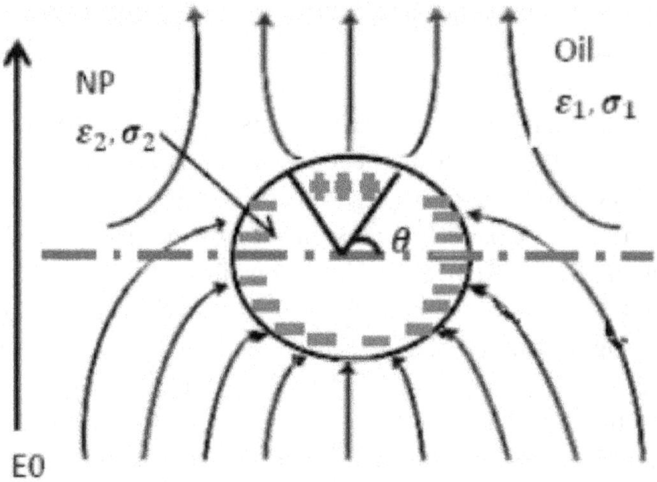

Figure 9. Charging dynamics of the two nanoparticles directed electric field is applied

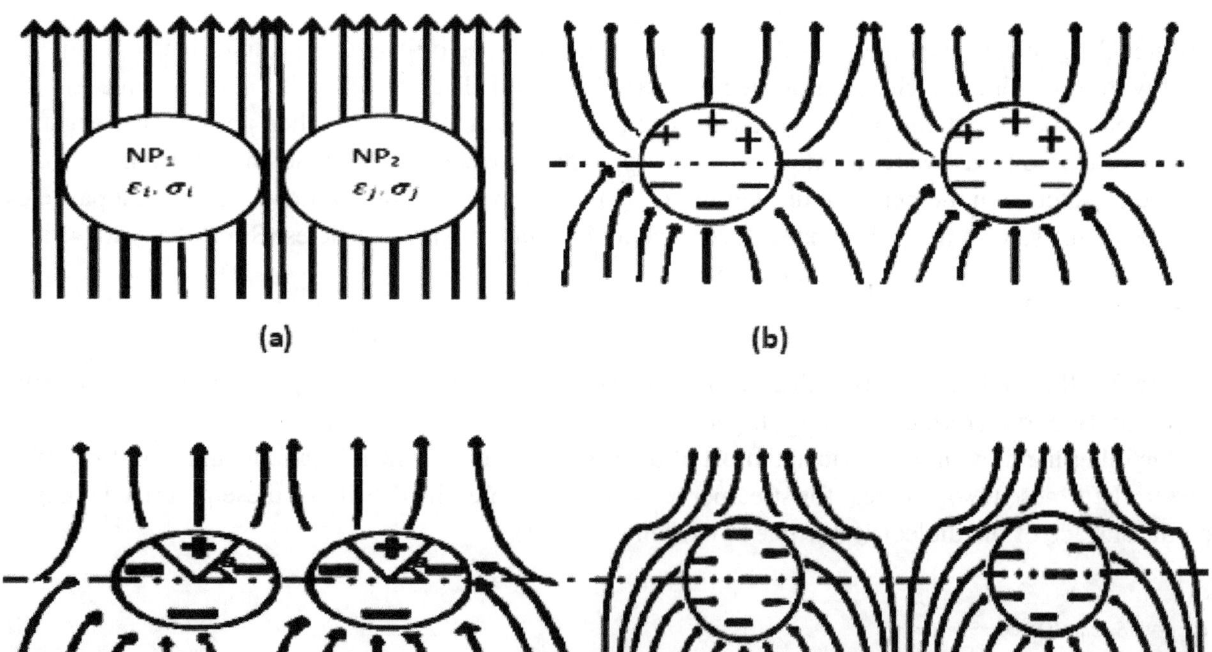

charging the long haul (time required for charging the nanoparticle in Diala b under applied field is (τ_{pc} = 0. 7791ns), as indicated Fig. (9 (a)), In any case, Fig. (9 (b)) positive and negative charges will create the easier and upper hemispheres of nanoparticle. The electrons that were prepared starting with ionization instantly are indeed these electrons that move rapidly on the positively accused hemisphere, as indicated in Fig. (9 (c)). Finally, Fig. (9 (d)) has indicated that the nanoparticle is totally secured with negative charges electrons. In this way, no more sure charges have showed up on nanoparticle's certain hemisphere and electrons were no more scavenged. Note that the certain and negative ions are not thought seriously about and will be viewed still throughout the charging process, to their portability ($\mu_n=\mu_p=1\times10^{-9}m^2.V^{-1}s^{-1}$), However, electron's mobility magnitude is ($\mu_e= 1 \times 10^{-4}$ $m^2{\cdot}V^{-1}{\cdot}s^{-1}$) and the volume fraction between nanoparticles and oil volume is neglected.

Surface charge density (σ_t) is an induced or polarized positive charge on nanoparticles and appears whenever the nanoparticle is exposed to field as a result of both dipole and replacement polarization on nanoparticle. Increasing density of the surface charge increases the effectiveness of nanoparticle in captures electrons. The density of surface charge of multi conductive nanoparticles is shown in Eq. (9) However, Eq. (10) gets thickness surface charge of dielectric or semi-conducive nanoparticle; then, Eq. (11) gets the strength thickness surface accuse at whatever point conductive and dielectric nanoparticles are utilized concerning illustration takes after (Thabet et al., 2017a; Thabet et al., 2017b).

$$\sigma t = 3n\varepsilon 1 E0 \cos\theta \tag{9}$$

$$\sigma_t = 3\varepsilon_1 E_0 \cos\theta \sum_{i=1}^{n} \frac{\varepsilon_{i+1}-\varepsilon_1}{2\varepsilon_1+\varepsilon_{i+1}} \tag{10}$$

$$\sigma_t = 3\varepsilon_1 E_0 \cos\theta (\sum_{i=1}^{n} (\frac{\varepsilon_{i+1}-\varepsilon_1}{2\varepsilon_1+\varepsilon_{i+1}}) + m) \tag{11}$$

Where $\varepsilon 1$ is permittivity of oil, n is the total number of used nanoparticles per unit volume, m is the total number of conductive nanoparticles per unit volume, θ is the angle of negative charges on positive hemi sphere, μ_{i+1} is permittivity of dielectric or semi-conducive nanoparticle

Saturation charge of multi-nanoparticles (Q_{st}) is the immersion accuse of multi-nanoparticles and which is the greatest electrons accuse that a nanoparticle might stand around its surface. In any case, it relies on the relative permittivity of oil, sort of nanoparticles; (magnetic or non-magnetic) and relative permittivity of non-magnetic nanoparticles (Thabet et al., 2017a; Thabet et al., 2017b). The rate of electrons' charge is deposited on nanoparticles; the extent of stored accuse expands rabidly from the beginning nanoseconds, then it slows down as the nanoparticles are near immersion. Thus, the multi-nanoparticle has short period to be accused for comparing with the long haul of streamer propagations as takes after (Thabet et al., 2017b).

$$Q_{tt} = \frac{t/\tau_{pc}}{1+t/\tau_{pc}}.(Q_{st}) \tag{12}$$

The charging flow of equal transformer oils nanofluids for individual and multi-nanoparticles has been predicted on Diala B, MIDEL7131and silicon oil in case the size of utilized nanoparticles is 25nm, E_0=10^8V/m and θ = 0. The impact of including individual sorts of individual and multi-nanoparticles (dielectric, conductive, and semi-conductive) with respect to surface charge density, immersion charge and charging element for mineral transformer oils has been exhibited in terms of effects.

4.3 Effect of Nanoparticles on Dynamic Charge in Nanofluids

A comparative study on the magnitude of deposited charge and attachment time constants of individual and multi-nanoparticles has been investigated on Diala B, Silicon and MIDEL7131 fluids. Individual and multi-nanoparticles expand extent of kept charge relying upon their polarizability and transformer oil permittivities. Transformer oils based-Fe_3O_4 might absorb bigger esteem of allowed electrons over other individual nanofluids and ($LiTaO_3$ + SiO_2) nanofluids. On the other hand, multi-nanoparticles (Fe_3O_4+ MgO) turn into right away rank for absorbing the biggest esteem for free electrons. According to attaching time that is time needed for nanoparticle to catch 80% of nanoparticle's charge (Sima et al., 2014), it's a critical issue for selecting effective nanoparticles that have thick, as small attachment time. Fe_3O_4 nanoparticles devour the smallest chance in getting significant accuse around chosen individual nanoparticles for chosen oils. In the event of utilizing multi-nanoparticles (Fe_3O_4+MgO), nanoparticles have the best chance around recommended distinct and multi-nanoparticle.

5 ELECTRICAL CONDUCTIVITY OF NANOFLUIDS

Electrical conductivity of transformer oils is profoundly influenced by including nanoparticles. Including nanoparticles to transformer oils increase both static and electrophoretic conductivity of resultant nanofluids. To solve this problem, this section displays new multi-nanoparticles method and specifies the impact of including individual and multi-nanoparticles for successful electrical conductivity transformer oils nanofluids. Furthermore, this section investigates the impact of volumetric concentration, temperature, applied field and zeta possibility with respect to successful electrical conductivity of the transformer oil nanofluids. A similar study has been introduced to point out the profit of including multi-nanoparticles as suspension multi-nano-metric inclusions inside power transformer oil. Transformer oil-based nanofluids (NFs) has pulled in incredible consideration for its change over insulating properties and heat exchange (Lewis, 2004; Pradhan et al., 2008). NFs are a colloid for unfaltering suspension of nanoparticles for transformer oil. Nanoparticles are included under transformer oil to enhance the cooling efficiency (Choi, 1995; Miao et al., 2012). It is additionally demonstrated that transformer oil-based nanofluids have higher AC breakdown voltage and positive drive voltage withstand compared with conventional oil (Lisunova et al., 2006).

The electrical conductivity of liquid suspensions of multiwall carbon nanotubes may be investigated. An exponential association between the electrical conductivity and the impostor concentration of the carbon nanotube in liquid suspension is investigated as well as the electrical conductivity for nanofluids in light of practical sulfonated carbon nanotubes. It is discovered that electrical conductivity increments 13 times in the nanofluid holding 0. 5%wt. nanotubes. The nanotube based on nanofluid, has been observed 3457. 1% extension in the electrical conductivity by including 1.44%wt. alumina nanoparticles for water (Ganguly et al., 2009; Liu et al., 2004). Around different electrical properties suspension, the compelling not main electrical conductivity reflects the insulation performance; moreover, it provides for the data on the level of scattering and stability of the nanofluid (Cruz et al., 2005; Cuquejo, 2006). The charge on the surface of nanoparticles pulls in a slim layer of ions for an opposite accuse of the nanoparticle surface. This twofold layer of particles dives for the nanoparticle, as it diffuses every last bit through the liquid. The electric potential at the limit of the twofold layer is known as the Zeta possibility of the particles and has values that commonly range between +100 mV to -100 mV (D'Souza, 2014; Smpson et al., 2012). In spite of the fact that the thermal conductivity of the nanofluid has been generally studied, the electrical properties of nanofluid are examined insufficiently. Only a couple of examinations have been carried out on the electrical conductivity of carbon nanotube-based nanofluids, and no precise numerous or major data is receptive to nanofluids holding nanoparticles as the dispersing materials.

5.1 Nanoparticles and Nanofluids Conductivity

Insulating oils that are used as transformer's liquids are unlikely to present any significant health or safety hazards when properly used in the recommended application, and when good standards of industrial and personal hygiene are maintained. The electrical conductivity of selected insulting oils and nanoparticles depicts as shown in Table 1. Nanoparticles are chosen to have extensive variety permittivities and conductivities (Fleming et al., 2005; Thabet et al., 2012; Wong & de Leon, 2010). Diala B oil is an uninhibited electrical dielectric liquid to be utilized where ordinary oxidation resistance is needed. Diala B Oil is a profoundly refined naphthenic mineral oil for standard low pour perspective characteristics. Diala B Oil is exceptionally managed to hold A high breakdown voltage as passed on. This empowers it to be used as A and only various applications without aid medicine (Product Data Sheet from Shell Lubricants, 2011). Kelamayi 25# oil is an exceedingly refined mineral oil and a chosen cell reinforcement. It can be planned for insulating and cooling transformers and its relative permittivity will be the same as Diala B Oil.

5.2 Estimation of Nanofluids Conductivity

The accumulation of nanoparticles in the build fluid has an ordinary measure of a few hundred nanometers. Brownian motion is accordingly dismissed as a model (Miao et al., 2012). The successful electrical conductivity of transformer oil-based NPs camwood can be displayed with excellent Maxwell conductivity and electrophoresis conductivity. Conductivity of nanofluids by classical Maxwell model σM can be expressed as follows:

$$\sigma_{M=}\sigma_f\left(1+\frac{3\varphi}{(\alpha+2)/(\alpha-1)-\varphi}\right) \tag{13}$$

Where σ_f and σ_p are the electrical conductivities of transformer oil and nanoparticle, φ is the volumetric fraction of nanoparticles, $\alpha=\sigma_p/\sigma_f$ is the ratio of conductivity.

The climbing of the electrical conductivity is mostly due to the electrical double layers (EDL) (Posner, 2009). The surface charge of nanoparticles, together with the ion cloud that constitutes the EDL, would actively increase the electrical conductivity of nanofluids. At applied electrical fields, the individual-charged nanoparticles might move when you quit offering on that one cathode to another, which forms the electrophoresis conductivity as follows:

$$\sigma_E = \frac{2\varphi\varepsilon_0^{\;2}\varepsilon_1^{\;2}\zeta^2}{\eta_0 r^2} e^{\lambda(T-T_0)} \tag{14}$$

Where ζ is the Zeta potentials of nanoparticles, T_0 initiates the temperature of nanofluids and T terminated temperature, η_0 is the dynamic viscosity at T_0, ε_1 is the relative permittivity of transformer oil, r is the radius of nanoparticle and λ is index viscosity. Hence, the modified electrical conductivity model of insulating oil-based nanofluids can be introduced based on Maxwell model and electrophoresis phenomenon, expressed as:

$$\sigma = \sigma_M + \sigma_E = \sigma_f\left[1 + \frac{3\varphi}{\frac{\alpha+2}{\alpha-1} - \varphi}\right] + \frac{2\varphi\varepsilon_0^{\;2}\varepsilon_1^{\;2}\zeta^2}{\eta_0 r^2} e^{\lambda(T-T_0)} \tag{15}$$

The static successful electrical conductivity of transformer oil nanofluids relies on the conductivity of the utilized oil and the conductivity of the chosen nanoparticles accepting the mono-dispersing nanoparticles of the transformer oil. Maxwell model will be utilized for foreseeing the static conductivity of multi-nanoparticles. The static conductivity of multi-nanofluids σ_{MT} can be expressed as follows:

$$\sigma_{MT} = \sigma_f\left(2 + \frac{3\varphi_1}{\frac{\alpha_1+2}{\alpha_1-1} - \varphi_1} + \frac{3\varphi_2}{\frac{\alpha_2+2}{\alpha_2-1} - \varphi_2}\right) \tag{16}$$

Where σ_{p1}, σ_{p2} and σ_f are electrical conductivity of the first type, the second type of nanoparticles respectively and the transformer oil. φ_1 and φ_2 are the volumetric fractions of first and second nanoparticles respectively. $\alpha_1=\sigma_{p1}/\sigma_f$ and $\alpha_2=\sigma_{p2}/\sigma_f$ are the ratios of first and second nanoparticles conductivity and fluid conductivity.

As shown by the hypothesis of colloid chemistry, nanofluids there have charges on the surface nanoparticles, and opposite charges might be pulled in around the particles, particularly the electrical double layer (EDL). Electrical double layer (EDL) might adequately grow the electrical conductivity of nanofluids. When an electrical field is applied, the individual accused nanoparticles might move from an individual cathode to another, which structures the electrophoresis conductivity. If the particles in colloid are exceptionally small, the electrophoresis speed of accused particles can be communicated by the accompanying comparison (Tang et al., 2011).

$$V_1 = 2E_0\varepsilon_0\varepsilon_1\zeta_1 / 3\eta \tag{17}$$

$$V_2 = 2E_0\varepsilon_0\varepsilon_1\zeta_2 / 3\eta \tag{18}$$

Where E is the electric intensity of applied field, ε_0 is the permittivity of vacuum, ε_r is the relative permittivity of transformer oil, ζ_1 and ζ_2 are the Zeta potentials of first and second nanoparticles respectively and η is the dynamic viscosity of nanofluid.

Similarly, as the transformer oil is a sort of feeble polarity liquid, the electrical possibility of nanoparticle surface is pretty nearly equivalent to the Zeta possibility (Eastman et al., 2001; Fathiah & Edyvean, 2015; Shen et al., 2011). Considering that the charged particles move with the uniform velocity, the electrophoretic conductivity caused by electrophoresis, i.e., the electrophoresis conductivity can be expressed as:

$$\sigma_{ET} = \frac{2\varepsilon_0^{\,2}\varepsilon_1^{\,2}}{\eta_0} e^{\lambda(T-T_0)} \left[\frac{\varphi_1\zeta_1^{\,2}}{r_1^2} + \frac{\varphi_2\zeta_2^{\,2}}{r_2^2} \right] \tag{19}$$

The model of Effective electrical conductivity of transformer oil-based multi-nanofluids can be introduced based on Maxwell model and electrophoresis phenomenon, as follows:

$$\sigma_T = \sigma_{MT} + \alpha_{ET} = \sigma_f \left[2 + \frac{3\varphi_1}{\frac{\alpha_1+2}{\alpha_1-1} - \varphi_1} + \frac{3\varphi_2}{\frac{\alpha_2+2}{\alpha_2-1} - \varphi_1} \right] + \frac{2\varepsilon_0^{\,2}\varepsilon_1^{\,2}}{\eta_0} e^{\lambda(T-T_0)} \left[\frac{\varphi_1\zeta_1^{\,2}}{r_1^2} + \frac{\varphi_2\zeta_2^{\,2}}{r_2^2} \right] \tag{20}$$

The effective electrical conductivity of nanofluid has main effect time of streamer propagation and breakdown voltage (Posner, 2009); therefore, it is necessary to study the factors affecting the effective electrical conductivity value.

5.3 Effect of Nanoparticles on Nanofluids Conductivity

The contrasts between the middle of the impacts of including individual and multi-nanoparticles can be communicated in emulating similar considerations. Including just the volumetric concentrations of the two individual sorts nanoparticles with chosen transformer oils, their size will be 15nm. The effective electrical conductivity contrasts from the compelling electrical conductivity by utilizing these nanoparticles independently. Including the same concentrations, the size of nanoparticle will be (15nm), the applied field (20MV/m) and temperature (40 °C concerning the illustration indicated for Table 3. For example, the effective electrical conductivity of (TiO_2 + ZnO) multi-nanofluids is more level than the successful electrical conductivity of TiO_2 distinct nanofluids and higher than the compelling electrical conductivity ZnO individual nanofluids.

Therefore, utilizing multi-nanoparticles provide for transformer nanofluids a characteristic of abating streamer propagation for sensible augment in successful electrical conductivity. In Table 4, however, the volumetric concentrations of individual nanoparticle and multi-nanoparticles oil are (0.1 wt. %) and the applied field (20MV/m), the impact of the electrical conductivity of multi-nanofluids contrasts with the individual-nanofluids. Multi-nanoparticles technology can effectively utilize different nanoparticles, similarly, since effective conductivity of multi-nanofluids increments with sensible strength to withstand the increasing temperature. Moreover, Figures (10, 11) indicate the effective electrical conductivity of power transformer oil-based nanofluids that have been expanded toward increasing volume rates of ZnO, and so; increasing viscosity list increments the successful electrical conductivity of power transformer oil.

Table 3. Effective Electrical Conductivity of Transformer Oil Nanofluids (20 °C)

Nanoparticles	Volumetric Concentrations of Nanoparticles in Transformer oils					
	Diala B			Kelamayi 25#		
	0.1%	0.2%	0.4%	0.1%	0.2%	0.4%
ZnO	6.967	13.932	27.864	6.430	12.859	25.718
TiO_2	30.378	60.755	121.510	28.039	56.077	112.15
TiO_2 + ZnO	18.602	37.201	74.397	17.170	34.336	68.668
$LiTaO_3$	25.453	50.905	101.809	23.493	46.985	93.968
SiO_2	2.346	4.691	9.382	2.166	4.330	8.659
$LiTaO_3$ + SiO_2	13.902	27.801	55.598	12.832	25.660	51.317
Graphite	15.288	30.575	61.149	14.111	28.220	56.440
MgO	9.937	19.872	39.743	9.171	18.342	36.682
Graphite+ MgO	12.615	25.226	50.449	68.668	51.317	46.564

Table 4. Effective Electrical Conductivity of Transformer Oil Nanofluids Under Thermal Conditions

Nanoparticles	Temperatures of Nanofluids					
	Diala B			Kelamayi 25#		
	40 °C	60 °C	80 °C	40 °C	60 °C	80 °C
ZnO	6.967	18.936	51.470	7.9376	17.663	39.309
TiO_2	30.378	82.575	224.460	39.416	87.721	195.226
TiO_2 + ZnO	18.602	50.559	137.428	23.588	52.492	116.818
$LiTaO_3$	25.453	69.187	188.067	32.517	72.366	161.052
SiO_2 (Silica)	2.3467	6.376	17.330	2.355	5.239	11.659
$LiTaO_3$ + SiO_2	13.902	37.784	102.701	17.439	38.806	86.358
Graphite	15.288	41.555	112.957	18.694	41.604	92.589
MgO	9.9370	27.008	73.415	11.701	26.040	57.951
Graphite +MgO	12.615	34.285	93.189	15.201	33.825	75.273

Figure 10. Effect of concentration of ZnO nanoparticles on effective conductivity of nanofluids

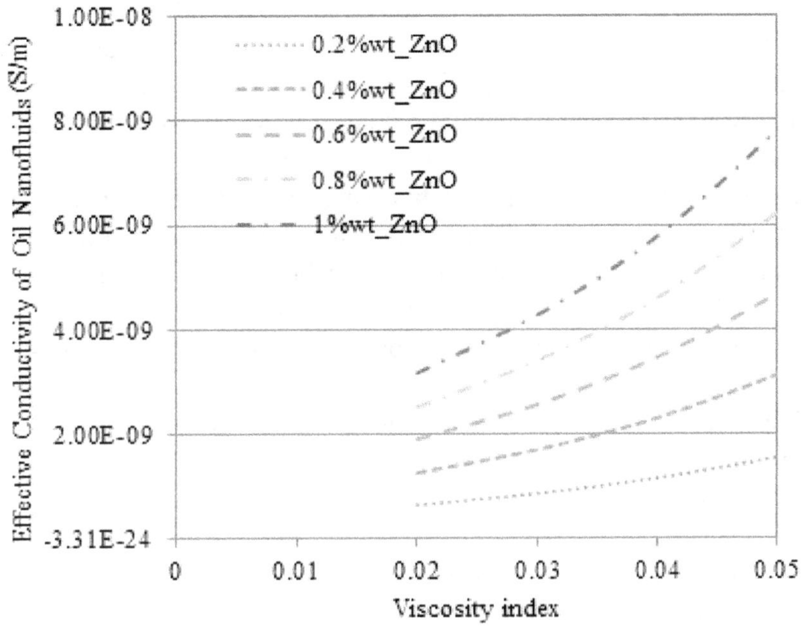

Figure 11. Effect of viscosity index on effective conductivity of transformer oil nanofluids

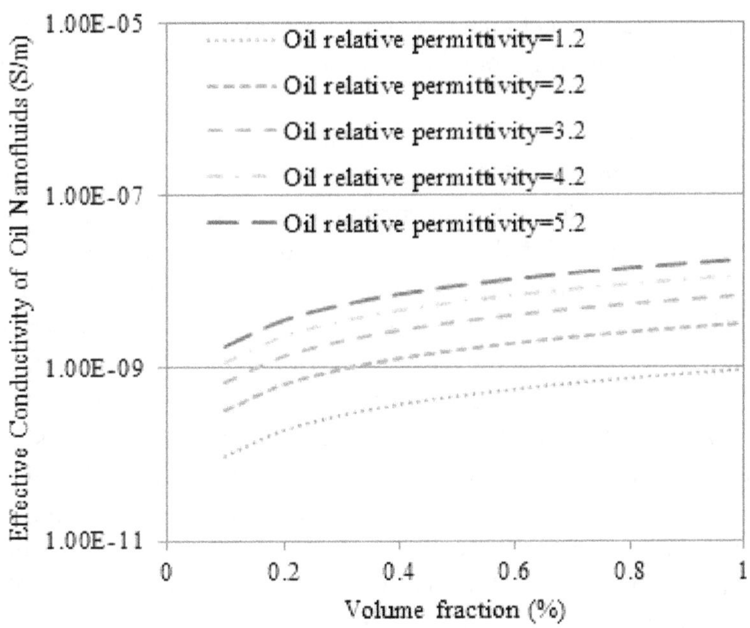

6 NANOFLUIDS IN LABORATORY

The dielectric strength of transformer oil nanofluids is an exceptionally imperative parameter for concentrating on the execution of transformer oils. Therefore, the electrical conductivity assumes an incredible role in increasing or decreasing the insulation strength of transformer oil nanofluids. Accordingly, this section displays lab-test estimations for breakdown voltage and effective relative conductivity of individual and multi-nanofluids. In this section, the lab-test examples have been readily created for standard procedures, devices, and materials to transform prominent nanofluids and multi-nanofluids transformer oil. Furthermore, this section examines the outcomes of breakdown voltage and successful relative conductivity of the applied lab-tests with respect to pure, nanofluids, and multi-nanofluid examples. Recently, considering the fill in expectation to improve both dielectric and electrical characteristics for high voltage insulation and power apparatus cooling transformer oils, a specific inventive case of such fill in can be the improvement of dielectric nanofluids. Nanofluids are readied by including nanoparticles suspension on transformer oil to move forward a few of the oil's insulating, electrical and thermal aspects (Feng et al., 2006).

Huge numbers of investigations have been conveyed to create the oil nanofluid. The scattered nano-sized particles in the extension of 1-100 nm under base fluid are known as nanofluids. The prerequisites for the generation of nanofluids are the much suspension, stable suspension, tough suspension without compound transforms and low agglomeration (Yu et al., 2010). However, the nanoparticles tend to aggravate under greater particles principally by the engaging constrains (Wei & Wang, 2010; Zhu & Yin, 2004), prompting the execution debasement of nanofluids. Thus, the long-haul soundness nanoparticles scattering in the group oil are at present undergoing lab-tests in this field. Substantially, worth of effort has been done to improve the scattering soundness of the nanoparticles in the base fluid (Lundgaard et al., 1998; Zhu & Yin, 2004).

Two-stage technique may be the most of comprehensively used method for delivering nanofluids. Nanoparticles, nanofibers, nanotubes, or different nanomaterials used to concern illustration of this system only are initially made as dry powders toward compound or physical techniques. Then, the nano-sized powder can be scattered under a fluid in the second preparing venture with the assistance of concentration escalated consideration magnetic energy agitation, ultra-nationalistic agitation, high-shear mixing, homogenizing, and ball processing. A two-stage strategy is the greatest investment technique on get ready nanofluids for expansive scales (Beroual & Aka-N'Gnui, 2002; Nguyen, 2013; Taylor et al., 2013; Yamashita et al., 1990). A one-stage strategy is similarly used to process nanofluids and decrease the agglomeration of nanoparticles, the one-stage transform comprises all and disperses the nanoparticles in the fluid. In this strategy, the methods for drying, storage, transportation, and scattering of nanoparticles are kept away from, with the goal that the agglomerated nanoparticles will be limited, and the reliability for fluids will be stretched (Nguyen, 2013; Technical Data Sheet from Ergon Refining, Inc., 2012). The group liquid utilized in this study will be HyVolt I transformer oil from Ergon Refining, Inc. organization. This uninhibited, electrical insulating oil is prepared from an extremely hydropower naphthenic oil to help the detailed prerequisite characterization in IEC 60296, ed. 4, 2012, and all determinations (User Manual of Elma Schmidbauer GmbH, 2007). Determinations of chosen nanoparticles are indicated in Table 5. All utilized nanoparticles are obtained by sigma Aldrich co (Thabet et al., 2012).

Table 5. Relative permittivity and conductivity of selected oils and nanoparticles

Nanoparticle	Relative Permittivity	Conductivity (*s/m*)
HyVolt I oil	2.3	$4*10^{-13}$
ZnO	7.4	10^{-6}
Al_2O_3	9.9	10^{-12}
SiO_2	4.5	10^{-11}

Regarding the prerequisites and precautions, nanoparticle camwood is unsafe for the human body since they gather in the organs, particularly the lungs. Quickly as the nanoparticles are scattered under base fluid, solvent may be not risky because of the holding between nanoparticles and build liquids. Precautions for further bolstering ought to be taken:

1. Preventing immediate contact between the middle of human body and nanoparticles.
2. Utilizing of a seethe hood, protective attempting clothing, impenetrable gloves and security glasses are essential throughout synthetization of nanofluids.
3. Avoiding breathing, hand, particular figure and eye contact to nanoparticles.

6.1 Methodology of Nanofluids Preparation

For the preparation of nanofluid, the two-stage strategy is utilized within this examination. Generally, the preparation of nanofluids begins with weighing nanoparticles in a glove box, nanoparticles are blended with mineral oil. Then, the mixture will be blended for AREC ceramic advanced thermal magnetic Stirrer, which provides for quick heating, high temperatures (up to 550 °C) and high imperviousness capable magnet to enable blending (up to 1500 rpm). At last, the mixture is placed under Elmasonic S 40 (H) for sonication of nanofluid to be greatly scattered. Elmasonic S 40 (H) has LED-display showing setting time of the cleaning period, turning knob for setting continuation and short-period operation from 1 to 30 min, temperature-controlled ultrasonic operation and sweep function for an optimized sound field distribution in the cleaning liquid by frequency modulation (User Manual of Power Electronical, n.d.; User Manual of Velp Scientifica, n.d.).

The magnetic mixing serves to scatter the nanoparticles uniformly in the build fluid, yet the vitality is not sufficient to break any agglomerated nanoparticles. With that goal, an ultra-nationalistic shower can be used to break the agglomerations nanoparticles. However, constant utilization of nanofluids will be ready for the same strategy; they require different times with respect to magnetic stirrer or ahead ultra-nationalistic gadget. 0. 05 wt. % of Al_2O_3 to HyVolt I transformer oil have two hours on hot stirrer (60 °C), two hours in ultra-nationalistic after you quit offering on that one hour around the high temp stirrer (60 °C) once more to get its nanofluid. HyVolt I based- 0.05 wt% ZnO nanofluid has two hours around heated stirrer (60°C) and three hours over ultra-nationalistic. HyVolt i based- 0. 05 wt. % SiO_2 nanofluid can be arranged by hot stirrer (60°C) for two hours with respect to 4 hours over ultra-nationalistic. HyVolt I based- (0. 025 wt. % SiO_2+ 0. 025 wt. % Al_2O_3) necessitates two hours with respect to hot stirrer (60 °C) and 5 hours in ultra-nationalistic at that point of particular case hour with respect to hot (60 °C) for preparation. To get ready HyVolt I based - (0.025 wt% SiO_2+ 0. 025 wt% Al_2O_3) nanofluid, two hours around high temp stirrer (60 °C) and 6 hours over ultra-nationalistic are necessary. HyVolt I based-(0. 025wt% for SiO_2+ 0. 025 wt%) nanofluid is placed for two hours of hot stirrer (60°C) and 6 hours in ultra-nationalistic.

The breakdown voltage test is done using Motorized Oil Breakdown Voltage Tester (PE-OBDV-M100). It has many advantages, such as giving smooth operation from zero potential to Flashover Start Voltage (FSV) with motorized operation, Digital display output, with display hold on trip Precise HV trip voltage measurement and voltage hold after Trip-Overload & Short-Circuit interlock for operator protection (User Manual of Velp Scientifica, n.d.). By clearing and drying the lab-test if repeated delicately in the recent past, each lab-test evades containments and water content. The vessel for breakdown voltage analyzer can be clean and dry, and the space between electrodes should be 2.5 mm. Following the spilling lab-test example regarding the vessel till oil level, the vessel is shut firmly and altered to mechanized oil breakdown Voltage analyzer (OBVT), as demonstrated in Fig. (12). Switch on (OBDV) analyzer concerning illustration is composed over its manual client at that point, noting the perusing of breakdown voltage ahead screen analyzer. Lab-test can be repeated ten times, and averages of the individual's readings will be made as break down voltage of lab-test. Electric conductivity of pure oil and nanofluids can be measured toward electrical conductivity measuring situated (ECMS).

Electrical conductivity measuring set comprises two electrodes, insulated starting with each other (The cathode has conductivity sensor working toward prompting an exchanging current to a shut circle for result and measuring its extent to determine conductivity. Each example is spilling in the cleaned cell of electric conductivity meter with k =0. 1 or k =1, and during the minute of electric conductivity meter equilibrium, the electric conductivity for oil or nanofluids will be recorded.

Figure 12. Motorized Oil Breakdown Voltage Lab-tester

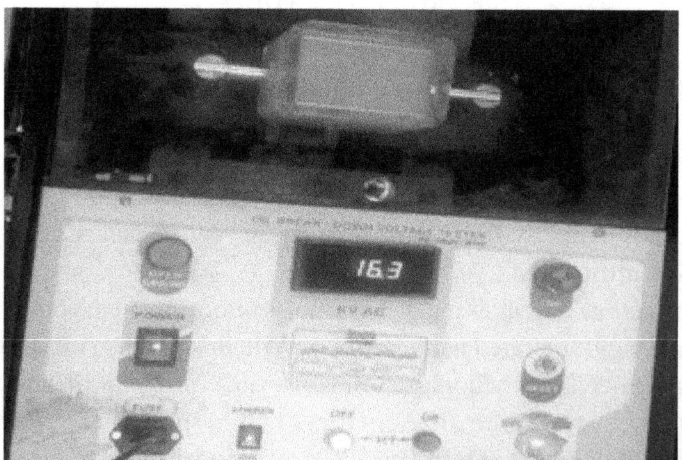

6.2 High Voltage Breakdown

Small concentrations of nanoparticles inside the transformer oil camwood transform electrical and dielectric properties for transformer oils. Breakdown voltage nanofluids progressions contrast with pure transformer oil. Effective electrical conductivity nanofluid has primary impact on figuring out the long run of streamer propagation and breakdown voltage. Figure (13) indicates breakdown voltages for pure Diala D oil transformer and the number of lab-tests conducted toward utilizing stirrer and without utilizing stirrer on uniform electric fields. On the other hand, Fig. 14 demonstrates the increasing rate of zinc oxide (ZnO) nanoparticles in the Diala D oil transformer dependent upon 1%wt builds dielectric strength of the modern materials separately. Volumetric rates of Zinc oxide nanoparticles are daintily raising dynamic thickness and effective electrical conductivity for energy transformer oil nanofluids.

Figure 13. Breakdown voltages in pure Diala D under uniform electric fields

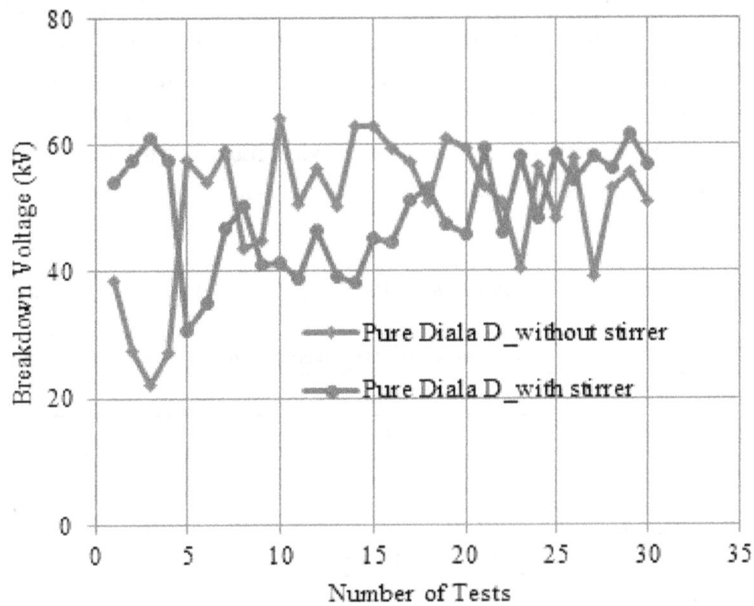

Figure 14. Breakdown voltages in (Diala D +1% wt ZnO) under uniform electric fields

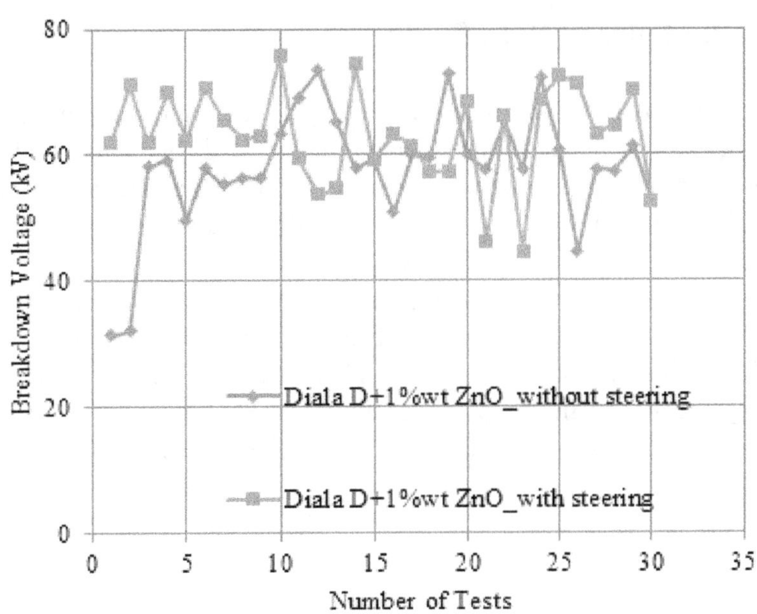

Moreover, increasing the electric applied field builds the extent of the kept accuse on the nanoparticles, since by increasing the space, the electrons can be stored. Thus, it has been investigated tentatively that ZnO/DialaD nanofluids have higher electrical and dielectric properties over pure insulating Fluids by including Zinc oxide nanoparticles.

Nanoparticles have been ionized and pulled in nothing electrons starting with transformer oil under an applied field, this movement raises the dielectric strength for nanofluids. Generally, it is known in the breakdown voltage that the tried nanofluids can be higher than breakdown voltage pure transformer oils, but all of this is for different values reliant on centralization and relative permittivity of chosen nanoparticles. 0. 05 wt. % Al_2O_3/ HyVolt i nanofluids have the most astounding breakdown voltage "around tried singular nanofluids reliant on relative permittivity for Al_2O_3 nanoparticles". 0.05wt.% SiO_2 / HyVolt i nanofluids have the smallest break down voltage because of small relative permittivity of SiO_2 nanoparticles. On the other hand, (0.025wt.% ZnO + 0.025wt.% Al_2O_3)/ HyVolt I nanofluid has the most noteworthy breakdown voltage "around chosen multi-nanofluids". Table 6 illustrates the breakdown voltage of pure oil, individual and multi-nanofluids.

Table 6. Breakdown Voltage of Pure Transformer Oils, Individual and Multi-nanofluids

Lab-Test Sample	Nanoparticles Concentrations	Breakdown Voltage (kV)
HyVolt I oil_ Com.	-	16
Al_2O_3/HyVolt I	0.05 wt. %	22.1
ZnO /HyVolt I	0.05 wt. %	18.57
SiO_2 /HyVolt I	0.05wt. %	16.3
(Al_2O_3+SiO_2)/ HyVolt I	(0.025wt. % SiO2+ 0.025wt. % Al_2O_3)	27.2
(ZnO + SiO_2) /HyVolt I	(0.025 wt. %SiO2+ 0.025wt. % ZnO)	26.1
(ZnO + Al_2O_3)/ HyVolt I	(0.025 wt. % Al2O3+ 0. 025 wt. % ZnO)	29.5

6.3 Effective Conductivity of Nanofluids

Compelling electrical conductivity is a standout amongst the principal purposes behind streamer propagation transformer oil-based nanofluids. Successful electrical conductivity will be reliant on huge numbers of variables, such as volume fraction, conductivity for utilized nanoparticles and conductivity for transformer oil, as shown in Table 7. The volume fraction is a primary variable for increasing the effective electrical conductivity because of increasing the number of nanoparticles into nanofluid. ZnO/ HyVolt I nanofluid has the most astounding compelling electrical conductivity around tried individual nanofluids. However, (ZnO + Al_2O_3)/HyVolt I nanofluid has the most astounding compelling electrical conductivity around chosen multi-nanofluids. Al_2O_3/HyVolt I nanofluids have the smallest successful electrical conductivity because of the small conductivity of Al_2O_3 nanoparticles. On the other hand, (ZnO + Al_2O_3)/HyVolt I nanofluid is the smallest successful electrical nanofluid.

Table 7. Static Electrical Conductivity of Pure Transformer Oils, Individual and Multi-nanofluids at 20°C

Lab-Test Sample	Nanoparticles Concentrations	Static Electrical Conductivity (S/m)
HyVolt I _ Commercial	-	$0.55*10^{-12}$
SiO_2 /HyVolt I	0.05wt. %	$11.5*10^{-12}$
ZnO/ HyVolt I	0.05 wt. %	$15.0*10^{-12}$
Al_2O_3/HyVolt I	0.05 wt. %	$11.0*10^{-12}$
(Al_2O_3+SiO_2)/ HyVolt I	(0.025wt. % SiO_2+ 0.025wt. % Al_2O_3)	$11.2*10^{-12}$
(ZnO + SiO_2)/ HyVolt I	(0.025wt. %SiO_2+ 0.025wt. % ZnO)	$13.1*10^{-12}$
(ZnO + Al_2O_3) /HyVolt I	(0.025 wt. % Al_2O_3+ 0. 025 wt. % ZnO)	$12.6*10^{-12}$

7 FORECAST AND RECOMMENDATIONS

Increasing nanoparticles' radii and volumetric concentration will increase the effective relative permittivity of nanofluids. Hence, the impact of nanoparticles radii will be better than the impact of volumetric concentrations for increasing the successful relative permittivity of nanofluids. Utilizing transformer oil–based multi-nanoparticles empower utilizing the individual kind of nanoparticles for high relative permittivity and the different kind for small relative permittivity for abating streamer propagation and keeping of the effective relative permittivity of nanofluids during sensible strength. Nanoparticles for (TiO_2+ZnO) aggravate the successful relative permittivity of nanofluids higher than the compelling relative permittivity of nanofluids that utilize nanoparticles for (SiO_2 +Graphite). High thermal conditions decline the dielectric constant nanofluids that utilize multi-nanoparticles; the rate of diminishment relies on the sort and concentration of nanoparticles. Thermal conditions and volume fraction are the primary factors for increasing the effective electrical conductivity of singular or multi-nanofluids. In any case, increasing the size of nanoparticle has no impact on effective conductivity; even in case of increasing Zeta possibility. Moreover, the applied field has restricted impact on increasing the effective conductivity due to a constrained effect of nanoparticle immersion charge. Magnetic nanoparticles provide for trapping electrons higher than non-magnetic nanoparticles because of its prompted charges to silicon or MIDEL7131.

Multi- nanoparticles have the capacity of moderating streamer and absorbing vast numbers of electrons during the exact restricted period contrasted with individual nanoparticles. Increasing size is to a greater extent compelling for increasing the extent of stored accuse on the nanoparticles over the electric applied field or relative permittivity. Determination of silicon oil is superior to MIDEL7131 and oil-dependent upon the intrinsic long haul necessary for the streamer propagation of both nanotechnology strategies. The profit of various nanoparticles technobabble relies on including the initial sort of nanoparticles (high relative permittivity) and the second sort of nanoparticles (small relative permittivity) for regulating the effective electrical conductivity of nanofluids. Therefore, including different nanoparticles (ZnO + TiO_2) to Diala B and Kelamayi 25# provide for multi-nanofluids with high characteristics of effective conductivity. Then, multi-nanofluids based-(SiO_2 +$LiTaO_3$) speak to the second catch up of high effective conductivity. Lab-test fill in checks that the compelling electrical conductivity multi-transformer oil nanofluids are superior to the electric conductivity of individual nanofluids. Suspension of multi-nanoparticles are more effective than individual nanoparticles for trapping electrons and increasing nanoparticle kept charges; however, the capacity of nanoparticles for polarization details the determination for gathering multi-nanoparticles inside nanofluids transformer oils. Charging progress for multi-nanoparticles is more proficient than the charging progress of individual nanoparticles in the same transformer oil nanofluids. Hence, magnitudes save charge by utilizing multi-nanoparticles in pairs regarding the utilization of individual nanoparticles at the same occasion. The privilege of the multi-nanoparticles' strategy is utilizing multi-nanoparticles in transformer oils which make a twofold activity of utilizing the same individual nanoparticles independently for absorption charges on the long run. Thus, utilizing multi-nanoparticles accumulation for (Fe_3O_4 + MgO) will be the best twines for charging progress.

A direct result of these twin nanoparticles is to absorb a vast number of electrons in a short time, after that multi-nanoparticle gathering of (TiO_2 +ZnO) hails next. Additionally, lab-test worth of effort victory on presumes that including both singular and multi-nanoparticles inside transformer oil build the breakdown voltage of resultant nanofluids for diverse characteristics subject to the kind of nanoparticles and their concentrations. Furthermore, utilizing multi-nanoparticles inside transformer oil is more effective than increasing the breakdown voltage resultant nanofluids in utilizing individual nanoparticles. The impact of nanoparticles radii is greater than the impact of volumetric concentrations upon increasing the compelling relative permittivity of nanofluids. Utilizing transformer oil-based multi-nanoparticles empowers utilizing of you quit offering on that one sort of nanoparticles with high relative permittivity and the different sort with small relative permittivity for abating streamer propagation and keeping of the effective relative permittivity for nanofluids toward sensible worth. The size of nanoparticles, the applied field and relative permittivity of nanoparticles or oil are the primary elements for increasing the saved charge for nanoparticles; however, the extent of the kept accuse in multi-nanoparticle is higher than the extent of stored charge which is consumed by individual nanoparticles. Accordingly, the multi-nanoparticles can lure to the same extent for the saved charge quicker than the singular nanoparticles. Effective electrical conductivity is subject to temperature, volume fraction, relative permittivity of transformer oil, relative permittivity of nanoparticles and the applied field.

The benefit of adding high relative permittivity nanoparticles sort with an alternate individual hosting a small relative permittivity empowers attracting a considerable measure of electrons and moderating propagation of streamer and decreasing effective analyzing of electrical conductivity of nanofluids with individual nanofluids. Including individual or multi-nanoparticles to transformer oil increments breakdown voltage for the resultant nanofluids with different characteristics subject to the kind of nanoparticles and their concentrations. The breakdown voltage of chosen multi-nanofluids is higher than the breakdown voltage of chosen individual nanoparticles. In any case, the static electrical compelling conductivity of chosen multi-nanofluids is superior than the static electrical compelling conductivity of chosen singular nanofluids.

REFERENCES

Beroual, A., & Aka-N'Gnui, T. (2002). Influence of Additives and Hydrostatic Pressure on Streamers Initiation and Dielectric Strength of Liquids. *The Proceeding of the International Conference on Electrical Insulation and Dielectric Phenomena*, 248-251.

Cavallini, A., Karthik, R., & Negri, F. (2015). The effect of magnetite, graphene oxide and silicone oxide nanoparticles on dielectric withstand characteristics of mineral oil. *IEEE Transactions on Dielectrics and Electrical Insulation*, *22*(No.5), 2592–2600. doi:10.1109/TDEI.2015.005016

Choi, S. U. S. (1995). Developments and applications of non-Newtonian flows. *International Mechanical Engineering Congress and Exposition, 66*, 99-105.

Cruz, R. C. D., Reinshagen, J., Oberacker, R., Segadães, A. M., & Hoffmann, M. J. (2005). Electrical conductivity and stability of concentrated aqueous alumina suspensions. *Journal of Colloid and Interface Science*, *286*(2), 579–588. doi:10.1016/j.jcis.2005.02.025 PMID:15897074

Cuquejo, J. (2006). Numerical and analytical studies of the electrical conductivity of a concentrated colloidal suspension. *The Journal of Physical Chemistry B, 110*(12), 6179–6189. doi:10.1021/jp057030e PMID:16553432

D'Souza, S. (2014). A Review of In Vitro Drug Release Test Methods for Nano-Sized Dosage Forms. Hindawi Publishing Corporation.

Eastman, J. A., Choi, S. U. S., Li, S., Yu, W., & Thompson, L. J. (2001). Anomalously increased effective thermal conductivities of ethylene glycol-based nanofluids containing copper nanoparticles. *Applied Physics Letters*, *78*(6), 718–720. doi:10.1063/1.1341218

Fathiah, M. Z., & Edyvean, R. G. (2015). The Role of Ionic Strength and Mineral Size to Zeta Potential for the Adhesion of P. putida to MineralSurfaces. *International Journal of Biological, Biomolecular, Agricultural, Food and Biotechnological Engineering*, *9*(7).

Feng, X., Ma, H., Huang, S., Pan, W., Zhang, X., Tian, F., Gao, C., Cheng, Y., & Luo, J. (2006). Aqueous-organic phase-transfer of highly stable gold, silver, and platinum nanoparticles and new route for fabrication of gold nanofilms at the oil/water interface and on solid supports. *The Journal of Physical Chemistry B, 110*(25), 12311–12317. doi:10.1021/jp0609885 PMID:16800553

Fleming, J., Pawlowski, T., Ammala, A., Casey, P. S., & Lawrence, K. A. (2005). Electrical conductivity and space charge in LDPE containing TiO/sub 2/ nanoparticles. *IEEE Transactions on Dielectrics and Electrical Insulation*, *12*(No.4), 745–753. doi:10.1109/TDEI.2005.1511100

Ganguly, S., Sikdar, S., & Basu, S. (2009). Experimental investigation of the effective electrical conductivity of aluminum oxide nanofluids. *Powder Technology*, *196*(3), 326–330. doi:10.1016/j.powtec.2009.08.010

Hwang, Zahn, O'Sullivan, Pettersson, Hjortstam, & Liu. (2010). Effects of nanoparticle charging on streamer development in transformer oil-based nanofluids. *J. Appl. Phys., 107.*

Hwang, J. G., Zahn, M., O'Sullivan, F. M., Pettersson, L. A. A., Hjortstam, O., & Liu, R. (2013). Effects of nanoparticle charging on streamer development in transformer oil-based nanofluids. *Journal of Applied Physics*, *114*, 702–711.

Hwang, Zahn, & Pettersson. (2011). Bipolar charging and discharging of a perfectly conducting spherein a lossy medium stressed by a uniform electric field. *J. Appl. Phys, 109.*

Jianlin, L., & Decai, L. (2010). Research on Magnetization Mechanism of Nano- magnetic Fluid. IEEE Transaction on Dielectrics and Electrical Insulation, 10, 633-637.

Jin, H. (2015). Dielectric Strength and Thermal Conductivity of Mineral Oil based Nanofluids. In *Master of Electrical Engineering* (pp. 1–170). Delft University of Technology.

Kasap, S. O. (2009). *Principles of Electronic Materials and Devices.* Xi'an Jiaotong University Press.

Kopcansky, P., Tomco, L., Marton, K., Koneracka, M., Timko, M., & Potocova, I. (2005). The influence of magnetic nanoparticles On Dc-and Acdielectric break down in Transformer oil-based magnetic fluid. *Journal of Magnetism and Magnetic Materials*, *289*, 613–624. doi:10.1016/j.jmmm.2004.11.117

Lesaint, O., & Massala, G. (1998). Positive streamer propagation in large oil gaps: Experimental characterization of propagation modes. *IEEE Transactions on Dielectrics and Electrical Insulation*, *5*(3), 360–370. doi:10.1109/94.689425

Lew. (2005). Laboratory Testing of Natural Ester Dielectric Liquids. Academic Press.

Lewis, J. (2004). Interfaces Are the Dominant Feature of Dielectrics at the Nanometric Level. *IEEE Transactions on Dielectrics and Electrical Insulation*, *11*(5), 739–753. doi:10.1109/TDEI.2004.1349779

Lisunova, M. O., Lebovka, N. I., Melezhyk, O. V., & Boiko, Y. P. (2006). Stability of the aqueous suspensions of nanotubes in the presence of nonionic surfactant. *Journal of Colloid and Interface Science*, *299*(2), 740–746. doi:10.1016/j.jcis.2006.03.012 PMID:16600279

Liu, L., Yang, Y., & Zhang, Y. A. (2004). Study on the electrical conductivity of multi-walled carbon nanotube aqueous solution. Physica E: Low-dimensional Systems and Nanostructures, 24(3), 343-348.

Lundgaard, L., Linhjell, D., Berg, G., & Sigmond, S. (1998). Propagation of Positive and Negative Streamers in Oil with and without Pressboard Interfaces. *IEEE Transactions on Dielectrics and Electrical Insulation*, *5*(3), 388–395. doi:10.1109/94.689428

Lv, Y., Ge, Y., Li, Ch., Wang, Q., Zhou, Y., Qi, B., Yi, K., Chen, X., & Yuan, J. (2016). Effect of TiO2 Nanoparticles on Streamer Propagation in Transformer Oil under Lightning Impulse Voltage. *IEEE Transactions on Dielectrics and Electrical Insulation*, *23*(4), 2110–2115. doi:10.1109/TDEI.2016.7556485

Lv, Y., Wang, Q., Zhou, Y., & Li, C. (2016). Effect of Fe3O4 nanoparticles on positive streamer propagation in transformer oil. AIP Advances Journal, 6(3).

Miao, J., Dong, M., Ren, M., Wu, X., Shen, L., & Wang, H. (2013). Effect of nanoparticle polarization on relative permittivity of transformer oil-based nanofluids. *Journal of Applied Physics*, *113*(20), 204103. doi:10.1063/1.4807297

Miao, J., Dong, M., & Shen, L. P. (2012). A modified electrical conductivity model for insulating oil-based nanofluids *IEEE in Condition Monitoring and Diagnosis (CMD) Conference*, *5*, 1126–1129.

Nguyen, D. V. (2013). *Experimental Studies of Streamer Phenomena in Long Oil Gaps* (PhD Thesis). Norwegian University of Science and Technology.

Posner, J. D. (2009). Properties and electrokinetic behavior of non-dilute colloidal suspensions. *Mechanics Research Communications*, *36*(1), 1–5. doi:10.1016/j.mechrescom.2008.08.008

Pradhan, D. K. (2008). Studies of Dielectric Relaxation and AC Conductivity Behavior of Plasticized Polymer Nanocomposite Electrolytes. *International Journal of Electrochemical Science*, *3*, 597–608.

Pradhan, D. K., Choudhary, R. N. P., & Samantaray, B. K. (2008). Studies of structural, thermal and electrical behavior of polymer nanocomposite electrolytes. *Express Polymer Letters*, *2*(9), 630–638. doi:10.3144/expresspolymlett.2008.76

Product Data Sheet from Shell Lubricants. (2011). Diala Oil B Electrical insulating oil (Vol. 3). Academic Press.

Rafiq, M., & Li, Ch. (2015). Preparation and Breakdown Properties of Mineral Oil Based Alumina Nanofluids. *IEEE International Conference Emerging Technologies (ICET)*, 575-578.

Randles, S. J. (2006). *Esters, in Synthetics, Mineral Oils, and Bio-Based Lubricants: Chemistry and Technology* (L. R. Rudnick, Ed.). CRC Press, Taylor & Francis Group.

Saad, X. Z., Khan, A., & Fedkiw, P. S. (2004). Nanocomposite Electrolytes Using Single-Ion Conducting Fumed Silica. *Electrochemical and Solid-State Letters*, *7*(10), 361–364. doi:10.1149/1.1792267

Sartoratto, Neto, Lima, Rodrigues de Sá, & Morais. (2005). Preparation and electrical properties of oil-based magnetic fluids. *J. Appl. Phys., 97.*

Segal, V., Hjortsberg, A., Rabinovich, A., Nattrass, D., & Raj, K. (1998). AC (60 Hz) and impulse breakdown strength of a colloidal fluid based on transformer oil and magnetite nanoparticles. IEEE Int'l. Sympos. Electr. Insul, 619-622.

Segal, V., Rabinovich, A., Nattrass, D., Raj, K., & Nunes, A. (2000). Experimental study of magnetic colloidal fluids behavior in power transformers. *Journal of Magnetism and Magnetic Materials*, *215*, 513–515. doi:10.1016/S0304-8853(00)00205-5

Shen, Zhao, & Kang. (2011). *Colloid and Surface Chemistry*. Bei Jing, Chemistry Industry Press.

Sima, W., Shi, J., Yang, Q., Huang, S., & Cao, X. (2015). Effects of conductivity and permittivity of nanoparticle on transformer oil insulation performance: Experiment and theory. *IEEE Transactions on Dielectrics and Electrical Insulation*, *22*(1), 380–390. doi:10.1109/TDEI.2014.004277

Sima, W. X., Cao, X. F., Yang, Q., Song, H., & Shi, J. (2014). Preparation of Three Transformer Oil-Based Nanofluids and Comparison of Their Impulse Breakdown Characteristics. *Nanoscience and Nanotechnology Letters*, *6*(3), 250–256. doi:10.1166/nnl.2014.1746

Smpson, L. Y., Luang, M. S., & Randles, J. (2012). *An Overview of the Zeta Potential* (Vol. 2). Technical Brief, Particle Sciences.

Tagami, N., Hyuga, M., & Ohk, Y. (2010). Comparison of Dielectric Properties Between Epoxy Composites With Nanosized Clay Fillers Modified by Primary Amine and Tertiary Amine. IEEE Transaction on Dielectrics and Electrical Insulation, 10, 214-220.

Takada, T., Hayase, Y., Takada, Y., & Okamoto, T. (2008). Space charge trapping in electrical potential well caused by permanent and induced dipoles for LDPE/MgO nanocomposite. *IEEE Transactions on Dielectrics and Electrical Insulation*, *15*(1), 152–160. doi:10.1109/T-DEI.2008.4446746

Tang, H., Zhang, J. P., Gu, F., Wang, F. C., & Wang, J. H. (2011). Study on Preparation of SiO2 Optical Fiber Preform by SG Method. *Advanced Materials Research*, *299-300*, 489–492. doi:10.4028/www.scientific.net/AMR.299-300.489

Taylor, Coulombe, Otanicar, Phelan, Gunawan, Lv, Rosengarten, Prasher, & Tyagi. (2013). Small particles, big impacts: A review of the diverse applications of nanofluids. *J. Appl. Phys., 113.*

Technical Data Sheet from Clearco Products Co, Inc. (2015). STO-50 Silicone Transformer Oil. Author.

Technical Data Sheet from Ergon Refining, Inc. (2012). HyVolt I Electrical Insulating Oil Marketing Specification. IEC 60296.

Technical Data Sheet from MIDEL. (2010). MIDEL 7131 Transformer Fluids (vol. 2). Author.

Thabet. (2020). *Design and Investment of High Voltage NanoDielectrics*. IGI Global. doi:10.4018/978-1-7998-3829-6

Thabet, A., Allam, M., & Shaaban, S. A. (2017a). Investigation on enhancing breakdown voltages of transformer oil nanofluids using multi-nanoparticles technique. *IET Generation Transmission and Distribution Journal*, (Nov). Advance online publication. doi:10.1049/iet-gtd.2017.1183

Thabet, A., Allam, M., & Shaaban, S. A. (2017b). Slowing Positive Streamer Propagation in Silicon and Ester Transformer Oil Using Multi-nanoparticles Technique. *International Journal of Applied Energy Systems*, *1*(1), 15–20. doi:10.21608/ijaes.2019.169957

Thabet, A., El-Dein, A. Z., & Youssef, N. M. (2012). Frequency Modulation Response for Effective Dielectric Response of New Nanocomposite Liquids. *International Review on Modelling and Simulations*, *5*(3), 1348–1355.

User Manual of Elma Schmidbauer GmbH. (2007). *Product profile of Elmasonic S 40 (H) Ultrasonic cleaning unit*. Academic Press.

User Manual of Power Electronical. (n.d.). *User Manual for Motorized Oil Breakdown Voltage Tester*. Academic Press.

User Manual of Velp Scientifica. (n.d.). *AREC Ceramic Digital Heating Magnetic Stirrer, F20500011*. Academic Press.

Wedin, P. (2014). Electrical breakdown in dielectric liquids-a short overview. *IEEE Electr. Insul. Mag*, *30*(6), 20–25. doi:10.1109/MEI.2014.6943430

Wei, X., & Wang, L. (2010). Synthesis and thermal conductivity of microfluidic copper nanofluids. *Journal of Particuology*, *8*(3), 262–271. doi:10.1016/j.partic.2010.03.001

Wong, K. V., & de Leon, O. (2010). Applications of Nanofluids: Current and Future. *Advances in Mechanical Engineering*, *2010*, 1–11.

Yamashita, H., Kawai, H., Stricklett, K. L., & Kelley, E. F. (1990). The Effect of High Pressure on Prebreakdown Phenomena in N-Hexane. *Proceeding of the 10th International Conference on Conduction and Breakdown in Dielectric Liquids*, 404-409. 10.1109/ICDL.1990.202933

Yang, Q., Yu, F., Sima, W., & Zahn, M. (2015). Space charge inhibition effect of nano-Fe_3O_4 on improvement of impulse breakdown Voltage of transformer oil based on improved Kerr optic measurements. *AIP Advances Journal*, *5*(9), 210–214. doi:10.1063/1.4931947

Yu, W., Xie, H., Chen, L., & Li, Y. (2010). Enhancement of thermal conductivity of kerosene-based Fe3O4 nanofluids prepared via phase-transfer method. *Journal of Colloids and Surfaces A*, *355*(1–3), 109–113. doi:10.1016/j.colsurfa.2009.11.044

Yuliastuti. (2010). *Analysis of dielectric properties Comparison between mineral oil and Synthetic ester oil* (Thesis). Faculty of Electrical Engineering, Mathematics, and Computer Science, Delft University of Technology.

Zhu, H. T., & Yin, Y. S. (2004). A novel one-step chemical method preparation of copper nanofluids. *Journal of Colloid and Interface Science*, *227*(1), 100–130. doi:10.1016/j.jcis.2004.04.026 PMID:15276044

Chapter 7
Nano-Magnetic Materials

ABSTRACT

The utilization of traditional magnetic materials and magnetic composite materials have been developing quickly in recent decades. The build of magnetic properties has been created, with pursuing the inclusion of a few diverse nanoparticles to formulate magnetic nanocomposite crystalline core. This chapter contains theories and characterization of magnetic nanocomposites that are compared with variant traditional magnetic materials. Finally, this chapter sheds light on regulations, forecast, and recommendations for manufactures for using new multi-nanocomposites magnetic materials in electrical applications.

1 DESIGN AND INVESTMENT

This chapter displays an investigation on new composite and nanocomposite magnetic modern materials. The effective conductivity and thermal effective conductivity have been predicted by including different sorts and rates of conductive particles (Al_2O_3, MgO, ZnO, Graphite…etc.) in the base matrices of Epoxy, Iron, or Silicon to formulate new composite and nanocomposite industrial materials. The characterization of successful conductivity for new polymeric composites has been investigated with different applied forces, incorporation sorts and their concentrations. In addition, the impact of consideration sorts and their concentrations on the effective thermal conductivities for thermal interface nanocomposite modern materials has been clarified and reported. Dielectric materials play a paramount role in expansive applications, such as electrical and electromagnetic applications. Moreover, this chapter offers the prediction for effective permeability of composite and nanocomposite dielectric materials dependent upon hypothetical investigation to detail the impacts of inserted magnetic inclusions in upgrading magnetic properties of dielectrics. Effective permeability of Plastics and Glass nanodielectrics has been predicted for including different sorts and rates of magnetic nanoparticles (Fe, Ni-Cu, Ni-Fe, MgZn_Ferrite, NiZn_Ferrite) for planning new nanodielectrics magnetic industrial materials. Soft nanoparticles powders have been utilized for new nanodielectrics regularly. These nanoparticles have the structure of a molecule size in the go of micrometer- on nano-sized grains and magnetic isotropy, e.g., an irregular conveyance of magnetic simple axes of the nano grains. Upgrading the characteristics of new nanodielectric magnetic streamlined materials has been successful. The effects have indicated a critical impact of inclusions appropriation

DOI: 10.4018/978-1-7998-8536-8.ch007

on the effective permeability of nanodielectric magnetic composites. Therefore, the impact of magnetic inclusions sorts and their concentration on the effective permeability of nanodielectric magnetic materials has become evident. Magnetic composite materials play a paramount part in expansive applications, such as transformers and electrical motors. In this chapter, the effective permeability of new magnetic composite materials dependent upon Maxwell-Garnett models have predicted the detailed impacts of installed inclusions. Therefore, it has been investigated on connected planning new magnetic composite streamlined materials toward including different sorts and rates of particles (Fe, silicon steels, Metglas, Co-Fe, Ni-Fe, MgZn_Ferrite, NiZn_Ferrite). All results have been investigated on the critical impact of inclusions distribution, incorporation sorts and their effect on the successful permeability of magnetic composite materials; along these lines, the characteristics of magnetic industrial materials toward utilizing the particles structure of an irregular conveyance for magnetic rather than difficult axes of the grains have been controlled and improved.

A wide array of research has been conducted about the preparation of nanocomposites, thus, planning and fabricating new nanocomposites for upgrading electrical and magnetic results has consequently been challenging. In this chapter, the prediction for effective permeability of new nanocomposites magnetic materials has been investigated in light of hypothetical methodologies at terahertz frequencies to show feeble electric and magnetic reactions. Hence, the impacts of installed magnetic multi-nanoparticles for upgrading magnetic properties of magnetic materials have been examined. In addition, it has been controlled to successful relative permeability of new nanocomposites magnetic materials dependent upon sorts and effects of the chosen magnetic multi-nanoparticles (Ferrite, silicon steel. NiZn_Ferrite, and Nikel_Ferrite). Additionally, the prediction results have indicated a critical impact ofinclusions radius and plan of inclusions on the successful permeability of new nanocomposites magnetic materials. Characterization of composite materials is a critical point in current electromagnetics. Thus, this chapter recommends new microstructure electromagnetics materials that are able to upgrade the electromagnetic properties for transformer core and its characterization. In addition, this chapter displays a theoretical analysis for the effective permeability prediction of the recommended magnetic composite materials for including different sorts and rates of particles (Fe, silicon steels, Metglas, Co-Fe, Ni-Fe, MnZn, MgZn, NiZn) to formulate new magnetic composite transformer core materials. A hypothetical investigation is acquainted to contemplate the impact of incorporation sorts and their effect on permeability. Numerical outcomes indicate critical aberration of consideration sorts and their effect on the effective permeability regarding accepted magnetic materials. Frequency Response Analysis (FRA) is used to research transformer core magnetic composites and specify upgrading magnetic effective intricate relative permeability of transformer core magnetic composites regarding variant frequencies (10^{-2}Hz – 10^{3}Hz).

2 TRADITIONAL MAGNETIC MATERIALS

For transformers cores, non-oriented silicon-iron (NO SiFe) is beyond question the most basic material for an around the world creation of roughly ten million tonsils annually, which may be 80% of the downright soft magnetic material business, including Ferrite, powder, amorphous, nickel Ferrite, NiZn_Ferrite (Krings et al., 2016; Mousavi & Agheb, 2011). The average material parameters which are taken under account throughout the transformer outline methodology are: • Magnetic saturation • Conductivity • Permeability• Iron losses (magnetic losses) • Magnetostriction (noise) • Delivery (forms, fully processed, semi-processed) • Market availability (supply, including geopolitical issues) •

Price. The utilized magnetic materials for transformers, such as Silicon-iron (SiFe) are the most broadly utilized materials. Iron is alloyed with a little amount of silicon from the transformer core. It may be accessible on a grain-oriented state (i.e. the material is anisotropic and has an alternate permeability in distinctive directions) or a non-oriented state (i.e. very near isotropic magnetic properties in all directions (Moses, 2012)). Electrical transformers generally utilize the grain-oriented magnetic materials because of their unidirectional flux change in the core legs and yoke. In contrast, rotational electrical machines primarily utilize non-oriented materials, since the flux bearing is for the most part not unidirectional and transforms heading (it rotates in the stator yoke and stator tooth rot tips).

The ferrite cores are more suitable for high frequency applications, and steel laminations are more suitable for low frequency applications. Both materials are accessible in a variety of primaries, each one of them is suitable for distinctive applicants. Ferrites are dense, homogeneous ceramic structures settled on toward blending iron oxide for oxides or carbonates of one alternately all the other metals, for example manganese, zinc, nickel, or alternately the first mass of the magnesium. The particular operating conditions and normal laminated core made of steel cover can be utilized just for the working frequency range of 50 - 70 Hz. For high frequency in the reach of kHz, the laminated core cannot be used. Hence, such frequency extent can best be produced for ferrite materials to be utilized. Hysteresis losses oversee upon 300 kHz. Toward high frequencies, eddy current losses will be incurred, since they have an inclination to change with frequency square. Therefore, on trouncing these losses, the ferrite core may be ready for laminated metal compound and powered metal cores are utilized. The amount of turns in the elementary and auxiliary may be restricted to the measure of the core. The number of turns in the primary and high side of the transformer is dictated by the standard wire gauge of the conductor which can be utilized. The core may be particularly planned for frequency capability. Ferrite is the best decision for high frequency transformer structure but for its mechanical toughness. Frequency transpires a vital variable. Switching power supplies has ended up, so mainstream due to their skill should maneuver at high frequencies; hence escalating their effectiveness. Since the prompted voltage of a transformer is dependent upon the adjusting magnetic flux, the flux (higher frequency) becomes more changing. With higher frequencies by different thought come into partake to minor frequencies, core material combination is driven toward core disseminated thought. Eddy current losses are low with the goal steel laminations. For higher frequencies, core material combination is determined toward core misfortune attention. Eddy currents and flows can be groundbreaking. In this crossroads, ferrite is normally utilized in light of its high electrical resistivity that minimizes eddy current losses. On the other hand, there is a cost to be paid for the conservative core losses, and with the expectation of this ferrite that has brought down saturation and permeability values (Lakshminaray & Balakrishnan, 2013).

3 MAGNETIC COMPOSITE MATERIALS

Composite materials are getting an expanding consideration on current advances because of their mechanical and electrical properties, configuration flexibility, and diminished expense and weight. Different composite materials are generally utilized within huge numbers of applications, e.g., mechanical support, high temperature exchange or isolation, electromagnetic filtering or shielding, or giving certain acoustic properties (Paulis et al., 2014). Composite materials are constructed by joining two of the materials, which have different electrical and physical properties. Usually, this may be a group material (matrix) for inserted inclusions. Inclusions might be in structure of crumbs, flakes, ðlings, whiskers, short or long fibers, and may have uncommonly intended shapes relying on huge numbers of factors, particularly mechanical quality and firmness (Koledintseva et al., 2006). Soft magnetic composites (SMC) challenge traditional materials, such as soft magnetic Ferrites and electrical steels of applications for exchanging magnetic ðelds. Soft magnetic composites, which are utilized within electromagnetic applications, can be portrayed as ferromagnetic powder particles encompassed by an electrical insulating film (Füzerova et al., 2012). New developments of magnetic materials and new processing systems technique SMC material are fascinating for provision of electrical machines. In later years, large fractions of scientists have attempted to move forward the magnetic properties execution of SMCs by selecting a suitableness material and applying a suitableness covering strategy (Périgo et al., 2011) or get ready composites dependent upon the amorphophallus titanium or nanocrystalline powders (Liu et al., 2009). Soft magnetic composite (SMC) materials and their provision over electromagnetic devices have undergone critical advancement because of their exceptional favorable circumstances, such as low eddy current loss, quasi-isotropy of mechanical and magnetic properties, low expense and low material utilization throughout the processing methodology. However, the magnetic properties of SMC materials are exactly unique in relation to the laminated SiFe materials, which are especially tested to the configuration and requisition of electrical machines.

The soft magnetic composite (SMC) materials, generating all the by powder metallurgical techniques, are made of iron powders (about 100 μm in diameter) for high purity and surface covering to electrical encasing and mechanical holding (Kollár et al., 2010; Weijie et al., 2016; Zhu et al., 2011). The covered iron powders, with high electrical resistivity, are pressed under a robust magnetic core after an arrangement of specialized foul steps, such as the pass on pressing, thermaling and curing medication. The SMC materials can be utilized for fabricating electromagnetic devices about uncommon structures. Due to the powdered nature, the magnetic property of SMC material is characteristically quasi-isotropic, despite slight magnetic anisotropic property that may be created because of the pressing and heat treatment in the preparation procedure of SMC cores (Liu et al., 2015). Extraordinary consideration ought to be given to soft magnetic composites, which comprise of a powder of really little disconnected iron particles. Machine parts settled on this material are not transformed as stacked cover sheets; however, they are specifically pressed together to a single part of alternately multi-part machine cores. The playing point of the powder is the plausibility for uncommon machine geometries for three-dimensional flux ways. Moreover, its high resistivity prompts low eddy current losses, which is a significant playing point to high frequency applications. The downside may be that its generally low permeability and flat B-H charge curve generally prompt most extreme flux thickness characteristics during average magnetic field characteristics of electrical machines. SMC material is utilized for electrical machines for a three-dimensional flux circulation. Further data around the compositions of SMC materials and workable future developments have been depicted (Li et al., 2014). Soft magnetic composite (SMC) is the structure which secures the surface of iron powder with the inorganic encasing skin film and adheres between iron powder particles for tar. For this reason, it can be incredibly reliant on the thickness, then afterward layering molding, permeability and saturation magnetic flux thickness for magnetic property of SMC of improving it so that thickness becomes high. The focal point and deficiency of a magnetic steel sheet and SMC are indicated in table 1(Krings et al., 2015).

For the improvement of present-day industry, an ever-increasing number of new substantial electric machines are generally provided to help in this advancement. Soft-magnetic composites for metallic inclusions are regularly used to build mechanical applications. Circular agglomerates with well-connected conductive particles successfully carry on substantial virtual molecule with an extra increment for effective

Table 1. Advantage and Fault of Magnetic Steel Sheet and SMC (Krings et al., 2015)

Category	Magnetic Steel Sheet	SMC
Advantages	High saturation magnetic flux	Small loss at high frequency
	High permeability	Three-dimensional structure is possible
	Low cost	Good recycling efficiency
Faults	Large loss at high frequency	Low saturation magnetic flux density
	Three-dimensional structure is difficult	Low permeability

permeability because of the flux effect in the agglomerate (Drnovšek et al., 2009). The new produced materials ended up especially paramount a direct result. Its physical properties change dramatically with size and for the nearby structure of the grains. Hence, regarding the built applications, these materials should be about low cost and ought to be simple to prepare, permitting them to be profoundly aggressive for the already-existing ones utilized within such building applications. Regarding soft magnetic materials that play a paramount role in wide applications, such as transformers and electrical motors, there is an enthusiasm toward soft magnetic composites materials due to the interest for scaling down of cores for energy electronic applications, transformers, stators and rotors of electro-motors. These materials are utilized within electromagnetic applications and can be depicted as ferromagnetic powder particles encompassed toward an electrical insulating film (Anhalt & Weidenfeller, 2012; Fuzerova et al., 2010; Yao et al., 2011). Along these lines in this research, it has been proposed new magnetic composite industrial materials whose magnetic characterization reaction have been improved regarding sorts and concentrations for chosen particles have been proposed. Therefore, a lot of distinctive micro-particles are reported as the best fillers fundamentally for building magnetic permeability.

3.1 Theory and Effective Parameters

The computation systems for estimating the successful permeability of magnetic composite modern materials which comprise micron/submicron-sized particles are inserted in distinctive matrices; recognizing irregular structures for magnetic particles in the base matrices. Hence, the past explanatory models (Balasubramanian & Venkatesh, 2013; Sihvola, 1999) have been used to plan hypothetical models for foreseeing the successful permeability of magnetic composite materials. Regarding Maxwell-Garnett recipe that plays a paramount part in 2D composite materials (Balasubramanian & Venkatesh, 2013), the effective permeability to irregular conveyance for molecule extent as 2-D round shapes essentially takes after:

$$\mu_{eff} = \mu_m + 2\mu_m \frac{\mu_i - \mu_m}{\mu_i + \mu_m - \varphi(\mu_i - \mu_m)} \tag{1}$$

Where,
μi is permeability of inclusions in the composites,
μm is permeability of main matrix of the composites,

φ is the volume fraction of inclusions inside the main matrix.

Consequently, the Bruggeman equation (Balasubramanian & Venkatesh, 2013) that is an alternate renowned homogenizing principle for foreseeing the effective permeability to symmetrically appropriation of molecule radius as 2-D hardware cylinders' shapes fundamentally takes after:

$$(1-\varphi)\frac{\mu_m-\mu_{eff}}{\mu_m+\mu_{eff}}+\varphi\frac{\mu_i-\mu_{eff}}{\mu_i+\mu_{eff}}=0 \tag{2}$$

Concerning the inclusions of the real composites that are not continuously comparative to spheres or cylinders, the over formulas are not fulfilled. Thus, the additional exact approach estimation (Sihvola, 1999) and the structure of magnetic composite material may be expected to be periodic, and the unit cell, which is the least volume, is viewed as a homogeneous magnetic substance, as indicated in Fig. (1). The effective permeability of magnetic composite materials can be characterized on the foundation of magnetic vitality offset in the unit cell, and it may be played, in this approach, that the first cell and the homogenized cell incorporate proportional magnetic vitality at both units inundated in equal magnetic field. Thus, this method can possibly be applied to any structure. The effective permeability of magnetic composite materials is given as follows:

$$\mu_{eff}=\frac{\int_{S_{homo}}\left|B_{homo}\cdot\right|dS}{\int_{S_{homo}}B_{org}.H_{orig}dS} \tag{3}$$

Where S_{homo} is the square of the original cell, B_{homo} is the magnetic flux density and H_{orig} is the magnetic field.

Figure 1. Scheme for homogenization of magnetic composite materials

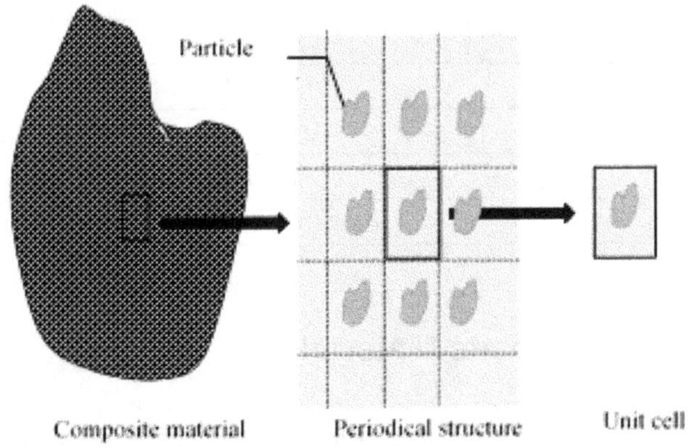

Magnetic materials, especially measured fillers hint at a critical change to physical, electrical, and magnetic properties, as opposed to their heft counterparts because of their high surface to volume proportion of the grains. Thus, impacts of types and effect of inclusions gran extent can be indicated on the execution of effective magnetic permeability of the composite. The effective permeability μeff of the suggested magnetic composite modern materials can be predicted by utilizing the above methodologies in light of inclusions permeability, weight rates of inclusions inside the matrix and primary matrix permeability of magnetic composite materials. Therefore, it can the successful permeability of magnetic composite that has been portrayed by including different rates of magnetic fillers irregular circulations on matrix material can be assessed (Lin et al., 2006).

3.2 Particles and Magnetic Materials

Magnetic composites can be depicted as ferromagnetic powder particles encompassed by the electrical insulating film. The decision for magnetic composite material is perplexing and relies on numerous factors in frequency, size of the component, physical quality and magnetic properties. This investigation illustrates the impacts of particles radius in iron high purity particles (single crystals to favored directions), silicon steels, Metglas, Co-Fe, Ni-Fe, MgZn_Ferrite, and NiZn_Ferrite on fabricating composite and nanocomposite streamlined materials. Table 2 depicts the principle electrical depiction properties for utilizing particles to upgrading magnetic properties of composite industrial materials.

Table 2. Electric and Magnetic Properties of Particles and Magnetic Industrial Materials

Magnetic Materials	Electric and Magnetic Properties	
	μ (H/m)	ρ (Ω.m)
Plastics	$1.25x10^{-6}$	$10^{16}-10^{22}$
Metglas	$125x10^{-2}$	$142x10^{-8}$
Silicon Steels	$93x10^{-3}$	$45x10^{-8}$
Fe "High Purity"	$1500x10^{-3}$	$10x10^{-8}$
Ni-Fe	$400x10^{-3}$	$55x10^{-8}$
Co-Fe	$70x10^{-3}$	$40x10^{-8}$
Ni	$1.25x10^{-4}$	69.9×10^{-9}
Isoperm Ni-Cu	$0.65x10^{-4}$	10^{-9}
Ferrite MgZn	$^{>}8x10^{-4}$	0.5
Ferrite NiZn	$2x10^{-5}$- $8x10^{-4}$	300

3.3 Characterization of Composite Materials

Most magnetic composite materials progress their magnetic properties because of the opposition of the new particles; however, general magnetic conduct technique of the composite materials can shift widely, contingent upon the structure of the material; especially for its electron setup. The accompanying outcomes clarify the characterization reaction of the suggested magnetic composite materials regarding different sorts and concentrations of chosen particles. Figure (2) indicates the effective permeability of silicon steel magnetic composites, noting that iron and Metglas particles are the best inclusions for upgrading the successful permeability of magnetic composite. Secondly, alloys particles have the second place in request for upgrading the effective permeability of magnetic composite industrial materials. Moreover, the effective permeability of silicon steels magnetic composites has been diminished toward including different rates of MgZn, And NiZn ferrite particles haphazard circulations of silicon steels matrix material.

Figure 2. Effective permeability of Silicon steels composites with various volume fractions of particles

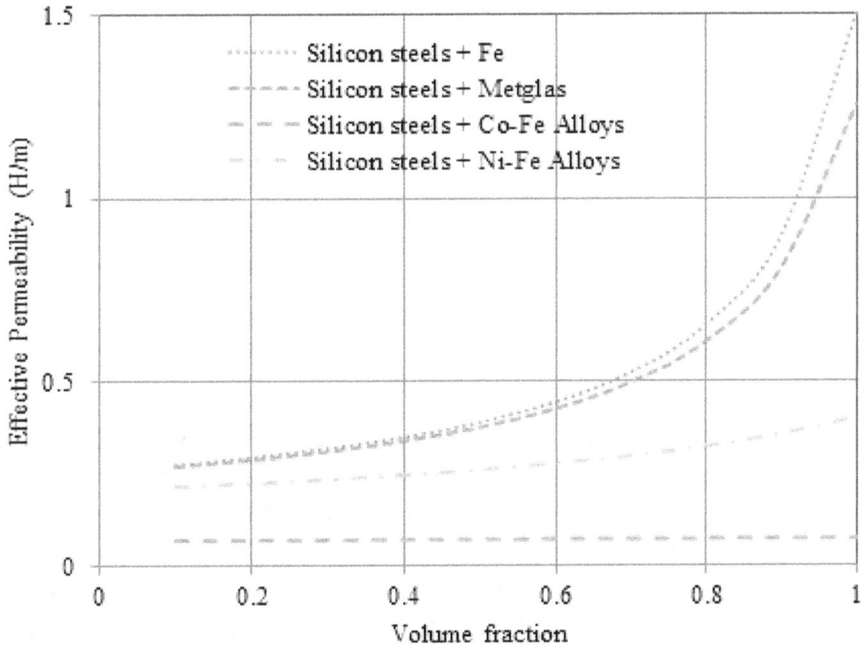

Figure (3) illustrates the effective permeability of Ni-Fe magnetic composites, it depicts that iron and Metglas particles are the best inclusions for upgrading the effective permeability of magnetic composite; then silicon steels particles occupy the second place in the request for upgrading the effective permeability of magnetic composite streamlined materials. However, Co-Fe Alloys particles help diminish the effective permeability of magnetic composite industrial materials. Examining impacts of ferrite particles is dependent upon Ferrites compositions that are compound exacerbates comprising ceramic materials with iron oxide as their central part. Large fractions of ferrite particles have magnetic materials and they are used to make changeless magnets, ferrite cores for transformers, and in different applications. In any case, the successful permeability of Ni-Fe magnetic composites is clear, noting that the effective permeability of NiZn Ferrite/ Ni-Fe Alloys magnetic composite is higher than the effective permeability of MgZn Ferrite/ Ni-Fe Alloys magnetic composite.

Figure (4) indicates the effective permeability of Co-Fe magnetic composites; it has been recognized that iron and Metglas particles record the best inclusions for upgrading the successful permeability of magnetic composite. Therefore, silicon steels and Ni-Fe Alloys particles help diminish the effective

Figure 3. Effective permeability of Ni-Fe Alloys composites with various volume fractions of magnetic particles

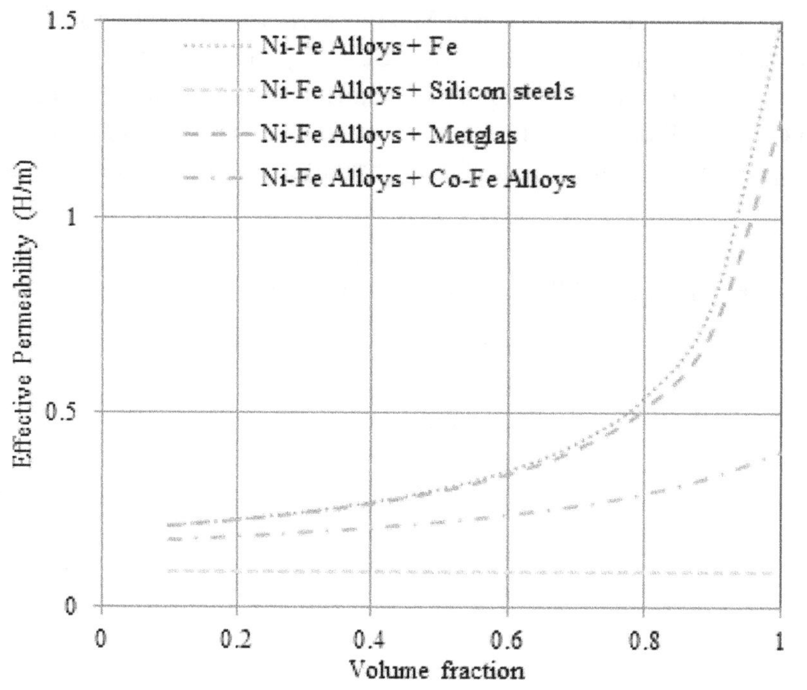

permeability of magnetic composite modern materials. Noteworthy, the successful permeability of NiZn Ferrite/ Co-Fe Alloys magnetic composite may be higher than the effective permeability of MgZn Ferrite/ Co-Fe Alloys magnetic composite. Figure (5) indicates the effective permeability of Metglas magnetic composites, it has been illustrated that expanding the rate of iron or Co-Fe Alloys particles enhances the effective permeability of magnetic composite. Furthermore, silicon steels and Ni-Fe Alloys particles diminish the effective permeability of magnetic composite industrial materials. In addition, the expanding magnetic particles of NiZn ferrite or MgZn ferrite to Metglas declines the effective permeability of the magnetic composite and the impact of these particles may be the same. Figure (6) indicates the effective permeability of iron magnetic composites; it has been delineated that Metglas and Co-Fe Alloys particles record the best inclusions for upgrading the effective permeability of magnetic composite with expanding their rates. Hence, silicon steels and Ni-Fe Alloys particles help diminish the effective permeability of magnetic composite modern materials. In any case, the effective permeability of NiZn Ferrite/ fe and MgZn Ferrite/ Fe magnetic composites has the same execution for the effective permeability of magnetic composite.

4 NANODIELECTRIC COMPOSITES BONDED BY SOFT MAGNETIC

In later years, Maxwell–Garnett has been altered and many people blend guidelines to foresee effective parameter computation for considering the perceptible magnetic properties of composite and nanocomposite streamlined materials in the reproduction strategies (Anhalt & Weidenfeller, 2007; Gonzo et al.,

Figure 4. Effective permeability of Co-Fe Alloys composites with various volume fractions of magnetic particles

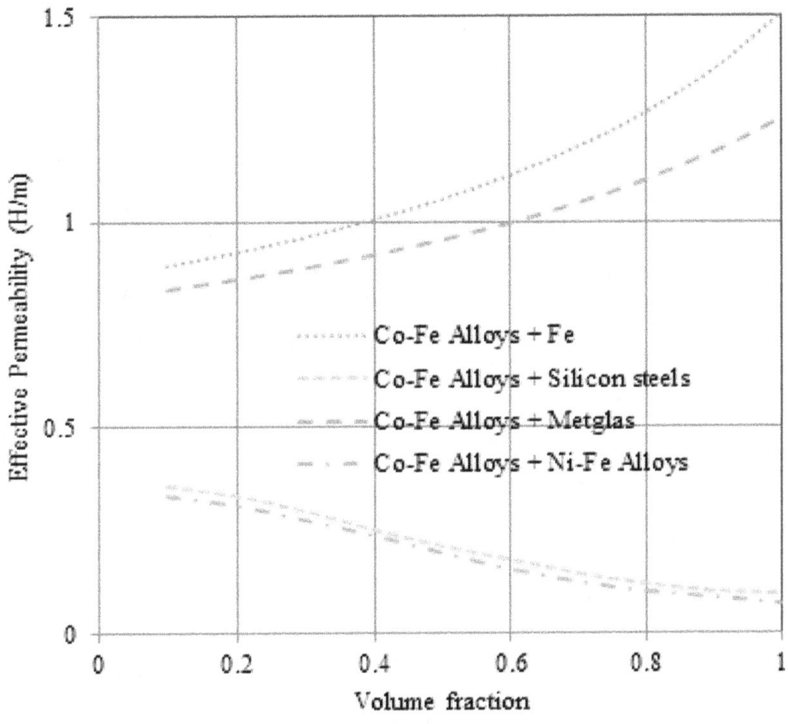

Figure 5. Effective permeability of Metglas composites with various volume fractions of magnetic particles

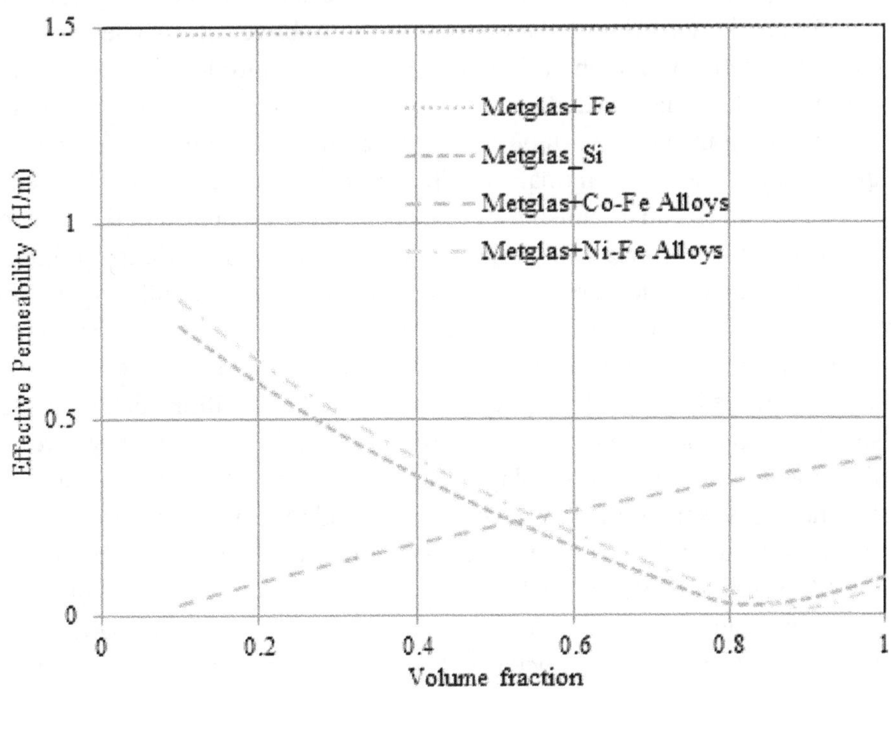

Figure 6. Effective permeability of Iron composites with various volume fractions of magnetic particles

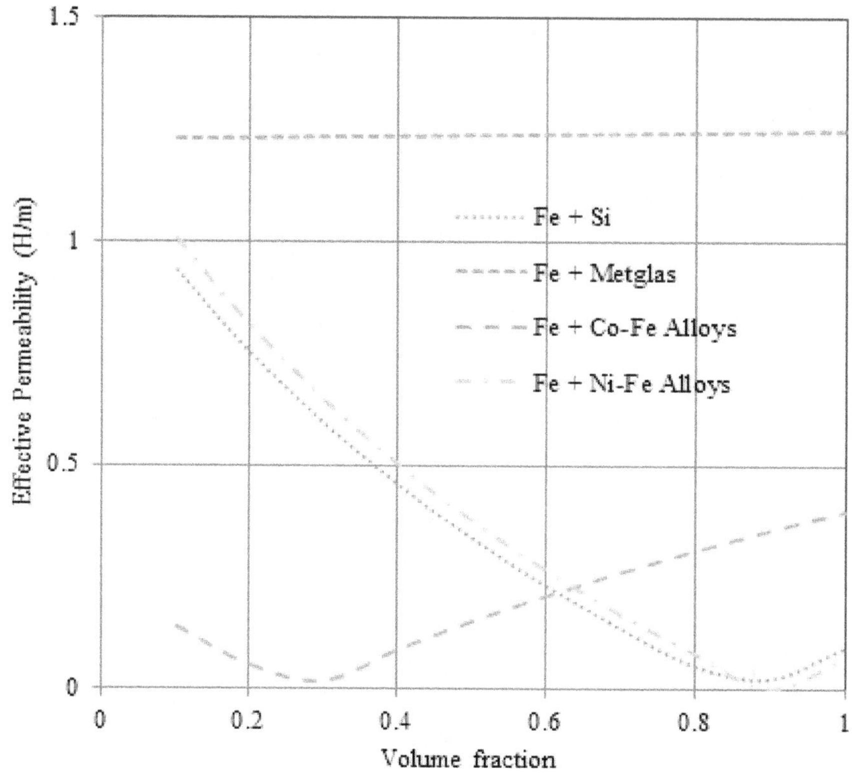

2006; Liu et al., 2009; Waki et al., 2005; Wu et al., 2006). Furthermore, the impacts of nano-size organic, inorganic and magnetic particles on polymers have been suggested (Thabet, 2020). The applications of soft magnetic materials in AC electrical and electronic devices oblige soft magnetic core materials that have high saturation magnetization, high curie temperature, high introductory permeability, low eddy current loss, low hysteresis loss, and low dielectric misfortune. Nanocomposite preparation offers another methodology for fabricating materials with novel magnetic and electric properties. On a magnetic/ceramic nanocomposite, the resistivity is dramatically increased. The trade coupling between neighboring magnetic nanoparticles can succeed the anisotropy and demagnetizing impact of the individual particles, having substantially superior soft magnetic properties than with accepted (large grain-size) materials. Nowadays, there are enormous endeavors continuously embraced toward outlining and manufacturing soft magnets with fantastic physical properties and high erosion safety for modern applications. Regarding building applications, these materials can be of low cost and can be simple to prepare, permitting them to be profoundly focused for the already- existing ones utilized within such applications. Straight and nonlinear simulated material in which the effective permeability may be negative in the low and low thickness lipoprotein extent can be displayed and assembled. At such frequencies, a dispersion mathematical statement drives the field under the medium, and this prompt exploring surprising effective – constitutive laws (Chiriac & Lupu, 2005; de Assis Olímpio Cabral et al., 2008; Lin et al., 2006; Sugimoto et al., 2005; Tellini & Bologna, 2010; Zhang et al., 2011; Zhang et al., 2001). In this chapter, new nanodielectric composites magnetic industrial materials whose magnetic characterization reaction has been improved in terms of sorts and concentrations of chosen nanoparticles have been proposed.

4.1 Theory and Effective Parameters

Our research will concentrate on figuring strategies for the frequency reliant permeability (complex intrinsic permeability μint and effective permeability μeff) of magnetic industrial materials composite comprising micron/submicron-sized nanocrystalline particles installed in distinctive matrix and acknowledging the nanocrystalline structure of magnetic particles and skin impact. This goal is dependent upon the taking after of explanatory models (Liu et al., 2009; Waki et al., 2005) that have been used to figure hypothetical hypotheses for foreseeing an effective permeably of composite and nanocomposite materials. Accordingly, this is a chance to ascertain the genuine intrinsic permeability (μi,true) of a magnetic circular molecule recognizing the skin impact, as shown in Figure 7. Generally, the molecule size can further bolster not surpass the skin profundity δ essentially to radar absorbing applications (Wu et al., 2006).

$$\delta = \sqrt{\frac{\rho}{\pi \mu_{int} f}} \tag{4}$$

Where ρ is the resistivity, μint is the intrinsic magnetic permeability, and f is the frequency.

Figure 7. Schematic for calculating true intrinsic permeability of a magnetic spherical particle

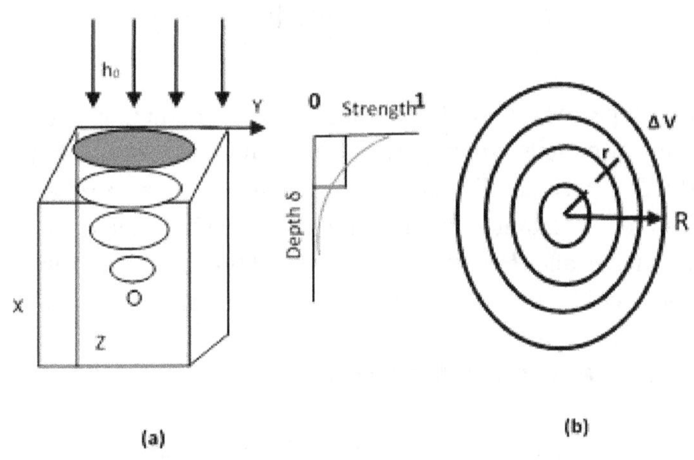

Furthermore, acknowledging a circular metallic molecule liable for an element field, the penetrated progressive magnetic field h abatements expand infiltration under the circular molecule size with d as indicated in Fig. (7), so that the element magnetic field can be obtained as takes after:

$$|h| = h_o e^{-(R-r)/\delta} \tag{5}$$

Where R is the radius of the spherical particle, the true intrinsic permeability of the spherical particle considering skin effect is as follows:

$$\mu_i = 1 + \frac{\Delta h_i \left(\mu_{int} - 1\right) dv / V}{h_o} \tag{6}$$

Regarding traditional blending decisions of particles for primary matrix for figuring out the effective permeability by Maxwell-Garnett's law, the effective permeability of composites regardless of the incorporation stage is scattered in the matrix stage irregular circulation (Liu et al., 2009), it can be as takes after:

$$\mu_{eff} = \mu_m + f\mu_m \frac{\mu_i - \mu_m}{\mu_m + (1-f)\left(N_v cos^2\theta + N_w sin^2\theta\right)\left(\mu_i - \mu_m\right)} \tag{7}$$

Where Θ stands for an acute angle between the direction of three main axes of flakelike inclusion and the external magnetic field H. Therefore, the inclusions are modeled as equivalent oblate spheroids to evaluate the demagnetizing factors N_v and N_w. Adding the effect of nonrandom distribution of inclusions on the effective permeability of the composite for the same model, there are two fitting parameters α and β as spatial position distribution factors as follows:

$$\mu_{eff} = \mu_m + f\mu_m \frac{\mu_i - \mu_m}{\mu_m + (1-f)\left(\alpha N_v cos^2\theta + \beta N_w sin^2\theta\right)\left(\mu_i - \mu_m\right)} \tag{8}$$

Moreover, the successful permeability of nanodielectric magnetic materials that has been described by including different rates of magnetic nanoparticles irregular circulations to polymer matrix material (Thabet, 2020). The effective permeability μeff of micro and nano iron particles/nonmagnetic matrix composite can at that point be ascertained by utilizing the above hypotheses in view of correct intrinsic permeability for magnetic materials. However, the impacts of types, concentration for micro-/nano-sized iron particle and frequency will be indicated in the execution of effective permeability in this chapter.

4.2 Dielectrics and Soft Magnetic Materials

Nanodielectrics are utilized within electromagnetic applications and can be depicted as ferromagnetic powder particles inserted to electrical insulating materials. The decision for nanodielectrics may be unpredictable and relies on looking into huge numbers of variables – the essential ones are constantly frequency, the size of the component, and the physical quality and magnetic properties. Nowadays, nanotechnology science can lead to gigantic leaps, primarily in the magnetic properties of the composite materials which will be influenced by the execution of the electromagnetic modern applications. Thus, this investigation illustrates the impacts of nanoparticle extent (50nm_diameter), iron high purity particles (single crystals to favored directions), Ni-Cu, Ni-Fe, MgZn, And NiZn. As shown in Table 2 that depicts the principle electrical portrayal properties regarding the utilization of magnetic nanoparticles, all these nanoparticles have been utilized for upgrading magnetic properties of dielectric materials.

4.3 Characterization of Nanodielectric Magnetic Materials

Particularly, nanodielectrics system is effective in yielding high execution composites. At whatever point, it has a great scattering of the fillers that have attained and required properties of the nanoscale filler to be considerably diverse or superior to the individuals of the matrix. Thus, this hypothetical contemplate of polymeric nanocomposites is recommended for utilizing different nanoparticles to define nanostructures. The proposed emulating new nanodielectric composites magnetic streamlined materials have improved their magnetic characterization reaction regarding the sorts and concentrations of chosen nanoparticles. The effective permeability of Plastics/Fe nanocomposite has been expanded toward including different rates of iron nanoparticles irregular and composed circulations in plastics matrix material. The effective permeability of Plastics/Fe nanocomposite expands for expanding volume portion of iron nanoparticles, particularly through volumetric fractions (90%wt: 100%wt) in the nanocomposite. In any case, expanding the intense point (0:90) doesn't have an adequacy figure in the effective permeability of nanocomposites. In recent researches, the successful permeability of Plastics/Ni and Plastics/Ni-Cu Alloys nanocomposite has been expanded toward including different rates of nickel and Ni-Cu Alloys nanoparticles irregular and composed circulations in silicon matrix material. Fig. (8) reveals how the expanding Ni-Cu Alloys nanoparticles increment the effective permeability of Plastics, but all the Ni-Cu Alloys nanoparticles have more level magnetic impacts over nickel nanoparticles in Plastics. Moreover, the intense point has a vitality variable for expanding the effective permeability execution toward utilizing Ni-Cu Alloys nanoparticles in association circulation inside plastics matrix. However, the effective permeability of Plastics/Ni nanocomposite that expands by expanding volume portion of nickel nanoparticles in the nanocomposite is clear. The expanding intense point (0:90) builds the effective permeability of nanocomposites. Nevertheless, the successful permeability of Plastics/Ni nanocomposite of irregular and composed circulations has indistinguishable twin characterization execution in a down intense point (0:30). The association dissemination for nickel nanoparticles inside plastics matrix has helped expand the effective permeability by expanding the intense point (60:90) regarding irregular dissemination. On the other hand, the effective permeability of Plastics/Ferrite MgZn and Plastics/Ferrite NiZn nanocomposites has been expanded by including different rates of ferrite nanoparticles irregular and composed circulations to plastics matrix materials. Fig. (9) demonstrates that expanding intense point (0:90) expands the effective permeability of nanocomposites. However, NiZn ferrite nanoparticles have easier magnetic impacts than MgZn ferrite nanoparticles in Plastics. Therefore, the intense point has a vitality component for expanding the successful permeability execution by utilizing Ni-Cu Alloys nanoparticles in association dissemination inside plastics matrix.

Glass does not hold any magnetic charges; it could be charged by different volume fractions of magnetic nanoparticles to be utilized within numerous electrical mechanical applications. Thus, the successful permeability of glass nanocomposite has been expanded by including different rates of magnetic nanoparticles irregular circulations for glass matrix material. Iron is the most elevated effective nanoparticle in the mentioned nanoparticles in expanding the effective permeability of glass nanocomposite. However, Ni-Fe Alloys are the most reduced effective nanoparticles for expanding the effective permeability of glass nanocomposites. On the other hand, the effective permeability of glass nanocomposites has been expanded by including different rates of ferrite magnetic nanoparticles irregular circulations to glass matrix material. Mg-ZnO ferrite can be the more successful nanoparticle over Mg-ZnO ferrite nanoparticles in the studied over glass nanocomposites for expanding the effective permeability of Glass nanodielectric composites. Soft magnetic nanoparticles provide for an up-primary for magnetic

Figure 8. Effective permeability of Plastics nanocomposite with various Ni-Cu Alloys nanoparticles volume fractions

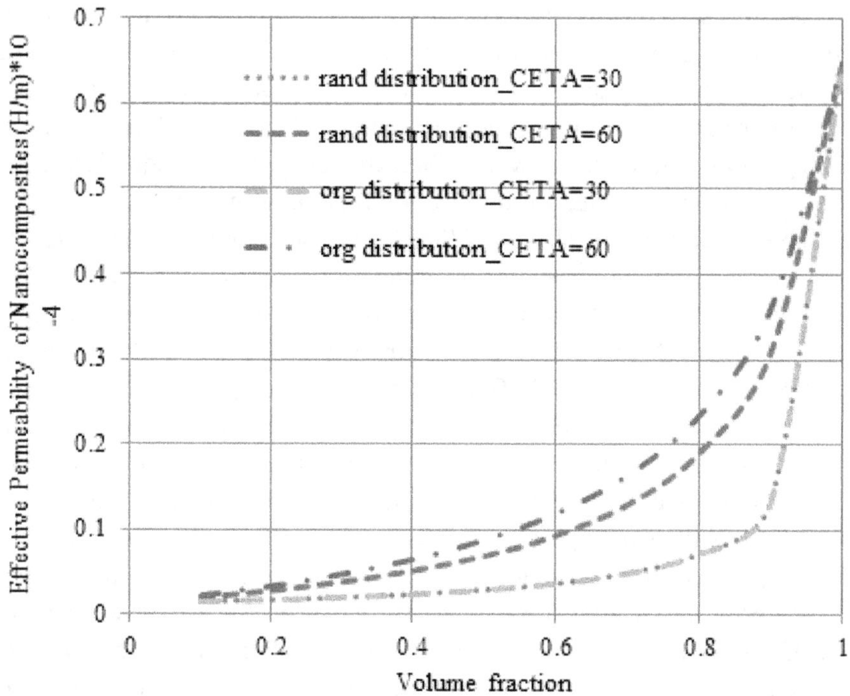

Figure 9. Effective permeability of Plastics nanocomposite with various MgZn Ferrite nanoparticles volume fractions

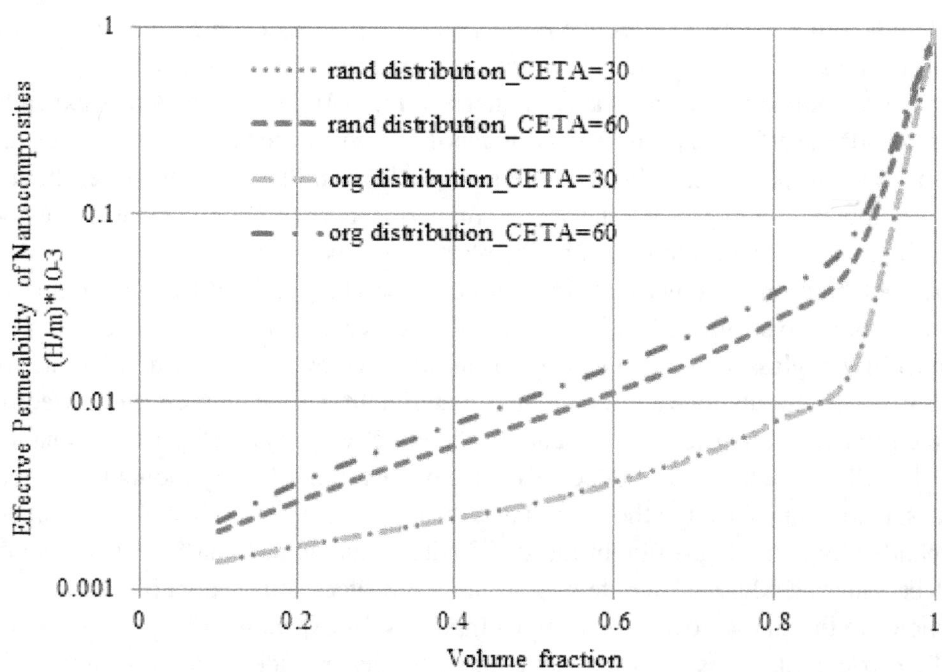

properties of the composites for considering the outcomes of nanodielectric composites. It is shown that nanoparticles provide for an incredible up-primary for magnetic properties of the composites. Nevertheless, dielectric composites can be the compass of the same outcomes of nanodielectric composites during specified extraordinary position conveyance factors (α, and β) and at specified indistinguishable twin flake-like scatters of the composite.

5 MAGNETIC COMPOSITE CRYSTALLINE CORE

Frequency response analysis (FRA) has been perceived as the vast majority dependable checking strategy for transformer winding relocation and deformity evaluation. It is secured upon the truth that the state of a winding frequency reaction in high frequencies is connected with winding geometry. The manifestation of clear shifts of thunder frequencies or new full concentrations for a reaction might portray broken states of windings (Ryder, 2003; Shintemirov et al., 2010). Investigation exercises have been embraced to use FRA in the improvement of suitableness lumped parameter scientific models of transformer windings (Abeywickrama et al., 2007; Abeywickrama et al., 2006; Abeywickrama et al., 2008a; Florkowski & Furgal, 2003; Rahimpour et al., 2003). This technique may be particularly advantageous to the point when the amounts of premium can't be measured straight forwardly. Such particular circumstances appear, for example at assessing inductances of transformer windings at their 1st thunder frequencies, which will be fundamental for understanding FRA information (Podoltsev, 2007; Podoltsev et al., 2007). The new formed materials end up especially paramount on account of their physical properties progressing dramatically for the size and for the nearby structure of the grains. Quitting offering on that one imperative provision is offered for transformer core materials, these materials ought to be of low cost and be simple to prepare, permitting them to be exceedingly focused with the routine existing ones. Regarding soft magnetic materials that play a paramount role in expansive applications, such as transformers. These materials utilized within electromagnetic applications might be depicted as ferromagnetic powder particles encompassed by an electrical insulating film (Anhalt & Weidenfeller, 2012; Fuzerova et al., 2010; Thabet & Repetto, 2013; Yao et al., 2011). The target of this effort is to portray the magnetic conduct of control transformer composite crystalline core utilizing frequency reaction investigation which is progressively prevalent for the appraisal of mechanical integument of control transformers. FRA strategy is made of the reality that the state of the reaction in higher frequencies is particularly dictated by the geometrical development of the transformer. Therefore, new magnetic composite transformer core materials have been recommended for upgrading force transformer magnetic characterization reaction identified with sorts and concentrations for chosen particles. Numerical effects anticipate that energy transformer core characterization is about the distinction between customary sorts over the use of variant sorts of particles for cores. This may be a chance for a beneficial concurrence for test effects. Therefore, large fractions of micro-particle composites with distinctive particles are investigated to achieve the best fillers' essential expansion of magnetic permeability.

5.1 Theory and Effective Parameters

This chapter highlights the figuring systems for the effective permeability of magnetic composite transformer core materials comprising micron/submicron-sized particles inserted in distinctive matrices. Hence, the past explanatory models (Lin et al., 2006; Wu et al., 2006) have been used to plan hypothetical models for foreseeing the effective permeability of magnetic composite materials. Regarding Maxwell-Garnett recipe that plays a paramount part in two extents of composite materials (Lin et al., 2006), the effective permeability to irregular dissemination of molecule measure as two measurements round shapes may be essentially provided. This is the exact approach that expects that the estimation of the magnetic composite material structure which is periodic, and the unit cell, which may be the least volume, are viewed as homogeneous magnetic substances (Yao et al., 2011). The effective permeability of magnetic composite materials is characterized according to the support of magnetic vitality harmony in the unit cell. It is also played in this approach that the first mobile and the homogenized cell incorporate equal magnetic vitality in both units that are inundated in proportional magnetic field as demonstrated in Fig. (10).

Figure 10. Homogenization of composite material

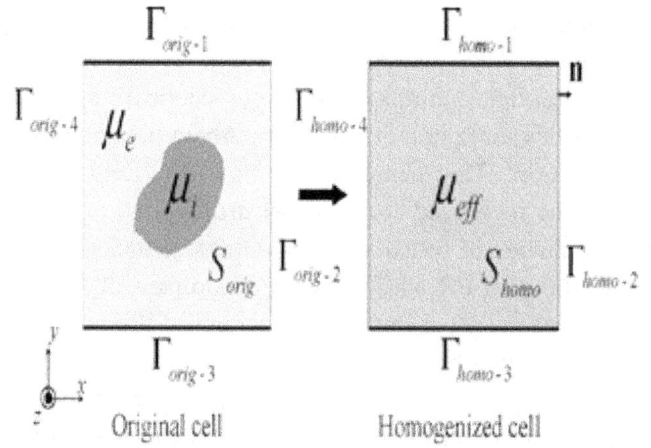

The schematic diagram and the equivalent magnetic circuit of the transformer core model (Waki et al., 2005) are established, as shown in Fig. (11). The magnetic core of the transformer is divided into three sections with respect to uniform fluxes φA, φB, and φC of each transformer limb. Each section of the magnetic core is represented by its reluctance Re, the reluctance of a section is determined by the magnetic parameters of lamination and the core geometry as follows (Lakshminaray & Balakrishnan, 2013; Moses, 2012):

$$Re = \frac{F}{\varnothing} = \frac{l}{\mu_{core_{eff}} \mu_o A} \tag{9}$$

Where, F is the magneto-motive force required for establishing the flux along the length of the section. ϕ is the produced magnetic flux. l is the length of the magnetic flux path along each section. A is the cross-section area of the core, $\mu_{core_{eff}}$ is the complex effective relative permeability of transformer core, and μo is the free space permeability.

Therefore, the effective nearby magnetic permeability of suggested magnetic composite transformer core materials can be predicted by utilizing the above methodologies in view of inclusions permeability, weight rates of inclusions inside the matrix and principle matrix permeability of magnetic transformer core composite materials. Thus, the unpredictable effective relative permeability of the transformer core is characterized as takes after:

$$\mu_{core_{eff}} = \mu'_{core_{eff}} - j\mu''_{core_{eff}} = k_{fe}\mu_r \frac{tanh\left(\left(1+j\right)b/\delta\right)}{\left(\left(1+j\right)b/\delta\right)} \tag{10}$$

Where, δ is the skin depth, kfe=2b/h represents the stacking factor, h and 2b are the thicknesses of a single lamination sheet of the core with and without insulation layer included respectively, and μr is the local magnetic permeability. The skin depth δ depends on the angular frequency ω of the magnetic field as follows:

$$\delta = \sqrt{2/\omega\sigma\mu_o\mu_r} \tag{11}$$

Where σ is the conductivity of lamination materials. Hence, using composite crystalline materials for transformer core laminations is required for deterministic effective conductivity (Oltean & Motoc, 2012; Salvadori et al.,). However, in this chapter, the influence of the types and concentration of inclusions particles is investigated on the performance of effective permeability of magnetic composite power transformer core.

5.2 Magnetic Particles and Core Materials

Nowadays, nanotechnology science can be settled on gigantic up-primary in the magnetic properties for transformer core materials; then the decision of magnetic composite material is mind complex and relies on numerous factors, the elementary ones are frequency, the size of the component, the physical quality and the magnetic properties. Thus, the suggested particles radius such as iron high purity particles (single crystals for favored directions), silicon steels, Metglas, Co-Fe, Ni-Fe, MgZn_Ferrite, and NiZn_Ferrite have portrayed the up-primary done execution of transformer core magnetic properties. Table i depicts the fundamental electrical depiction properties of the use of particles which have been utilized for up-grading magnetic properties of transformer core materials. Magnetic transformer core has tremendous concentrations in terms of the quality and increments the impact of magnetic fields handled by electric currents and flows and lasting magnets. The impact of core laminations is to keep eddy currents and flows on profoundly curved ways that encase little flux and along these lines decrease their magnitude. The slim core laminations can be utilized in High-frequency transformers, for a portion of really dainty steel laminations fit to work up to 10 kHz. In practice, the parameters of core transformer, such as the extent of winding, core lamination…etc. as a rule dead set utilize a cover test (Podoltsev, 2007; Podoltsev et al., 2007). The accompanying results are conveyed on the unpredictable effective relative permeability and reluctance of the new transformer core composite magnetic materials that have been proposed for upgrading the magnetic characterization reaction regarding the sorts and concentrations of chosen particles. The specifications of Suggested Transformer Core Materials have been tabulated in Table 2.

5.3 Characterization of Magnetic Crystalline Core

Utilizing the explanatory model portrayed by MATLAB project for the reproduction of the transformer core model composites materials has been investigated regarding the characterization of effective permeability of the composites for different sorts and concentrations to acquire cost-fewer composite materials. The outcomes are checked for the capacity of the proposed composites on changing the core cover parameters regarding the reference values. Regarding the developing enthusiasm toward researching high-frequency phenomena in transformers, this chapter displays recommended composites for parameter identification of a laminated core for force transformers dependent upon reference Frequency Response Analysis. Fig. (11) indicates mind complex successful relative permeability for different magnetic composites toward different rates of iron, Metglas, silicon steels, and ferrite MgZn particles haphazardly distributed, then Fig. (11) reveals the upgrading effective mind complex relative permeability of magnetic composite materials, including magnetic particles, which have higher permeability over the base matrix and through. It is to be noted that this sort of particles has the primary variable for specifying the execution of the effective permeability of the composites indicated in Fig. (11). Metglas particles are effective particles for upgrading successful permeability of the composites. MgZn and silicon steels particles have terrible impact on the successful permeability of the composites. Fig. (12) indicates effective mind complex relative permeability for Different transformer core magnetic composites that have been shifted by including certain appropriation rate (30%wt) of Metglas, silicon steels, ferrite MnZn, and ferrite MgZn particles for iron building matrix. Frequency response Analysis (FRA) for cost-fewer transformer core

Figure 11. Effective Complex Relative Permeability with various Volume Fractions of Soft Magnetic Particles

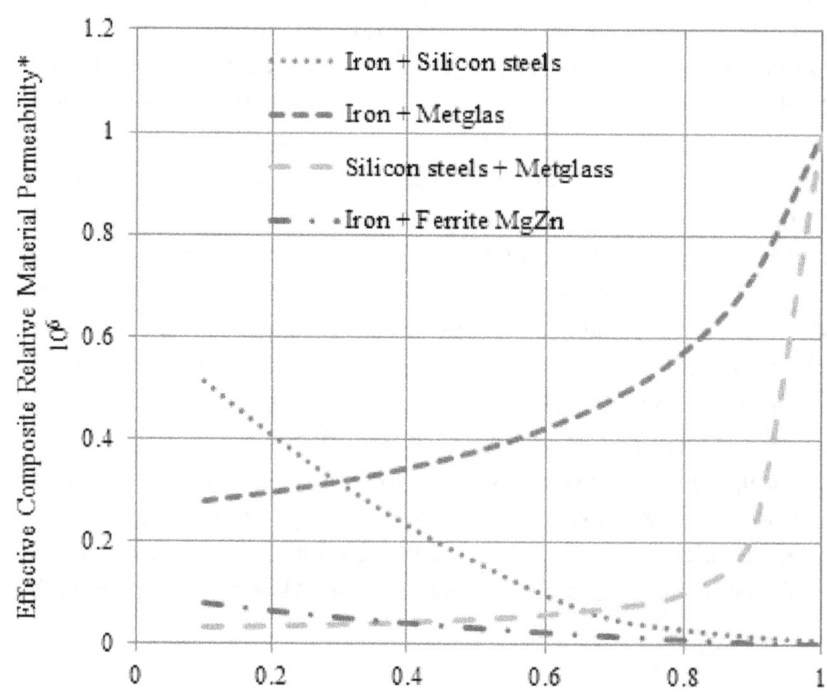

magnetic composites has been investigated in the range of (10^{-2}Hz – 10^{3}Hz) for specifying and upgrading the magnetic effective complex relative permeability of the transformer core magnetic composites that rely on the higher permeability of magnetic particles of the base matrix.

Figure 12. FRA for effective complex transformer core relative permeability with various magnetic particles

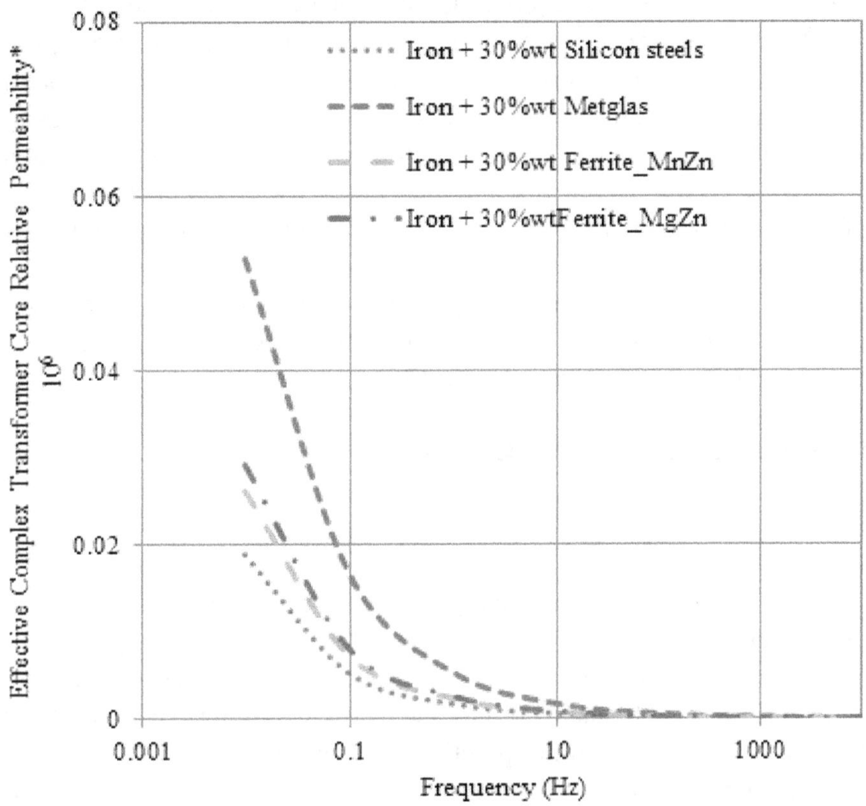

Nowadays, the execution of effective complex transformer core relative permeability for different volume fractions of Ferrite_MgZn particles haphazardly scattered in iron matrix material are investigated. In any case, the obtained results are over end for that gotten toward (Oltean & Motoc, 2012; Salvadori et al.,). Moreover, the transformer core reluctance of cost-fewer magnetic composites has been investigated (10^{-2}Hz – 10^{3}Hz) regarding fluctuating sorts toward sure concentration of magnetic particles. It illustrates the reluctance of different transformer core magnetic composites that have been shifted by including certain rate (30%wt) of Metglas, silicon steels, ferrite MnZn, and ferrite MgZn particles of haphazard circulations with iron build matrix. On the other hand, the execution of transformer core reluctance for different volume fractions of Ferrite_MgZn particles haphazardly scattered on iron matrix material can be prevailing. Magnetic particles in the composite evolve in magnetic reluctance of the composite about evolving effective complex transformer core relative permeability. The obtained outcomes are in line with that understanding (Oltean & Motoc, 2012; Salvadori et al.,).

6 NANOCOMPOSITES MAGNETIC MATERIALS

The most relatable nanostructured magnetic materials are nanoparticles, dainty nano-layers, nanowires, nanocrystalline materials and nanocomposites. The magnetic properties of these materials vary in few approaches from the individuals of mass magnetic materials. The usage of low-dimensional structures turns into an ever-increasing significant amount because of the scaling down of devices due to the novel emerging nanoscopic properties. Magnetic properties of a material transform signiðcantly for diminished extents. At the same time, magnetic dainty ðlms still represent the framework of most gadget applications, making nanoscale magnetic structures that enable breakthroughs. By lessening dimensions, materials can be customized to enhance the fancied properties of a speciðc provision. This provides another system for controllably evolving magnetic properties. With decreasing dimensions, an extensive variety of nanostructures, such as slim ðlms, nanowires, particles, junctions, tubes could be made. When conventional strategies for magnetic property control (doping and alloying) are joined with an expansive extent of nanoscale geometries, control of intrinsic magnetic properties which aggregate magnetization, curie point, and anisotropy is workable. In addition, in joining materials with different magnetic properties, new functionalities, such as titan magneto imperviousness (Ikeda & Dohmeki, 2005), trade bias, multi-ferromagnetism, and magnetoelectricity can be attained.

6.1 Preparation of Nanocomposite Cores

Inert-Gas Condensation

The inert-gas buildup may be a standout amongst a large portion of generally utilized preparation strategies for nano-size particles. The material is dissipated under low weight idle gas, such as immaculate helium. Vapor condenses under nanometer measure particles that are gathered on the surface of a fluid nitrogen-cooled pivoting barrel. The radius of particles turns into littler when the thickness of the gas is more diminutive and when the softening point of the metal is higher. The almost round particles are gathered by scratching the surface under a pipe and finally in a compacting assembly, the last continuously in the same vacuum cell. The compaction may be generally led in two stages, the second constantly over 1GPa. The trademark shape is a mm-thick circle with the greatest breadth of 15 mm. The last relative thickness ranges between 70% And 97%, contingent upon the metal's pliability. The molecule measure ranges start with 2 on 100nm. Since this procedure prevents pollution, it is suitable for basic investigations about physical properties. As stated by the irregular anisotropy model (RAM), the effective anisotropy is decreased if the grain measure is more diminutive than the anisotropy trade length (Mazaleyrat & Varg, 2000).

High-Energy Milling

Numerous metastable alloys might be obtained toward mechanical alloying with nano-sized crystallite-structure; the powder molecule size remaining over the micrometer level. In contrast, the ball processing of oxides such as magnetite or hematite prompts nanoparticles. Starting with these, immaculate iron nanoparticles are obtained by hydrogen diminishment.

Cryogenic Melting

When a liquid metal may be in contact with contact a cryogenic liquid, vaporous layer manifestations can appear on the surface. In this calefaction layer, the metallic vapor condenses under ultra-fine particles whose radius depends basically on the vapor weight (Ikeda & Dohmeki, 2005; Mazaleyrat & Varg, 2000). The gas prepared from the cryogenic fluid (argon or nitrogen) transports the particles under a canvas channel. In order to have additional vapor pressure, the metal must make the thermal up by a few hundred degrees in its dissolving temperature (over 20003C for Fe, Ni, Co). Consequently, cryogenic dissolving of ferromagnetic metals can't be conducted over a ceramic crucible, which reacts under these states. The levitation liquefying strategy employs an RF inductor that induces eddy currents and flows in the metal and both suspend the high temperature of the metal. A picture of the reactor is seen in Fig. (13 left). Practically, a pole of metal can be slid down under the reactor. Drop for liquid metal structures at the edge and tumbles onto the inductors can be levitated in a complete change under nanoparticles. The generation is constant at a rate of something like 150g/h. The as-obtained iron nanopowders have a spheroidal state for normal breadth of about 32nm (78% of particles are short of 20nm) over the ideal case, their size is superbly circular in viewing Fig (13 right hand). Then after passivation by give or take 1nm oxide layer, the particles are no more pyrophoric and may have a chance to be manipulated over air.

Figure 13. Cryogenic levitation melting experimental apparatus (left hand), TEM micrograph of the as-obtained nanoparticles (Abeywickrama et al., 2008b; Szabó et al., 2010)

An accepted methodology to elucidate the properties of composites utilizes blending rules, equations that relate the intrinsic properties of inclusions and the host matrix with the effective properties of composite frequencies in a basic glorified model recognized as a round structured Incorporation (Marina et al., 2012). Nowadays, Maxwell–Garnett and parts of blending decides have been changed to anticipate a successful parameter count for discovering the crazy perceptible magnetic properties of composite and nanocomposites magnetic materials in the reenactment strategies (Liu et al., 2009; Waki et al., 2005). The frequency response analysis (FRA) system pioneered by dick and Erven has become a progressively prominent method for the outside evaluation of the mechanical integument of control transformers, toward searching for the allowed deformations of the windings and the core (Abeywickrama et al., 2008b). This chapter exhibits the frequency dependencies of effective material parameter (permeability) of various sorts of magnetic nanoparticles; therefore, the magnetic characterization reaction of the new nanocomposites materials regarding conventional magnetic materials has been improved.

6.2 Theory of Magnetic Nanocomposites Using Individual Nanoparticles

The effective permittivity and permeability of the nanocomposites' materials are ascertained in distinctive approaches (Szabó & Füzi, 2014; Szabó et al., 2010). However, the optical wavelength of the occurrence electromagnetic wave as takes after are as follows:

The size parameter: $$x = \frac{\sqrt{\mu_r^h \varepsilon_r^h}\,\omega r}{c_0} \tag{12}$$

Where,
ε_r^h is the relative permittivity of the base matrix material.
μ_r^h is the relative permeability of the base matrix material.
r is the radius of spherical nanoparticles.
ω is the angular frequency of incident electromagnetic wave.
c_0 is the speed of light in vacuum.

Moreover, Mie theory explains exactly and offers analytic resolution in form of infinite series (Bohren & Huffman, 2004) as follows:

$$a_n = \frac{m\gamma_n(mx)\widehat{\gamma_n}(x) - \gamma_n(x)\widehat{\gamma_n}(mx)}{m\gamma_n(mx)\widehat{\varphi_n}(x) - \varphi_n(x)\widehat{\gamma_n}(mx)}$$

$$b_n = \frac{\gamma_n(mx)\widehat{\gamma_n}(x) - m\gamma_n(x)\widehat{\gamma_n}(mx)}{\gamma_n(mx)\widehat{\varphi_n}(x) - m\varphi_n(x)\widehat{\gamma_n}(mx)} \tag{13}$$

Where,

$m = \sqrt{\varepsilon_r^i \mu_r^i} \Big/ \sqrt{\varepsilon_r^h \mu_r^h}$ is the contrast of the refractive index.

ε_r^i is permittivity of spherical inclusions.
μ_r^i is permeability of the spherical inclusions.
φn and γn $_a$re Riccati - Bessel functions.

Thus, the effective electrical permittivity ε_r^{eff} and the effective magnetic permeability μ_r^{eff} are as follows:

$$\varepsilon_r^{eff} = \varepsilon_r^h \frac{x^3 + 3i\zeta a_1(mx)}{x^3 - \frac{3}{2}i\zeta a_1(mx)} \tag{14}$$

$$\mu_r^{eff} = \mu_r^h \frac{x^3 + 3i\zeta b_1(mx)}{x^3 - \frac{3}{2}i\zeta b_1(mx)} \tag{15}$$

Where ζ is volume fraction.

6.3 Theory of Magnetic Nanocomposites Using Multiple Nanoparticles

A hypothesis of magnetic nanocomposites utilizing numerous nanoparticles, blending solids and composites has prepared new mixtures and composites (Gouda et al., 2004; Tuncer, 2002; Tuncer et al., 2001); thus, it might be getting a greater amount of enhancement in magnetic characterization and can predict a close estimation count technique for deciding the successful relative permeability of nanocomposites magnetic materials that utilizes multi-nanoparticles. Thus, it supposes three concentric magnetic permeability plates for permeabilities $\mu_{h,}$ μ_i, and μ_j which are embedded inside the effective medium with the magnetic permeability μ, as presented in Fig. (14): the effective complex. The magnetic permeabilities are assessed with respect to the close estimation count of multi nanoparticles with the principal matrix. The effective magnetic permeability of new nanocomposites, whether the inclusions are multi nanoparticles which are conveyed inside the matrix, will have a chance to be as takes after:

$$\varepsilon_r^{effi} = \varepsilon_r^h \frac{x_i^3 + 3i\zeta a_{1i}(m_i x_i)}{x_i^3 - \frac{3}{2}i\zeta a_{1i}(m_i x_i)} \tag{16}$$

$$\mu_r^{effi} = \mu_r^h \frac{x_i^3 + 3i\zeta b_{1i}(m_i x_i)}{x_i^3 - \frac{3}{2}i\zeta b_{1i}(m_i x_i)} \tag{17}$$

Where,

$$a_{1i} = \frac{m_i \gamma_n(m_i x_i)\widehat{\gamma_n}(x_i) - \gamma_n(x_i)\widehat{\gamma_n}(m_i x_i)}{m_i \gamma_n(m_i x_i)\widehat{\varphi_n}(x_i) - \varphi_n(x_i)\widehat{\gamma_n}(m_i x_i)} \tag{17.i}$$

$$b_{1i} = \frac{\gamma_n(m_i x_i)\widehat{\gamma_n}(x_i) - m_i \gamma_n(x_i)\widehat{\gamma_n}(m_i x_i)}{\gamma_n(m_i x_i)\widehat{\varphi_n}(x_i) - m_i \varphi_n(x_i)\widehat{\gamma_n}(m_i x_i)} \tag{17.ii}$$

Where the size parameter of the first type nanoparticles is

$$x_i = \sqrt{\varepsilon_r^h \mu_r^h}\, \omega r_i / c_0 \tag{17.iii}$$

Figure 14. Schematic diagram for new nanocomposites using multi-nanoparticles

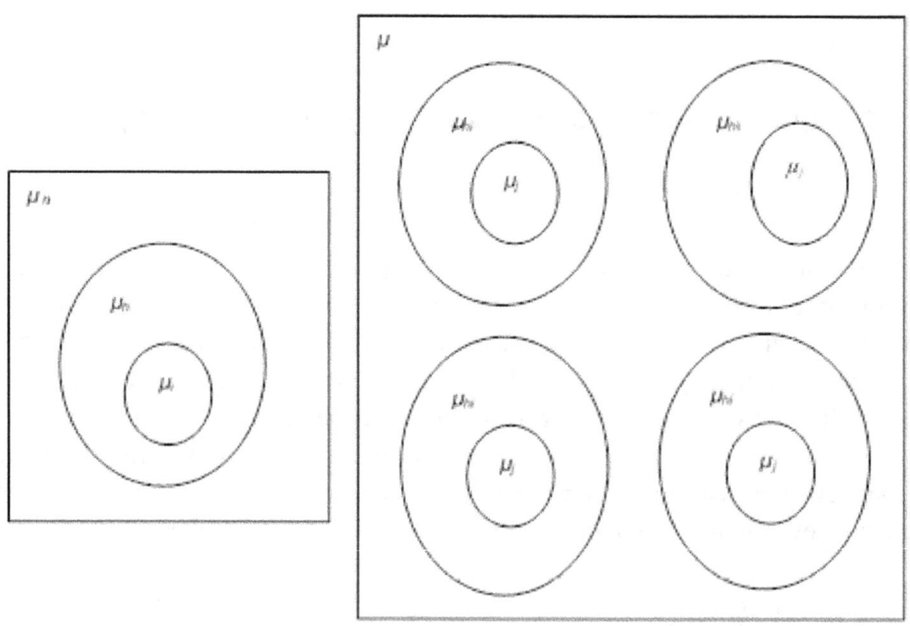

Moreover, the contrast of the refractive index first type nanoparticles is

$$m_i = \frac{\sqrt{\varepsilon_r^i \mu_r^i}}{\sqrt{\varepsilon_r^h \mu_r^h}} \tag{17.iv}$$

Where,

ε_r^i is the relative permittivity of the first type of inclusion material.
μ_r^i is the relative permeability of the first type of inclusion material.
ε_r^h is the effective relative permittivity of the base matrix.
μ_r^h is the effective relative permeability of the base matrix.
r_i is the radius of the first type of inclusion material.

$$\mu_r^{effj} = \mu_r^{effi} \frac{x_j^3 + 3i\zeta b_{ij}\left(m_j x_j\right)}{x_j^3 - \frac{3}{2} i\zeta b_{1j}\left(m_j x_j\right)} \tag{18}$$

Where,

$$b_{1j} = \frac{\gamma_n\left(m_j x_j\right)\widehat{\gamma_n}\left(x_j\right) - m_j \gamma_n\left(x_j\right)\widehat{\gamma_n}\left(m_j x_j\right)}{\gamma_n\left(m_j x_j\right)\widehat{\varphi_n}\left(x_j\right) - m_j \varphi_n\left(x_j\right)\widehat{\gamma_n}\left(m_j x_j\right)} \tag{18.i}$$

Where, the size parameter of multi nanoparticles is

$$x_j = \sqrt{\varepsilon_r^{effi} \mu_r^{effi}}\, \omega r_j \Big/ c_0 \tag{18.ii}$$

Additionally, the contrast of the refractive index of multi nanoparticles is

$$m_j = \frac{\sqrt{\varepsilon_r^j \mu_r^j}}{\sqrt{\varepsilon_r^{effi} \mu_r^{effi}}} \tag{18.iii}$$

Where,

ε_r^j is the relative permittivity for the second inclusion material.
μ_r^j is the relative permeability for the second inclusion material.
ε_r^{effi} is the effective relative permittivity for unique nanoparticles.
μ_r^{effi} is the effective relative permeability for unique nanoparticles.
r_j is the radius of the second inclusion material.

6.4 Characterization of Magnetic Nanocomposites

Terahertz (THz) or the T-ray radiation is attracting significant consideration for quite some time with the improvement of biomedical imaging, security screening, manufacturing personal satisfaction control and sensors. Terahertz applications are further increased by the advancement in the field of nanocomposites magnetic materials, since regular materials best show feeble electric and magnetic reactions at terahertz frequencies. Nanocomposites magnetic materials, dissimilar to regular materials, can be artificially custom-made to show solid full conduct at terahertz frequencies (Khodasevych et al., 2012). Moreover, nanoparticles have the capacity for regulating magnetic and electric properties of traditional magnetic materials. Hence, the general magnetic conduct technique of the nanocomposites materials has been changed for electric and magnetic mechanical applications. With respect to Table 2, it depicts that the primary electrical and magnetic depiction properties of using magnetic nanoparticles (Dia.: 10nm) which have been utilized for upgrading magnetic properties of materials.

As shown in Fig. (15), the effective relative permeability of Ferrite/Silicon steel nanocomposites during frequency 1THz and separate consideration radius have been investigated. Expanding silicon steel nanoparticles increments the effective relative permeability of ferrite regarding the volume fractions but expanding the consideration radius has been diminished by the up-primary in the effective relative permeability of ferrite nanocomposites. However, Fig. (16) indicates the effective relative permeability of Ferrite/Silicon steel nanocomposites for expanding frequency toward different volume portion. Volume portion is an imperativeness variable for moving forward the successful relative permeability execution toward utilizing silicon steel nanoparticles over frequency reaction investigation. Expanding volume portion has expanded the effective relative permeability of ferrite nanocomposites suddenness inside the extent of (10 – 40THz); then, effective relative permeability dives can be stable.

Figure 15. Effect of inclusion radius on effective relative permeability of Ferrite/ Silicon Steel nanocomposites

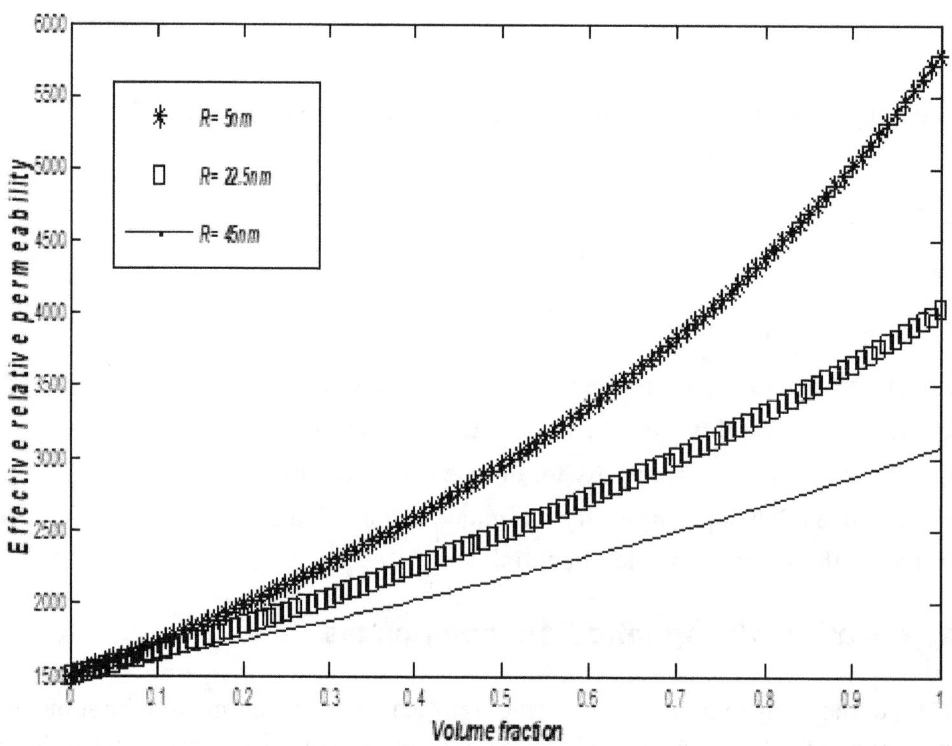

Figure 16. Frequency response analysis for effective relative permeability of Ferrite/ Silicon Steel

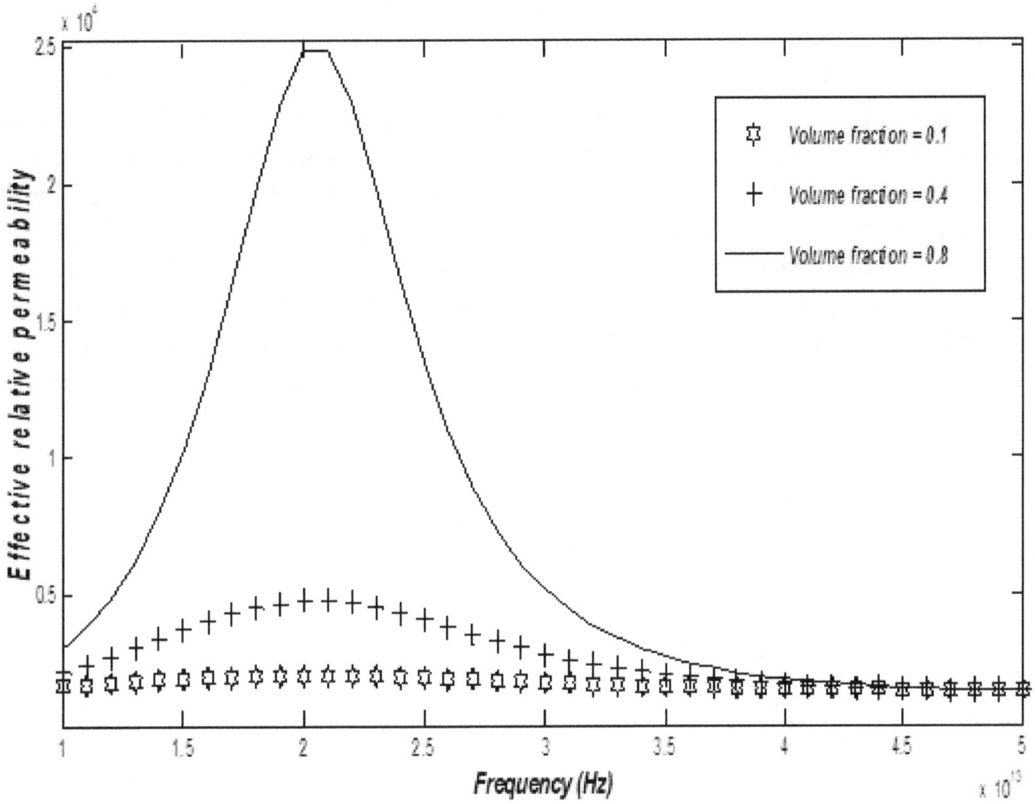

Nowadays, using multi-nanoparticles has improved the magnetic characterization at 1THz. It has been self-evident that the expansion in the effective relative permeability of ferrite nanocomposites happens with expanding the volume portion. Utilizing multi-nanoparticles is higher than the successful relative permeability of ferrite nanocomposites with expanding volume portion by utilizing an absolute kind of nanoparticles main. On the other side, vitality gives the idea in plan for nanoparticles inside figuring of the successful relative permeability to new nanocomposites that are utilizing multi-nanoparticles. The impact of multi-nanoparticles on the successful relative permeability of ferrite nanocomposites during high frequency extent (10–50THz) for diverse specified volume fractions to multi-nanoparticles is indicated. The enhancement in the effective relative permeability of ferrite nanocomposites happens by utilizing multi-nanoparticles. Utilizing multi-nanoparticles has expanded the effective relative permeability of ferrite nanocomposites suddenness inside the range of 10–30THz. Multi-nanoparticles (Nickel ferrite and silicon Steel) is checked for upgrading the magnetic properties concerning the utilization of individual nanoparticles. The evolving of host magnetic materials illustrates the effective relative permeability of Ni_Ferrite nanocomposites for different volume fractions of multi nanoparticles in frequency 1THz. The effective relative permeability of Ni_Ferrite nanocomposites that has been expanded toward expanding the effect of NiZn_ Ferrite, silicon steel nanoparticles in the Ni_Ferrite at effective relative permeability may be not improved by including ferrite nanoparticles due to the decrease of relative permeability of nanoparticles compared with base matrix materials. Thus, multi-nanoparticles have improved the effective relative permeability of Ni_Ferrite magnetic materials. The recurrence reaction analysis has demonstrated the impact of multi nanoparticles magnetic materials on the effective relative permeability of Ni_Ferrite nanocomposites for expanding frequency range (10-50THz) for different rates of nanoparticles 30wt% and 40wt%. The effective relative permeability of Ni_Ferrite/Ferrite nanocomposites may be diminishing inside the frequency extent (10-30THz); moreover, a concentration of 40wt% of the chosen magnetic materials has improved the successful relative permeability toward the same extent. Ferrite and silicon steel nanoparticles have struck the best enhancement on the successful relative permeability of Ni_Ferrite nanocomposites. Fig. (17) indicates the effective relative permeability of NiZn_Ferrite nanocomposites with different volume fractions of multi nanoparticles at frequency 1THz. The successful relative permeability of NiZn_Ferrite nanocomposite has diminished toward expanding the effect of ferrite or nickel ferrite nanoparticles so that utilizing silicon steel nanoparticles upgrades the effective relative permeability of NiZn_Ferrite regarding the plan in nanocomposites show. In any case, the individual Silicon steel nanoparticles struck them a slight expanding in the effective relative permeability of NiZn_Ferrite nanocomposites.

Figure 17. Effective relative permeability of NiZn_ Ferrite nanocomposites using individual and multi-nanoparticles

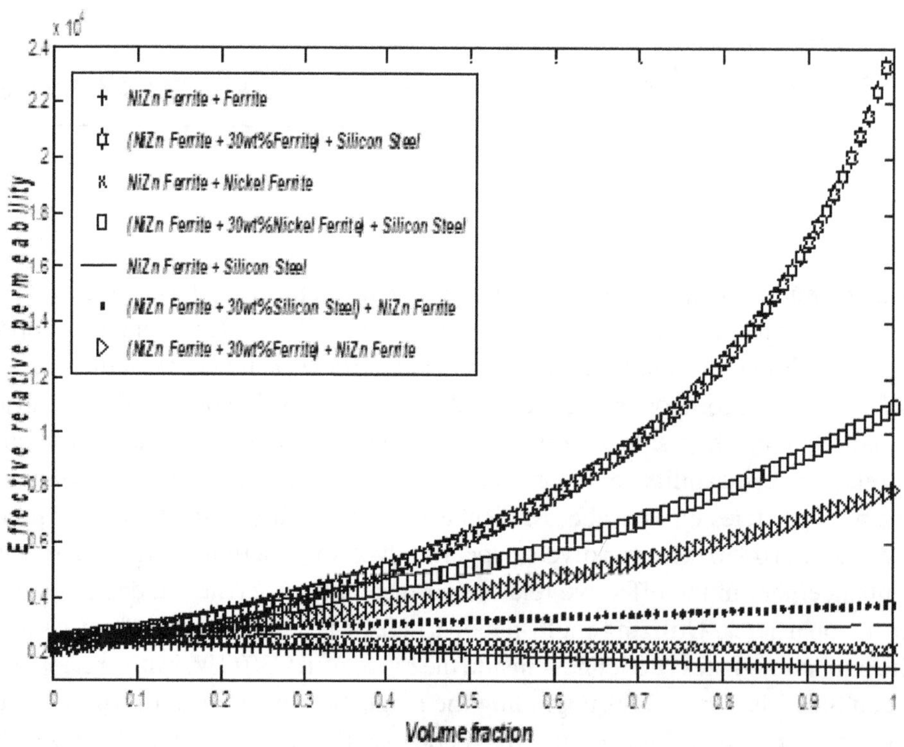

Additionally, Fig. (18) indicates the impact of multi-nanoparticles magnetic materials on the effective relative permeability of NiZn_Ferrite nanocomposites inside the frequency extent of (10–50THz) during distinctive volume fractions. Utilizing multi-nanoparticles has expanded the effective relative permeability of NiZn_ferrite nanocomposites; moreover, the course of action of multi-nanoparticles

(Ferrite, silicon steel and Ni_Ferrite) inside the group base matrix is a critical figure for upgrading the effective permeability of NiZn_Ferrite nanocomposites.

Figure 18. Effective relative permeability of NiZn_Ferrite nanocomposites using individual and multi-nanoparticles

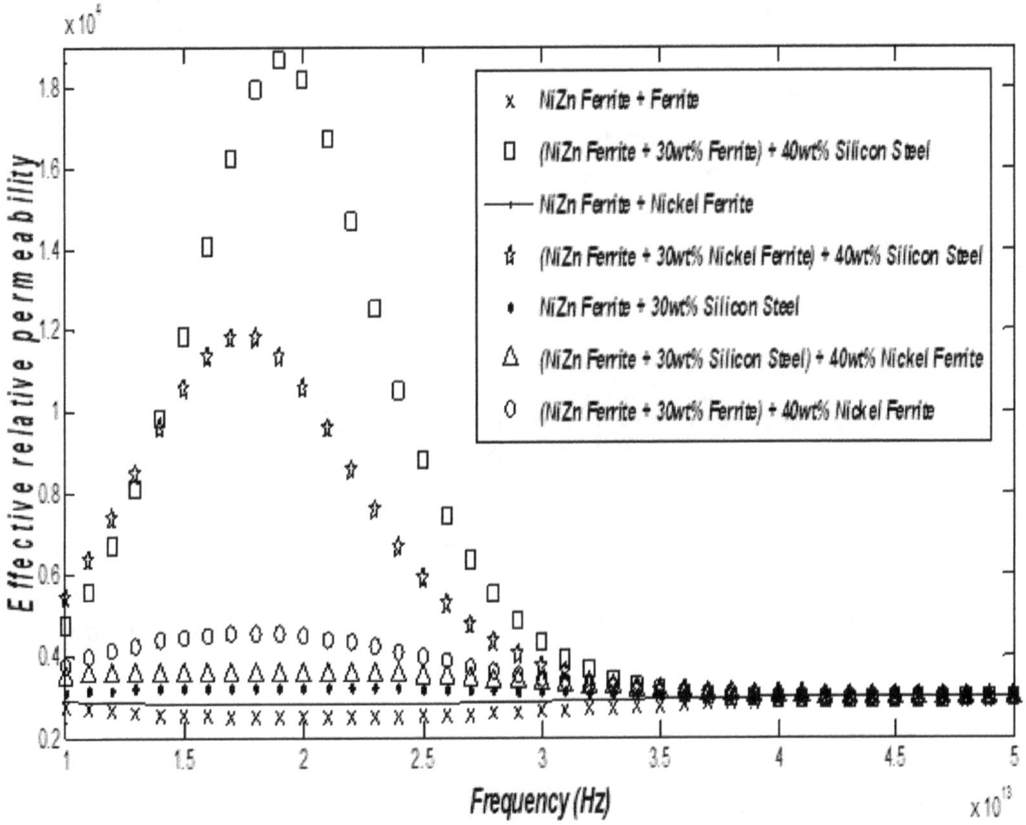

For a high permeability of host matrix, Fig. (19) indicates the effective relative permeability of silicon steel nanocomposites for different volume fractions at 1THz. Ni_Ferrite, alternately ferrite nanoparticles, diminishes the successful relative permeability for silicon steel nanocomposite with expanding the volume portion, but the expansion of NiZn_Ferrite nanoparticles to silicon Steel/Ferrite or silicon Steel/Ni_Ferrite has expanded the effective relative permeability of silicon steel nanocomposites. Expanding the effect of NiZn_Ferrite nanoparticles expanded the effective relative permeability of silicon Steel/NiZn_Ferrite nanocomposites. Hence, the addition of Ni_Ferrite and 30wt% of NiZn_Ferrite nanoparticles to silicon steel nanocomposite decreased the effective relative permeability of silicon steel nanocomposites. However, Fig. (20) indicates the impact of distinct and multi-nanoparticles magnetic inclusions on the effective relative permeability of silicon steel nanocomposites inside the frequency extent of (10 -50THz). The successful relative permeability of silicon steel nanocomposites has been progressed by expanding frequency through utilizing multi-nanoparticles of (Ni_ferrite or Ferrite) and NiZn_Ferrite nanocomposites. The course of action of multi nanoparticles (NiZn_ferrite, and (Ni_Ferrite or Ferrite) inside the group base matrix may be a critical element for upgrading magnetic properties of

Figure 19. Effective relative permeability of Silicon Steel nanocomposites using individual and multi-nanoparticles

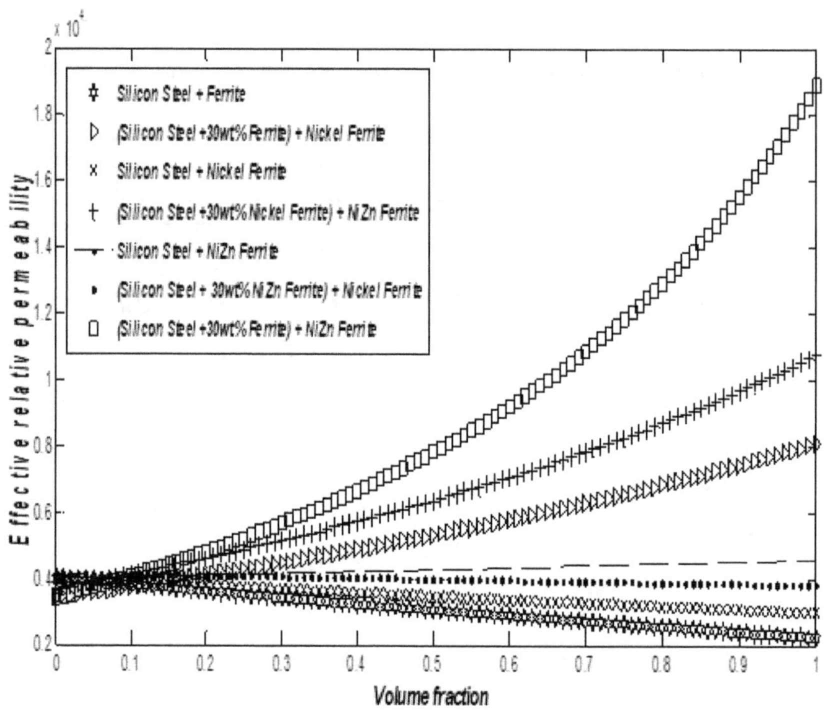

Figure 20. Effective relative permeability of Silicon Steel nanocomposites using individual and multi-nanoparticles

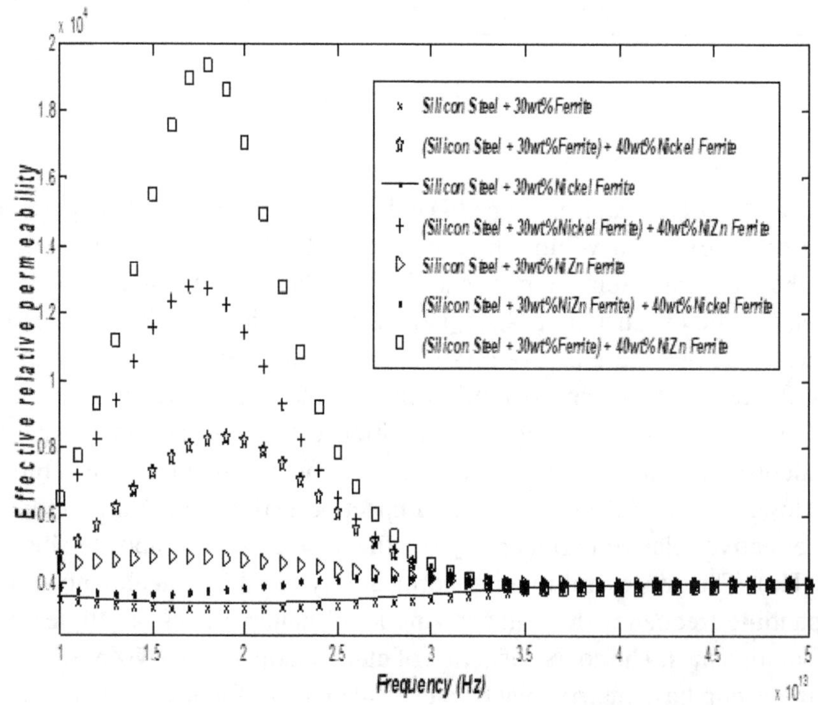

silicon steel materials. The enhancement in the effective relative permeability of silicon Steel/ (Ferrite and NiZn_Ferrite) nanocomposites proved to be the most ideal approach for upgrading the magnetic properties of silicon steel materials regarding the utilization of distinct nanoparticles.

7 FORECAST AND RECOMMENDATIONS

Percolation tunneling element is a critical element to evidence the stream for current through irregular resistor networks, particularly the effective conductivity of polymeric composites. Compressive powers connected to the polymeric composite material examples improve the effective conductivity of polymeric composites materials. The effective conductivity and thermal conductivity of modern polymeric composite materials can be controlled by utilizing different conductive sorts and rates of particles in irregular distributions, relying on the sorts and rates for cost-fewer particles. Al_2O_3 particles are more effective for expanding effective conductivity of the composite materials than Iron, SiO_2, and MgO particles. Foreseen to be an aid for business manufactures, it can show that graphite nanoparticles are additionally effective for diminishing the thermal conductivity of nanocomposites regarding silicon nanoparticles. Nevertheless, MgO nanoparticles are more effective for expanding thermal conductivity of nanocomposites than Al_2O_3 and ZnO nanoparticles. Plastic sorted out nanodielectric composites circulation can be done by fabricating units that help expand effective permeability nanocomposites more than plastic irregular dissemination nanodielectrics in intense point (30:90) over dielectric matrix. Characterization of the effective permeability of Glass nanodielectric composite has been improved by including different rates of magnetic nanoparticles irregular circulations, as stated by the sorts and volume portion of nanoparticles inside the dielectric matrix.

Dielectric Composites can achieve the same results about nanodielectric composites in specified exceptional position conveyance variables (α, and β) and in specified indistinguishable twin flake-like scatters of the composite. Iron, Metglas, and Co-Fe Alloys particles help upgrade the effective permeability of silicon steels magnetic composites. Nevertheless, Alloys particles are in the second request for upgrading the effective permeability of silicon steels magnetic composite materials. However, ferrite particles and silicon steels particles have the property to decrease the effective permeability of magnetic composite streamlined materials. Incorporation radii are the vitality variables for upgrading the permeability of magnetic nanocomposites materials; moreover, the diminish in the consideration radii expand the successful relative permeability of nanocomposites magnetic materials. In addition, the relative permeability of nanoparticles requires higher polarizing properties than the group base matrix to obtain a high change over magnetic characterization. Using multi-nanoparticles has investigated the best new patterns for upgrading the effective relative permeability about utilizing unique nanoparticles; therefore, the new magnetic nanocomposites can make Hosting multi-nanoparticles upgrade the permeability for building matrix magnetic material. The course of action about multi-nanoparticles inside the host base matrix is a critical component for upgrading the effective permeability of nanocomposites; that means, the position of nanoparticles inside the group matrix defines the last dream for upgrading effective relative permeability of base matrix. Upgrading the effective intricate relative permeability for magnetic composite materials obliges including magnetic particles Hosting higher permeability over the base matrix. Therefore, cost-fewer transformer core magnetic composites rely on the utilized magnetic particles sort. Magnetic reluctance of the magnetic composite materials varies regarding the evolving effective complex transformer core relative permeability. Accordingly, it will make evolving magnetic proportional out reluctances. Blending magnetic composite to transformer core magnetic composites results in evolving the effective permeability of magnetic composite regarding the progress of their different electrical properties toward easier and higher frequencies concerning the points out.

REFERENCES

Abeywickrama, N., Podoltsev, A. D., Serdyuk, Y. V., & Gubanski, S. M. (2007, May). Computation of parameters of power transfromer windings for use in frequency response analysis (FRA). *IEEE Transactions on Magnetics*, *43*(5), 1983–1990. doi:10.1109/TMAG.2007.891672

Abeywickrama, N., Serdyuk, Y. V., & Gubanski, S. M. (2006). Exploring possibilities for characterisation of power transfromer insulation by frequency response analysis (FRA). *IEEE Transactions on Power Delivery*, *21*(3), 1375–1382. doi:10.1109/TPWRD.2006.875855

Abeywickrama, N., Serdyuk, Y. V., & Gubanski, S. M. (2008a). High-frequency modeling of power transfromers for use in frequency response analysis (FRA). *IEEE Transactions on Power Delivery*, *23*(4), 2042–2049. doi:10.1109/TPWRD.2008.917896

Abeywickrama, N., Serdyuk, Y. V., & Gubanski, S. M. (2008b, July). Effect of Core Magnetization on Frequency Response Analysis (FRA) of Power Transformers. *IEEE Transactions on Power Delivery*, *23*(3), 1432–1438. doi:10.1109/TPWRD.2007.909032

Anhalt, M., & Weidenfeller, B. (2007). Magnetic properties of polymer bonded soft magnetic particles for various filler fractions. *Journal of Applied Physics*, *101*(2), 023907. doi:10.1063/1.2424395

Anhalt, M., & Weidenfeller, B. (2012, April). Permeability of Soft Magnetic FeCoV-Composites for Varying Filler Fractions. *IEEE Transactions on Magnetics*, *48*(4), 1545–1548.

Balasubramanian, M., & Venkatesh, S. (2013). Partial Discharge and Conduction Current Signature Analysis for the Assessment of Transformer Oil Behavior under Accelerated Stress Condition. *International Journal on Electrical Engineering and Informatics*, *5*(4), 411–421. doi:10.15676/ijeei.2013.5.4.2

Bohren, C. F., & Huffman, D. R. (2004). *Absorption and Scattering of Light by Small Particles*. Wiley.

Chiriac, H., & Lupu, N. (2005, October). New FeNbB-Based Bulk Amorphous and Nanocomposite Soft Magnetic Alloys. *IEEE Transactions on Magnetics*, *41*(10), 3289–3291. doi:10.1109/TMAG.2005.854721

de Assis Olímpio Cabral, F., Luis de Araujo Machado, F., Humberto de Araujo, J., Maria Soares, J., Ricalde Rodrigues, A., & Araujo, A. (2008, November). Preparation and Magnetic Study of the CoFe2O4-CoFe2 Nanocomposite Powders. *IEEE Transactions on Magnetics*, *44*(11), 4235–4238. doi:10.1109/TMAG.2008.2001545

Drnovšek, Bregar, & Pavlin. (2009). Numerical study of effective permeability of soft-magnetic composites with conductive inclusions. *Journal of Applied Physics, 105*.

Florkowski, M., & Furgal, J. (2003). Detection of transformer winding deformations based on the transfer function-measurements and simulations. *Measurement Science & Technology*, *14*(11), 1986–1992. doi:10.1088/0957-0233/14/11/017

Füzerova, Fuzer, Kollar, Hegedüs, & Bure. (2012). Analysis of the complex permeability versus frequency of soft magnetic composites consisting of iron and fe73cu1nb3si16b7. *IEEE Transaction on Magnetics, 48*, 1545-1548.

Fuzerova, J., Fuzer, J., Kollar, P., Hegedus, L., Bures, R., & Faberova, M. (2010, February). Analysis of the Complex Permeability Versus Frequency of Soft Magnetic Composites Consisting of Iron and Fe73Cu1Nb3Si16B7. *IEEE Transactions on Magnetics, 46*(2), 440–442.

Gonzo, E. E., Parentis, M. L., & Gottifredi, J. C. (2006). Estimating models for predicting effective permeability of mixed matrix membranes. *Journal of Membrane Science, 277*(1-2), 46–54. doi:10.1016/j.memsci.2005.10.007

Gouda, O. E., Thabet, A., & El-Tamaly, H. H. (2004). How to Get Low Dielectric Losses in Binary and Multi-Mixture Dielectric at High Frequency (Vol. 4). Academic Press.

Ikeda, K., & Dohmeki, H. (2005). *Study on improvement in the motor property by the difference in the fabrication density of the soft magnetic composite*. IEEE.

Khodasevych, I. E., Shah, C. M., Sriram, S., & Bhaskaran, M. W. (2012). Elastomeric Silicone Substrates for Terahertz Fishnet Metamaterials. Appl. Phys. Lett., 100(6), 061101-1–061101-3.

Koledintseva, M., Chandra, S., Dubroff, R., & Schwartz, R. (2006). Modeling of dielectric mixtures containing conducting inclusions with statistically distributed aspect ratio. *Progr. Electromagn. Res, 66*, 213–228. doi:10.2528/PIER06110903

Kollár, P., Hegedüs, L., Füzer, J., Bureš, R., & Fáberová, M. (2010). R. Bureˇs and M. Fáberová, "Ac magnetic properties of vitroperm based composite materials. *Acta Physica Polonica A, 118*(5), 787–789. doi:10.12693/APhysPolA.118.787

Krings, A., Boglietti, A., Cavagni, A., & Sprague, S. (2016). *Soft magnetic material status and trends in electric machines* (Vol. 10). IEEE Transaction on Industrial Electronic.

Krings, A., Cossale, M., Tenconi, A., Soulard, J., & Cavagni, A. (2015). *Characteristics comparison and selection guide for magnetic materials used in electrical machines* (Vol. 6). IEEE. doi:10.1109/IEMDC.2015.7409206

Lakshminaray, V., & Balakrishnan, P. (2013). Analysis and design of ferrite core transformer for high voltage, high frequency which is used inozonators. Academic Press.

Li, Y., Liu, F., Yang, Q., & Ren, P. (2014, August). Magnetic anisotropic properties measurement and analysis of the soft magnetic composite materials. *IEEE Transactions on Applied Superconductivity, 24*(5), 1–4. doi:10.1109/TASC.2014.2347317

Lin, G. Q., Wu, Y. P., & Li, Z. W. (2006, October). Improvement of the Electromagnetic Properties in Composites With Flake-Like Co2Z Powders by Molten-Salt Synthesis. *IEEE Transactions on Magnetics, 42*(10), 3326–3328. doi:10.1109/TMAG.2006.879626

Liu, C., Zhu, J., Wang, Y., Guo, Y., & Lei, G. (2015, June). Comparison of claw pole machines with different rotor structures. *IEEE Transactions on Magnetics, 51*(11), 1–4. doi:10.1109/TMAG.2015.2443022

Liu, T., Zhou, P. H., Deng, L. J., & Tang, W. (2009). Spatial orientation and position distribution effect on the effective permeability of composites consisting of aligned flakes. *Journal of Applied Physics*, *106*(11), 114904. doi:10.1063/1.3264641

Marina, Koledintseva, & Yelsukov. (2012). *Frequency-Dependent Effective Material Parameters of Composites as a Function of Inclusion Shape*. inTech.

Mazaleyrat, F., & Varg, L. (2000). Ferromagnetic nanocomposites. *Journal of Magnetism and Magnetic Materials*, *7*, 253–259. doi:10.1016/S0304-8853(00)00128-1

Moses, A. (2012). Energy efficient electrical steels: Magnetic performance prediction and optimization. *Scripta Materialia*, *67*(6), 560–565. doi:10.1016/j.scriptamat.2012.02.027

Mousavi, S. E., & Agheb, G. (2011). Investigation of effects on core losses in single phase power transformers. *Arch. Electr. Eng*, *60*(1), 35–47. doi:10.2478/v10171-011-0004-9

Oltean, I. D., & Motoc, D. L. (2012). Factors influencing the electrical conductivity of composites with iron particles. *IEEE International Conference on Optimization of Electrical and Electronic Equipment (OPTIM),* 124-129.

Paulis, F., Nisanci, M., Orlandi, A., Koledintseva, Y., & Drewniak, L. (2014). Design of homogeneous and composite materials from shielding effectiveness speciðcations. *IEEE Transactions on Electromagnetic Compatibility*, *56*(2), 343–351. doi:10.1109/TEMC.2013.2280463

Périgo, E., Nakahara, S., Pittini, Y., Hazan, Y., & Graule, T. (2011). Magnetic properties of soft magnetic composites prepared with crystalline and amorphous powders. *J. Magn. Mater.*, *323*(15), 1938–1944. doi:10.1016/j.jmmm.2011.02.025

Podoltsev, A. D. (2007, December). Multiscale Computations of Parameters of Power Transformer Windings at High Frequencies. Part II: Small-Scale Level. *IEEE Transactions on Magnetics*, *43*(12), 4076–4082. doi:10.1109/TMAG.2007.908371

Podoltsev, A. D., Abeywickrama, K. G. N. B., Serdyuk, Y. V., & Gubanski, S. M. (2007, November). Multiscale Computations of Parameters of Power Transformer Windings at High Frequencies. Part I: Small-Scale Level. *IEEE Transactions on Magnetics*, *43*(11), 3991–3998. doi:10.1109/TMAG.2007.906220

Rahimpour, E., Christian, J., Feser, K., & Mohseni, H. (2003). Transfer function method to diagnose axial displacement and radial deformation of transformer windings. *IEEE Transactions on Power Delivery*, *18*(2), 493–505. doi:10.1109/TPWRD.2003.809692

Ryder, S. A. (2003). Diagnosing transformer faults using frequency response analysis. *IEEE Electrical Insulation Magazine*, *19*(2), 16–22. doi:10.1109/MEI.2003.1192032

Salvadori, Teixeira, Cattani, Nikolaev, Savkin, Oks, Park, Phillips, Yu, & Brown. (2012). On the electrical conductivity of Ti-implanted alumina. *Journal of Applied Physics, 111*(6).

Shintemirov, A., Tang, W. H., & Wu, Q. H. (2010, January). Transformer Core Parameter Identification Using Frequency Response Analysis. *IEEE Transactions on Magnetics*, *46*(1), 141–149. doi:10.1109/TMAG.2009.2026423

Sihvola, A. (1999). Electromagnetic mixing formulas and applications. *IEE Electromagnetic Waves Series, 47*, 63–68.

Sugimoto, S., Haga, K., Kagotani, T., & Inomata, K. (2005, October). Magnetic Properties of Co/(CoNi) Fe2O4 Nanocomposite Magnet Powders. *IEEE Transactions on Magnetics, VOL., 41*(10), 3971–3873.

Szabó, Z., & Füzi, J. (2014, April). Equivalence of Magnetic Metamaterials and Composites in the View of Effective Medium Theories. *IEEE Transactions on Magnetics, 50*(4), 1–4. doi:10.1109/TMAG.2013.2288297

Szabó, Z., Park, G.-H., Hedge, R., & Li, E.-P. (2010, October). A unique Extraction of Met material Parameters Based on Kramers–Kronig Relationship. *IEEE Transactions on Microwave Theory and Techniques, 58*(10), 2646–2653. doi:10.1109/TMTT.2010.2065310

Tellini, B., & Bologna, M. (2010, December). Magnetic Composite Materials and Arbitrary B-H Relationships. *IEEE Transactions on Magnetics, 46*(12), 3967–3972. doi:10.1109/TMAG.2010.2079332

Thabet. (2020). *Design and Investment of High Voltage NanoDielectrics*. IGI Global. . doi:10.4018/978-1-7998-3829-6

Thabet & Repetto. (2013). Predicting Effective Permeability of Nanodielectric Composites Bonded by Soft Magnetic Nanoparticles. *International Journal of Chemical, Materials Science and Engineering, 7*(11), 354-359.

Tuncer. (2002). How Round is Round on Accuracy in Complex Dielectric Permittivity Calculations a Finite Size Scaling Approach. *Tr. J. of Physics, 15.*

Tuncer, Serdyuk, & Gubanski. (2001). Dielectric Mixtures-Electrical Properties and Modeling. *IEEE Trans. El. Insul. And Dielec., 2*(40).

Waki, H., Igarashi, H., & Honma, T. (2005, May). Estimation of Effective Permeability of Magnetic Composite Materials. *IEEE Transactions on Magnetics, VOL., 41*(5), 1520–1523. doi:10.1109/TMAG.2005.845071

Weijie, X., Duan, N., Wang, S., Guo, Y., & Zhu, J. (2016). *Modeling and measurement of magnetic hysteresis of soft magnetic composite materials under different magnetizations* (Vol. 10). IEEE Transaction on Industrial Electronics.

Wu, L. Z., Ding, J., Jiang, H. B., Neo, C. P., Chen, L. F., & Ong, C. K. (2006). a_ H. B. Jiang, and C. P. Neo, L. F. Chen, and C. K. Ong, "High frequency complex permeability of iron particles in a nonmagnetic matrix. *Journal of Applied Physics, 99*(8), 083905. doi:10.1063/1.2190719

Yao, C., Shimada, Y., Qin, G. W., Pei, W. L., Okamoto, S., Kitakami, O., Endo, Y., & Yamaguchi, M. (2011, October). Study of Permeability for Composites Including Fe, NiZn Ferrite and Fe-B-P Particles. *IEEE Transactions on Magnetics, 47*(10), 3160–3162. doi:10.1109/TMAG.2011.2148698

Zhang, Ohnuma, & Masumoto. (2011). Soft Magnetic Co-(TiN) Composite Films Realized within a Wide-Range of Cobalt Content. *IEEE Transactions on Magnetics, 47*(10), 3795-3798.

Zhang, Y. D., Wang, S. H., Xiao, D. T., Budnick, J. I., & Hines, W. A. (2001, July). Nanocomposite Co/SiO2 Soft Magnetic Materials. *IEEE Transactions on Magnetics*, *37*(4), 2275–2277. doi:10.1109/20.951146

Zhu, J., Guo, Y., Lin, Z., & Huang, Y. (2011). Development of pm transverse flux motors with soft magnetic composite cores. *IEEE Transactions on Magnetics*, *47*(10), 4376–4383. doi:10.1109/TMAG.2011.2157320

Chapter 8
Nano-Tech Transformer Cores

ABSTRACT

This chapter describes optimal characterization of composites and nanocomposites materials that handle traditional transformer cores to design transformer core materials. It attempts to offer new designs for new transformer cores to control the magnetization parameters. This chapter draws attention to theories and effective nanoparticle structure that tackle the characteristics of transformer cores. It further sheds light on the effects of nanoparticles on magnetization loss of transformer core by using individual and multiple magnetic nanocomposites. The forecasting and recommendations of the magnetic characterization are also presented for 1-phase and 3-phase transformer cores.

1 NANO-TECH VISION

The structure of magnetic materials is a fundamental parameter for specifying the magnetic characterization of the transformer core. This chapter displays upgrading magnetic characterization of transformer cores toward utilizing new nanotechnology techniques. The effective magnetic parameters of new magnetic nanocomposites materials for the transformer cores (single phase and three-phase) have been predicted, taking into account late hypothetical methodologies. In recent designs, the impacts of variant sorts and concentrations of magnetic multi-nanoparticles on charge loss of transformers cores have been mulled over regarding universal transformer cores. Ideal sorts and concentrations of nanoparticles have been characterized to controlling the reluctance and magnetization losses of transformer cores utilizing the multi-nanoparticles technique. A similar investigation has portrayed the industrial features for utilizing multi-nanoparticles against differentiate nanoparticles for transformers cores. The fundamental development of a transformer is moderately straightforward and delicately known; however, an incredible level of intricacy is included in the close subtle elements of operation, especially transformers for lower frequency rating that have magnetic core material. The core configuration permits tight coupling of the middle of elementary and secondary windings for voltages prompted in the windings, as stated by Faraday's law. Although an appropriately stacked transformer might show high efficiency, few sorts of loss can be manifested, with two specifically figuring the vast majority prominently: (a) winding loss because of the resistance in the coils, and (b) core loss because of eddy current and hysteresis. The last sort of loss has

DOI: 10.4018/978-1-7998-8536-8.ch008

verifiably been alleviated toward utilizing dainty electrical steel laminations as opposed to a solid core, for a metallurgical cosmetic of the electrical steel hosting a significantly expanded electrical resistance. Such laminations serve to decrease eddy current loss significantly (Martin, 2008). Hence, in order to decline the eddy current loss and total core loss, multi-layer silicon steel lamination for thinner sheets is broadly applied in electromagnetic equipment, such as power transformers and electrical machines. Therefore, examining magnetic properties of the silicon steels is signiðcant in order to decrease the core loss for electrical apparatus (Buswell, 2001; Li et al., 2016; Sievert, 2000).

Many researchers have investigated the demonstrating of core loss under the non-sinusoidal waveforms by utilizing Fourier series. The Fourier series aided in a qualitative investigation of the core loss under non- sinusoidal excitations, in any case it is not quantitatively exact (Albach & Brockmeyer, 1996). An alternate approach of demonstrating is by relating the core loss of the rate of progress for flux (dB/dt). As stated by this principle, the summed-up core loss model for non- sinusoidal excitation can be named altered Steinmetz equation (MSE) (Reinert et al., 1999). At present, the conventional core loss model is a popular technique in machine outline programming instruments. The conventional model solves Maxwell's equations for the static B–H bend. After acquiring the numerical result by utilizing the static curve, core loss will be ascertained in view of the core loss equation utilizing coefficients that are concentrated starting with trials (Ionel et al., 2006; Zhang et al., 2011). The existing macroscopic core loss calculation strategies could be sorted under two main groups, loss separation and empirical technique. In the ðrst method, the magnetic loss is subdivided under three primary components: Hysteresis loss, traditional eddy current loss and excessive loss. The second bunch that is engaging for designers are the experimental techniques which initiated in 1892 by Steinmetz presenting a comparison to determine the magnetic loss, which can have just been subject to the top esteem of the magnetic induction (Albach & Brockmeyer, 1996). Afterwards, a more all mathematical statement may have been utilized, which likewise acknowledges the frequency reliance of loss. It is known as original Steinmetz equation (OSE), which is basically about the bend ðtting of a few estimations under sinusoidal excitations during different frequencies and peak magnetic inductions (Reinert et al., 1999). The current ideas of the physical beginning and system of loss in magnetic materials can be reviewed under three universal categories: Hysteresis losses, eddy current or dielectric losses and the remaining losses. Core (iron) losses are basically identified by polarizing or stimulating the core and display the whole chance if the transformer is powered on, in any case if there's any load or not (Agheb & Høidalen, 2013; Ionel et al., 2006; Zhang et al., 2011).

2 TRADITIONAL TRANSFORMER CORE

A transformer comprises a coil and an iron core. Generally, there are two coils; in particular, the elementary core and the auxiliary curl. The two coils are not related physically but rather associated by a magnetic field. Transformers for use during power or sound frequencies commonly have cores constructed of high permeability silicon steel. The steel has a permeability much more than that for allowed space, and the core consequently serves to significantly lessen the polarizing present and keep the flux with a way which nearly couples the windings. Early transformer developers quickly acknowledged that cores constructed from solid iron brought restrictive eddy current losses, and their plans alleviated this impact for cores comprising packs of insulated iron wires. Recent designs have constructed the core by stacking layers of slim steel laminations, a standard that remained in use. Every lamination is insulated starting with its neighbors by a dainty non-conducting layer for insulation. One basic outline of laminated core is committed starting with interleaved stacks for E-shaped steel sheets topped with I-shaped pieces, prompting its name of 'E-I transformer's. Such a configuration has a tendency to show additional losses but is extremely prudent to fabricate. The cut-core or C-core sort is constructed toward winding a steel strip around a rectangular structure holding the layers together. It is that point cut in two, shaping two c shapes, and the core amassed by tying the two c halves together for a steel strap. They have the preference for the flux being continuously situated parallel to the metal grains, thus reducing reluctance.

The core of transformers is constructed 'core form' or 'shell form', as shown in Fig. (1). When windings encompass the core, the transformer is core form. At the point when windings are encompassed by the core, the transformer is shell manifestation. Shell structure outline is more common than the core type plan for dissemination transformer requisitions because of the relative simplicity for stacking the core around winding core. The core type outline has a tendency to, as a general rule, be additionally economical, and in this manner more prevalent than shell manifestation outline for high voltage power transformer provisions toward the more level conclusion of their voltage and control rating ranges (less than or equivalent to, nominally, 230 kV or 75 MVA). At higher voltage and power ratings, shell manifestation transformers have a tendency to be more pervasive. The shell type plan has a tendency to be favored for extra-high voltage and higher MVA provisions because, if that's only the tip of the iceberg work required for manufacture, shell structure transformers will be portrayed as hosting naturally finer kVA-to-weight ratio, preferred short circuit strength characteristics and higher resistance to travel damage (Agheb & Høidalen, 2013; Ionel et al., 2006).

Figure 1. Types of transformer core (Kulkarni & Khaparde, 2004)

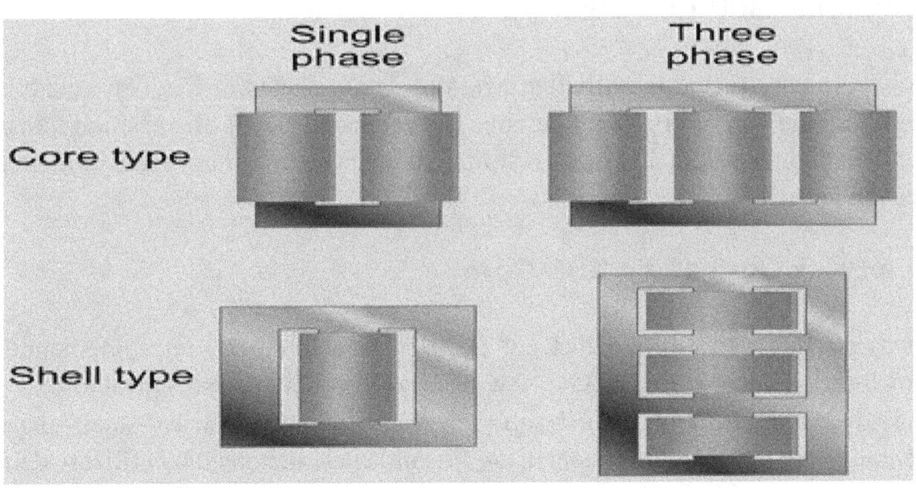

Saturation is a condition for which the magnets are fully transformer cores and produce greatest magnetic flux. This, as a rule, happens though the transformer is not enormously sufficient for the application, while a percentage of the parts influence the flexibility and make distortions that could mischief the core. Core immersion camwood additionally increase the temperature and charge the current transformers. In saturation, some of the transformer core and present sounds era sources stream straightforwardly to the essential and secondary windings. Some of the past investigations indicate that the symphonious substance builds together to the point when the current no-load and high magnetic flux density injects. Today a lot of individual transformers are broadly exasperating toward sounds. Symphonious sections have brought on losses in the transformer which permits fizzles to work. The effectiveness of transformer cores for a considerable length of time is a mix of two principal factors: Handling of predominant magnetic materials and advanced transformer core geometries. From the solid, non-silicon iron cores of the late 1880's of the routine butt- lap and mitered core outlines of late times, the no-load losses of transformer cores have been decreased by nearly 70%. This reduction has been attained because of the preferred comprehension of the flux conduct technique inside the core, and the better use of the properties of present-day electrical steels. The reduction of core losses in transformer cores has been a real region for looking into many large fractions across time (Dermawan et al., 2016; Mclyman & Colonel, 2004).

3 TRANSFORMER CORE DESIGN

A magnetic core is a piece of attractive material with a high attractive porousness used to limit and guide attractive fields in transformers attractive gatherings. It is made of ferromagnetic metal like iron, or ferrimagnetic mixtures like ferrites. The high porousness, comparative with the encompassing air, causes the attractive field lines to be moved in the center material. The attractive field is frequently made by a current-hefting loop of wire around the center. There are various types of transformer core design as follows: (Collett & Buswell, 2011).

3.1 A Conventional Butt-Lap Core

Customary cores are constructed by utilizing layers of dainty sheets of magnetic material with each layer lamination lest you quit offering on that one. The extensive extent of a fraction transformer cores prompts moderately long development times resulting in noteworthy commitment of the manufacturing fetches of such devices.

3.2 Design a Brick Core and a Sub-Core

Configuration transformer cores utilize stacks of electrical steel to decrease development times without resulting in any increment in the core losses. New plan exhibits have down control loss over the routine core by taking advantage of the high anisotropy for grain-oriented electrical steels and reducing the flux along the transverse heading of the steel. Core loss has been alleviated by utilizing dainty electrical steel laminations as opposed to a robust core for metallurgical cosmetics of the electrical steel hosting of a significantly expanded electrical resistance. Such laminations serve to decrease eddy current losses considerably. Recently, consideration has been provided for the advancement of wire core engineering to further decrease core losses and to move forward the effectiveness of the manufacturing framework. A standout amongst the magic characteristics of wire core engineering organization is the adaptability for which a transformer core might be arranged. It is accepted that exceedingly unpredictable core shapes camwood is the earlier constructed shapes which can decrease the amount of core "hot spots".

Figure 2. A concept of a power transformer employing wire core material, present in the eight core sectors positioned in the transformer windings (Jieli et al., 2001)

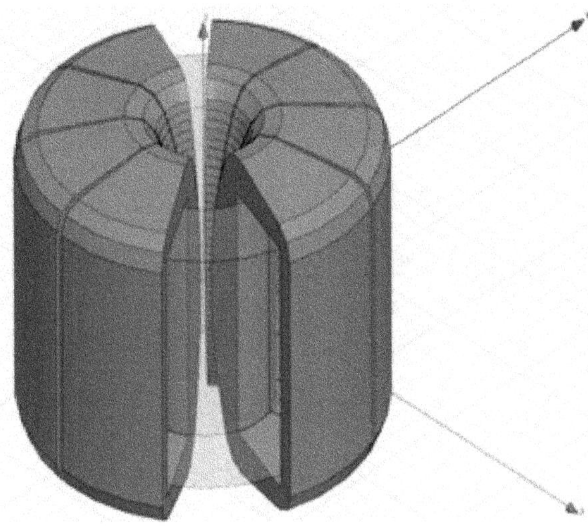

As indicated in Fig. (2), the core is really broken dependent upon eight parts, and the high- and low-voltage windings are concentric around the z-axis. A configuration of this kind is generally shielded, possibly upgrading its materialness particularly in electrically loud situations (Jieli et al., 2001).

4 TRANSFORMER CORE COMPOSITES

An amount of new transformer topologies, novel core joint types, and moved forward electrical steels have been formed so as to enhance the transformer effectiveness (Mu & Lee, 2014). Composite magnetic cores have been recommended for the reduction of the manufacturing and operational expense of conveyance transformers without the change of the transformer topology or magnetic materials (Kefalas & Kladas, 2012). A composite wound core is a magnetic core constructed of a mix of customary grain-oriented and high charge grain-oriented electrical steel laminations. Its operating standard is the flux thickness non-uniformity for wound cores. A three-phase five-legged wound transformer core is constructed of two high permeability Si-Fe wound cores and two routine Si-Fe wound cores. The three-phase magnetic core is amassed on two expansive inward wound cores made of high permeability grain-oriented electrical steel and two littler external wound cores made of accepted grain-oriented electrical steel. This arrangement is in light of late test series of the flux thickness conveyance of the three-phase five-legged wound core transformer. As stated in ref. (Chang et al., 2010), the top flux thickness of the two external cores is more level than the two inward cores as shown in Fig. 3. Therefore, low cost, customary grain-oriented electrical steel can be utilized for the external cores.

Figure 3. Simplified scheme of wound core and its geometrical parameters (Chang et al., 2010)

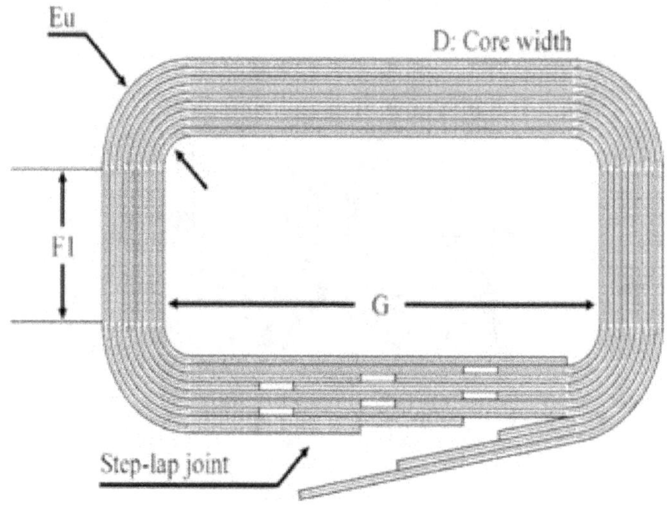

5 NANOCOMPOSITES TRANSFORMER CORES

Few previous and late investigations have examined the loss thickness and permeability of nanocrystalline cores and other magnetic cores, such as Ferrites, amorphophallus titanium metal alloys, and so forth (Shin et al., 2010). Fe-based Nanocrystalline (FE-N), which has properties like high immersion flux, low or high permanence, high permeability, and low loss density, is frequently decided as the material of the transformer core, magnetic switch, and sifting inductor, and so forth. FE-N has particular provisions for its distinctive permanence. For example, FE-N with high permanence is thought to be handy as the material of a magnetic switch in magnetic pulse layering. Regularly, the coercive drive of general magnetic material increments as the Br/Bs builds (Br is the permanence, and Bs is the immersed induction). However, the coercive drive of FE-N declines as the Br/Bs increases, and it also has some impact on the loss densities of FE-N for diverse permanence. It is adaptable to get high or low permanence FE-N toward longitudinal magnetic processing, or transverse magnetic processing, separately. Thus, it is vital to concentrate on the pulse magnetic properties of FE-N for diverse permanence and to decide on the suitableness FE-N cores, as stated by the particular provision. Currently, the pulse magnetic properties of FE-N cores under high charge rate are measured in the same or comparable working states. The strategies are isolated under two classifications. One is in view of the magnetic switches, and the pulse magnetic properties will be obtained under the cosmos of the senior pulse excitation (Hernández et al., 2010). The opposite is in view of pulse magnetic parts with consistent charge rate, and the pulse magnetic properties are measured under the square wave pulse excitation (Liu et al., 2013). The pulse magnetic properties break down starting with the B-H curves, and the measuring standards of the strategies are comparable. However, the courses of action of the tried windings around the magnetic core are not the same. Double-winding test circuit is traditional and generally utilized at the same time as testing the hysteresis loops of magnetic cores (Muhlethaler et al., 2012). However, in some cases, including the examination of magnetic properties and the outline of magnetic switches, single-winding test circuit is additionally applied to test the B-H bend and loss characterization. The decision of the test circuit will finally impact the trying effects.

In the design-oriented demonstrating and reproduction for control hardware and pulsed power systems, exact and useful models of the magnetic sections are of noteworthy viability. The Recreation models of magnetic components, including reluctance models, permeable alloy models, inductance matrix models, T-circuit models, and gyrator capacitor models have been investigated. A progressive magnetic out model including iron immersion and losses is recommended (Radun, 2014). Dynamic high-fidelity magnetic equivalent circuits are exact and material science-built models of magnetic components; however shockingly, the reduced-order ones are of high intricacy (Davoudi et al., 2009). DC Bias Restraining Structure, at the same time the HVDC run for monopolar ground circuit operation mode present in the earth, can achieve different possibility dc bias racial inclination of the transformer. When there is dc streaming under the transformer winding, the dc magnetic flux m and the ac magnetic flux superimpose which makes person A large fraction cycle magnetic flux density builds significantly and the other half cycle magnetic flux density diminishes. Measures ought to be made to smother the transformer dc bias racial inclination. A sort of magnet recompense system has been made for the dc bias racial inclination in view of the nanocomposite material core. The nanocomposite core may have been charged to neutralize the dc inclination in the transformer core at the dc bias racial inclination occurring. The magnetic material can be demagnetized at the dc bias racial inclination disappearing, and the transformer is resumed of the ordinary fill in. A sort of magnet payment system can be taken to the dc bias racial inclination dependent upon a nanocomposite material core. The nanocomposite core can be charged to neutralize the dc inclination in the transformer core at the point when the DC bias racial inclination happens. The magnetic material can be demagnetized at the point when the dc inclination disappears, and the transformer is resumed of the typical fill in (Chen et al., 2016).

5.1 Theory and Effective Parameters

The demonstrating of a transformer concentrates with respect to finding circulations for all the electromagnetic fields components on an obliged frequency extend upon knowing its geometry (structure and dimensions). Moreover, the magnetic electric properties of the materials utilized for windings, magnetic core and insulation and parameters of its working mode (open/short out test, and also through this way, observing and stock arrangement of all instrumentation) is enhanced. The effective parameters of the windings can be further confirmed starting with the obtained result of the field issue. A transformer's core is regularly constructed of grain turned silicon steel laminations. Eddy current losses are minimized toward expanding resistivity through s for the silicon of the steel and additionally by decreasing the lamination thickness. Average lamination thicknesses are utilized within control transformer assembling extent starting with 0. 23 to 0. 35 mm. The fabrication of up-to-date transformer's core out of thin, naturally resistive laminations for anisotropic properties has a magnetic circuit with high permeability at mains frequency. A standout amongst the vast majority essential sections in the control transformer is the transformer core, which serves similarly as low reluctance for the magnetic flux linking grade and auxiliary windings to create the pro-fractional magnetic circuit of the core (Kulkarni & Khaparde, 2004; Sweetser, 2002). In order to decrease the reluctance of transformer core, the point of this single section is to consider the impact of nanocomposites core material on the transformer core reduction. Consider a cross area of an ordinary three-limb wound transformer core, as demonstrated in Fig (4.a) where the period of winding is eager for frequency reaction estimations. As shown in Fig. 4, the magnetic core of the transformer can be partitioned under three sections regarding uniform fluxes Φ_A, Φ_B, Φ_C and of each of the transformer's period (Shintemirov et al., 2010).

Figure 4. Three-limb transformer core (Shintemirov et al., 2010)

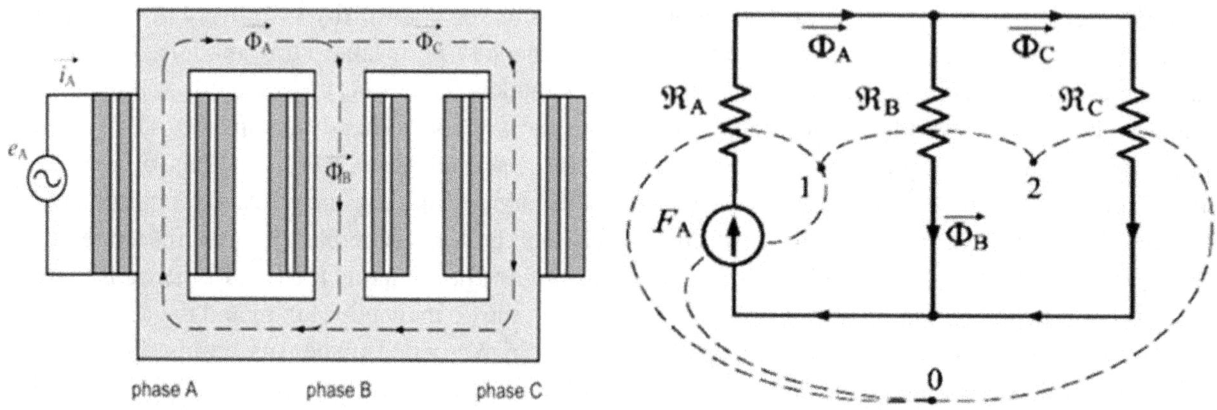

a-Cross section of a three-limb transformer b- Equivalent magnetic circuit of a three-limb transformer

In order to establish the pro-fractional magnetic out of the core, each area of the magnetic core is spoken to toward its reluctance $\mathscr{R}$ which gives a connection to the middle of the comparing ñux Φ and the magneto motive power f is obliged to secure that ñux along the length of the area. The reluctance of a section is dead set toward the magnetic parameters of lamination and the core geometry (Shintemirov et al., 2010). The complex effective relative permeability $\hat{\mu}_{clef}$ is deðned by using the following expression (Abeywickrama et al., 2007; Abeywickrama et al., 2008):

$$\hat{\mu}_{clef} = \mu'_{eff} - j\mu''_{eff} = \mathcal{K}_{fe}\mu_r \frac{\tanh\left(\left(1+j\right)\ddagger/_{2\delta}\right)}{\left(\left(1+j\right)\ddagger/_{2\delta}\right)} \tag{1}$$

Where, $\mathcal{K}_{fe}$ is stacking factor which is defined by the following expression:

$$\mathcal{K}_{fe} = \ddagger/_{h} \tag{1.i}$$

h is the fraction of steel in the laminated core
$\ddagger$ is the thicknesses of a lamination sheet of the core
μr is the local relative magnetic permeability

The skin depth δ depends on the angular frequency *É* of the magnetic ðeld, and is deðned as follows:

$$\delta = \sqrt{2/\left(\bar{\omega}\sigma\mu_r\mu_0\right)} \tag{1.ii}$$

Where, σ is the electrical conductivity of the core lamination.

The equivalent magnetic circuit of the core is presented in Fig (4.b), where $\mathcal{R}_A$ & $\mathcal{R}_B$ and $\mathcal{R}_C$ represent the reluctances of the three paths in the magnetic core carrying magnetic fluxes Φ_A and Φ_B and Φ_C respectively. Magnetic characteristics of the transformer core using individual nanoparticles has been described by the reluctance $\mathcal{R}_i$ which depends on the complex effective relative permeability $\hat{\mu}_{clefi}$ of core lamination with individual nanoparticles embedded in transformer core material and deðned by using the following expression:

$$\mathcal{R}_i = \frac{l}{A\mu_0\hat{\mu}_{clefi}} \tag{2}$$

Where, A is the cross section area of the core

$$\hat{\mu}_{clefi} = \mu'_{clefi} - j\mu''_{clefi} = \mathcal{K}_{fe}\mu_{reffi}\frac{\tanh\left((1+j)\ {}^{\ddagger}\!/_{2\delta_{effi}}\right)}{\left((1+j)\ {}^{\ddagger}\!/_{2\delta_{effi}}\right)} \tag{2.i}$$

The effective relative magnetic permeability μ_{reffi} denotes magnetic nanocomposites materials (Bohren & Huffman, 2004; Pleite et al., 2006; Thabet et al., 2016; Thabet et al., 2018). However, the effective skin depth by using individual nanoparticles $\delta ef_{fi\ d}$epends on the angular frequency ω of the magnetic ðeld and is deðned as follows:

$$\delta_{effi} = \sqrt{{}^{2}\!/\!\left(\hat{\omega}_{effi}\mu_{reffi}\mu_0\right)} \tag{2.ii}$$

The effective electrical conductivity of the magnetic nanocomposites σe_{ffi} is defined as follows:

$$\sigma_{effi} = (1-\zeta)\sigma_h + \zeta\sigma_i. \tag{3}$$

Where,

σ_h is the electrical conductivity of base matrix

σ_i is the electrical conductivity of the first type of inclusion material

Similarly, in case of magnetic characteristics of transformer core using multi-nanoparticles, the magnetic characteristics of core transformer by using multi-nanoparticles has been described by the magnetic reluctance $\mathcal{R}_j$ which depends on the complex effective relative permeability $\hat{\mu}_{effj}$ of core lamination with multi nanopatics embedded in transformer core material.

5.2 Nanoparticles Structure in Transformer Core

In practice, the parameters of core lamination, such as neighborhood relative magnetic permeability μ_r and the electrical conductivity of the core lamination σ are controlled toward utilizing Table (1) which indicates the principal electrical and magnetic parameters of Single-phase transformer and Three-phase transformer. Moreover, the magnetic nanoparticles (Dia.: 10 nm) are utilized for upgrading the magnetic properties of materials.

Table 1. Specifications of Transformer Models (Easwarlal & Palanisamy, 2013)

Transformers Specifications	1-φ transformer	3-φ Transformer
KVA	5	500
Voltage rating	11000/400 V	11000/433 V
Stacking factor	0.9	0.9
Length of flux path	1.028 m	4.816 m
Length of right section flux path	-	2.0315 m
Length of left section flux path	-	2.0315 m
Length of central section flux path	-	0.753 m
Cross section area of core transformer	0.00687 m^2	0.0378 m^2
Thickness of a single lamination sheet of the core(ŧ)	0.35mm	0.35mm
Power frequency	50Hz	50Hz
Number of primary turns	6560	1100(per phase)
Maximum magnetic flux density	1.1(Wb/m^2)	1.2(Wb/m^2)

5.3 Effect of Nanoparticles on the Characteristics of Transformer Core

Recently, the effective relative permeability of ferrite nanocomposites of transformer core for different volume fractions of multi-nanoparticles at power frequency 50 Hz have shown a slight expansion in the effective relative permeability of ferrite nanocomposites for transformer core lamination that has been struck by expanding volume fraction for nickel ferrite. However, high increment in the effective relative permeability of ferrite + 30wt. % nickel ferrite nanocomposites transformer core lamination has been struck by including different rates of NiZn_Ferrite or silicon steel nanoparticles. On the other side, the vitality seems a course of action for nanoparticles involved in the computation of successful relative permeability for new nanocomposites by utilizing multi-nanoparticles. The effective relative permeability of ferrite +30wt. %NiZn_Ferrite nanocomposites to transformer core lamination has been expanded by including different rates of silicon steel nanoparticles. It has been manifested that the expansion in the effective relative permeability of ferrite nanocomposites to transformer core lamination that occurs to expand the volume fraction by utilizing multi-nanoparticles is higher than the upgrade in the successful relative permeability of ferrite nanocomposites for transformer core lamination and aid the expanding volume fraction by utilizing a solitary kind of nanoparticles. Figure (5) illustrates the effective relative permeability of nickel ferrite nanocomposites of transformer core for different volume fractions of multi-nanoparticles at power frequency 50 Hz. It has become self-evident that the effective relative permeability of nickel ferrite nanocomposites transformer core has been reduced by expanding the volume fraction of ferrite nanoparticles. However, including NiZn_Ferrite or silicon steel as a second sort of nanoparticles into nickel ferrite + 30wt. % ferrite nanocomposite transformer core lamination increase the effective relative permeability of nickel ferrite nanocomposites transformer core for expanding the volume fraction. Furthermore, the sort of nanoparticles magnetic materials that have been utilized for expanding the viable relative permeability of nanocomposites transformer core is a significant element for upgrading the successful relative permeability of nanocomposites transformer core by using multi-sort nanoparticles. The upgrade in the effective relative permeability of nickel ferrite nanocomposites for transformer core has occurred toward utilizing multi- nanoparticles for expanding the volume fraction.

Additionally, the viable relative permeability of NiZn_Ferrite nanocomposites to transformer core has been reduced by expanding the concentration of ferrite nanoparticles. Moreover, silicon steel or nickel ferrite are used as second sort nanoparticles under NiZn_Ferrite + 30wt. % ferrite nanocomposites transformer core lamination has expanded the effective relative permeability of NiZn_Ferrite nanocomposites of the transformer core for expanding the volume fraction of the second sort nanoparticles. Including silicon steel with NiZn_Ferrite + 30 wt. % nickel ferrite nanocomposites for transformer core has improved the expansion in the successful relative permeability of NiZn_Ferrite nanocomposites for transformer core. However, including nickel ferrite under NiZn_Ferrite +30wt. % silicon steel nanocomposites to transformer core of the second sort nanoparticles has reduced the effective relative permeability of NiZn_Ferrite nanocomposites of the transformer core for expanding the volume fraction. Figure (6) indicates the effective relative permeability of silicon steel nanocomposites to transformer core with different volume fractions of multi-nanoparticles during the power frequency 50Hz. The viable relative permeability of silicon steel nanocomposites has been expanded by expanding the concentration of NiZn_Ferrite, ferrite or nickel ferrite nanoparticles in the silicon steel transformer core lamination. The upgrade in the effective relative permeability of silicon Steel nanocomposites has occurred by utilizing multi-nanoparticles. Multi-nanoparticles utilizing silicon steel + 30wt. % ferrite + NiZn_Ferrite as multi-nanoparticles are the best inclusions utilized for upgrading the effective permeability of silicon steel nanocomposites to transformer core lamination. Therefore, Silicon steel + 30wt. % nickel ferrite + NiZn_Ferrite multi-nanoparticles are in the second request for upgrading the successful relative permeability of silicon Steel nanocomposites transformer core lamination. Utilizing singular ferrite nanoparticles has been the slightest incorporation for upgrading the effective relative permeability of silicon Steel nanocomposites transformer core lamination.

The reluctance of ferrite nanocomposites to transformer core has been reduced by expanding the rate of NiZn_Ferrite, nickel ferrite or silicon steel nanoparticles in the ferrite transformer core. Including silicon steel or NiZn_Ferrite nanoparticles with ferrite + 30wt. % nickel ferrite has improved the diminish in the reluctance of ferrite nanocomposites for single phase transformer core with expanding the rate of silicon steel nanoparticles. Utilizing multi-nanoparticles can enhance the magnetic characteristics of ferrite nanocomposites for transformer core by upgrading the decline toward upgrading the diminish in the reluctance of ferrite nanocomposites for single phase transformer core to become higher than the diminish that has struck them by utilizing individual nanoparticles. As stated by evolving the host magnetic materials, Fig. (7) indicates the reluctance of nickel ferrite nanocomposites to single phase transformer core with different volume fractions during power frequency 50 Hz. The reluctance of nickel ferrite nanocomposites for single phase transformer core has been expanded by expanding the fixation of ferrite nanoparticles. However, utilizing nickel ferrite + 30wt. % ferrite + NiZn_Fer-

Figure 5. Effective relative permeability of Nickel Ferrite nanocomposites for transformer core lamination with various volume fractions of multi-nanoparticles

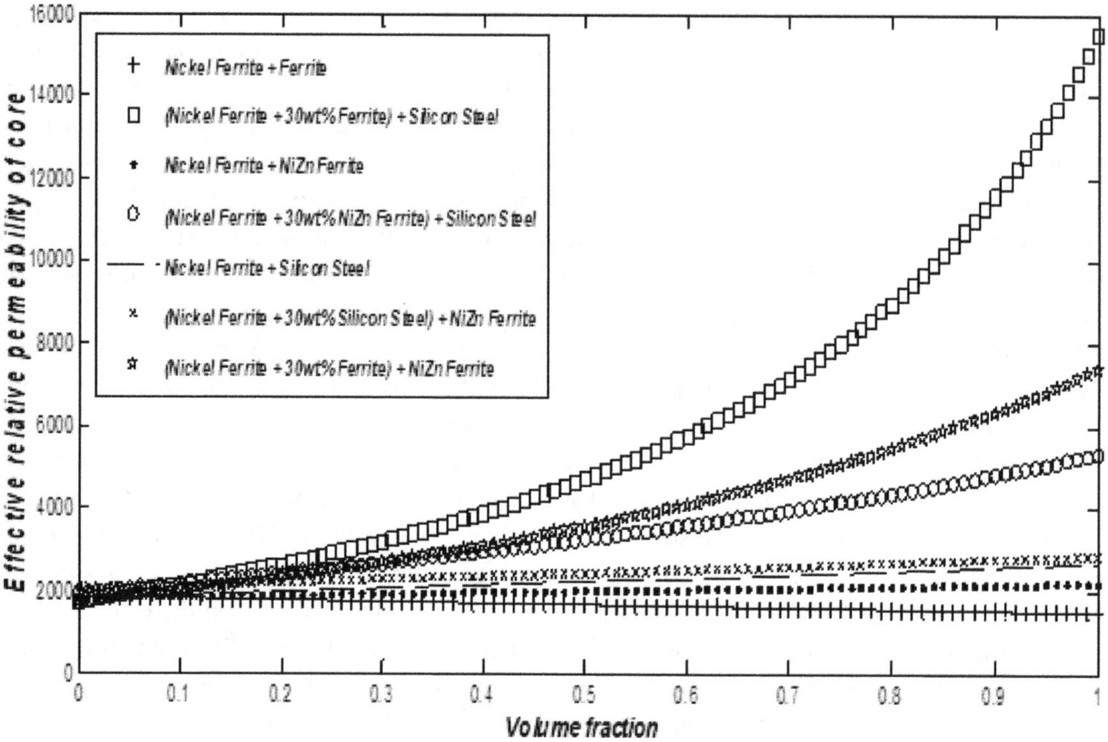

Figure 6. Effective relative permeability of Silicon Steel nanocomposites for transformer core lamination with various volume fractions of multi-nanoparticles

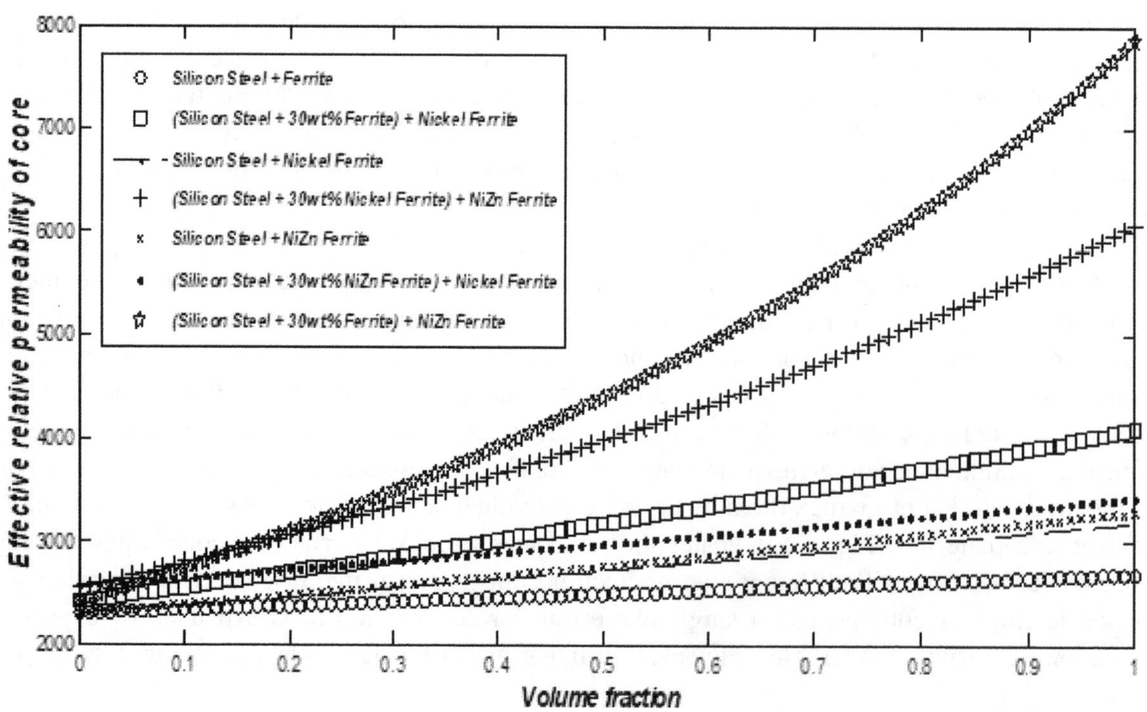

rite or nickel ferrite + 30wt. % ferrite + silicon steel as multi-nanoparticles has reduced the reluctance of nickel ferrite nanocomposites core for single phase transformers. On the other hand, expanding the rate of NiZn_Ferrite nanoparticles in Nickel ferrite for single phase transformer core has reduced the reluctance of nickel ferrite + 30wt. % NiZn_Ferrite nanocomposites corE for single phase transformer. Utilizing 30wt. %NiZn_Ferrite and expanding the fixation of silicon Steel nanoparticles has improved the diminish of the reluctance for nickel ferrite core of single-phase transformers. On the other side, a slight decline in the reluctance of nickel ferrite + 30wt. % silicon steel nanocomposites for single phase transformer core has been struck toward expanding the concentration of NiZn_Ferrite in the nickel Ferrite/Silicon steel nanocomposites core of single-phase transformers.

Utilizing ferrite individual nanoparticles has expanded the reluctance of NiZn_Ferrite nanocomposites to single phase transformer core for expanding the volume fraction. However, the addition of nickel ferrite or silicon steel nanoparticles to NiZn_Ferrite + 30wt. % ferrite has reduced the reluctance of NiZn_Ferrite + 30wt. % Ferrite nanocomposites for single phase transformer core with expanding the volume fraction. Expanding the concentration of silicon steel singular nanoparticles reduced the reluctance of NiZn_Ferrite nanocomposites to the single-phase transformer core. However, the addition of nickel ferrite and 30wt. % of silicon steel nanoparticles to NiZn_Ferrite nanocomposites expanded the reluctance of NiZn_Ferrite nanocomposites for the single-phase transformer core. Utilizing 30wt. % nickel ferrite and expanding the concentration of silicon steel nanoparticles in the NiZn_Ferrite nanocomposites for the single- phase transformer core improved the reduction in the reluctance of NiZn_Ferrite + 30wt. % nickel ferrite nanocomposites core of single-phase transformer cores. Figure (8) indicates the reluctance of silicon steel nanocomposites for single phase transformer core lamination with different volume fractions for individual and multi-nanoparticles at power frequency 50Hz. Nickel ferrite, ferrite, or NiZn_Ferrite nanoparticles reduced the reluctance of silicon steel nanocomposites for the single-phase transformer core with expanding the volume fraction by utilizing distinctive nano particles. Moreover, the addition of NiZn_Ferrite nanoparticles to silicon steel + 30wt. % ferrite or silicon steel + 30wt. % nickel ferrite has improved the reduction of the reluctance of silicon steel nanocomposites single phase transformer core. Expanding concentration of NiZn_Ferrite nanoparticles reduced the reluctance of silicon steel + 30wt. % NiZn_Ferrite nanocomposites. Addition of nickel ferrite and 30wt. % of NiZn_Ferrite nanoparticles to silicon steel nanocomposites improved the reduction in the reluctance of silicon steel nanocomposites core. Utilizing multi-nanoparticles has expanded the decrease in the reluctance of silicon steel nanocomposites transformer core lamination.

Figure (9) indicates the reluctance of ferrite nanocomposites to straight area and the total reluctance of three phase transformer core for different volume fractions in power frequency 50 Hz. Utilizing individual and multi-nanoparticles, the reluctance values of right section ferrite nanocomposites core and the total reluctance have been higher than the reluctance of vital section ferrite nanocomposites core for three-phase transformers with expanding the volume fraction. Figure (10) depicts the total reluctance of nickel ferrite nanocomposites for three-phase transformer core lamination with different volume fractions at power frequency 50 Hz. The total reluctance of nickel ferrite nanocomposites for three-phase transformer cores has been reduced by expanding the fixation of chosen multi-nanoparticles in the nickel ferrite nanocomposites core of the three-phase transformers.

Figure (11) portrays the total reluctance of NiZn_Ferrite nanocomposites for three-phase transformer cores with different volume fractions toward power frequency 50 Hz. In any case, Fig. (12) indicates the total reluctance of silicon steel nanocomposites of three-phase transformer core for different volume fractions in power frequency 50 Hz. Utilizing multi-nanoparticles, the performance of the total

Figure 7. Reluctance of Nickel Ferrite nanocomposites for single phase transformer core with various volume fractions of multi-nanoparticles

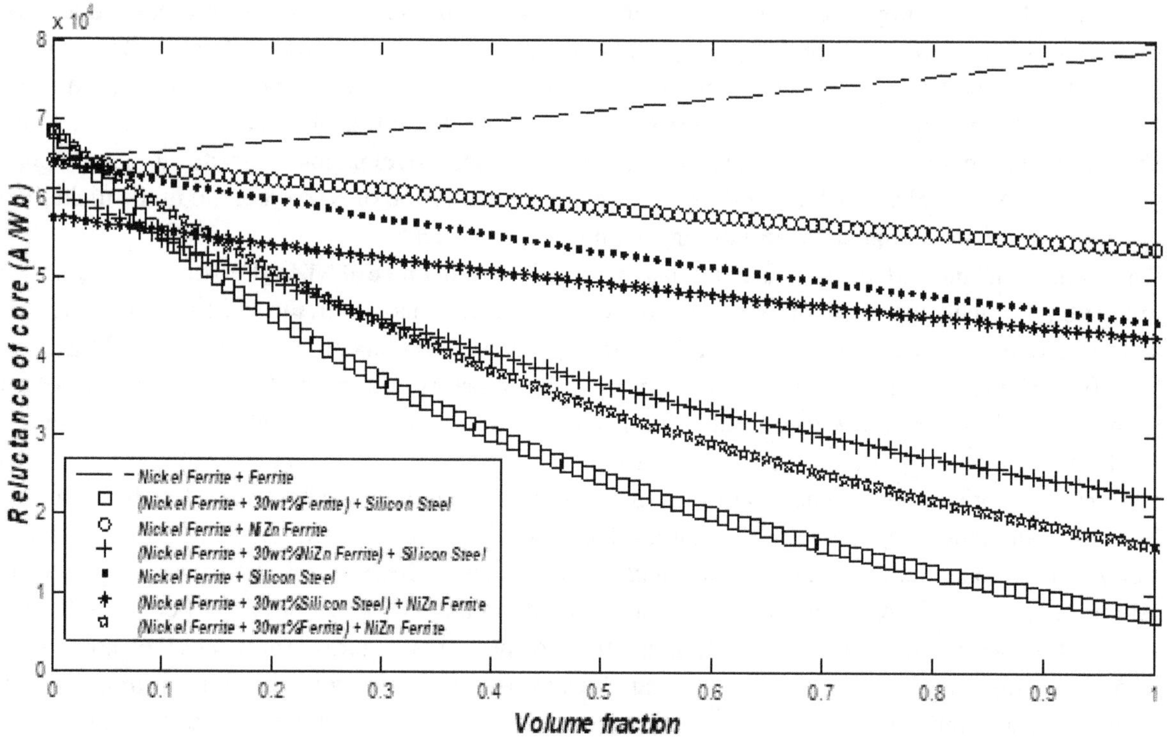

Figure 8. Reluctance of Silicon Steel nanocomposites for single phase transformer core with various volume fractions of multi-nanoparticles

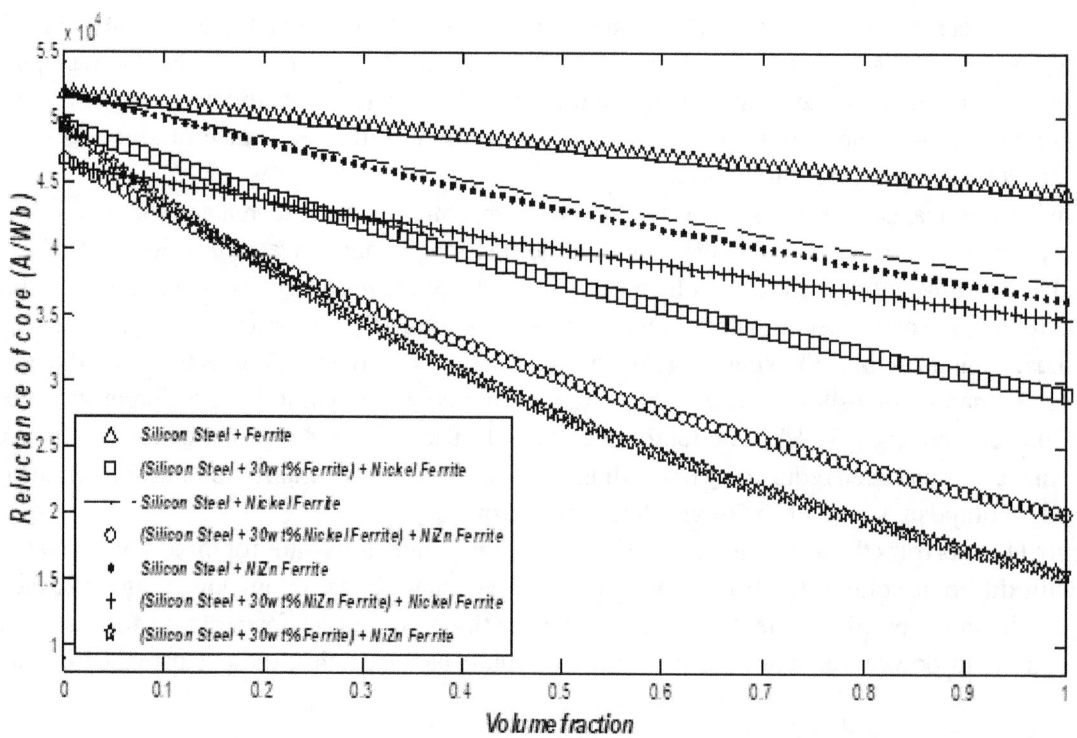

Figure 9. Total reluctance of Ferrite nanocomposites for three-phase transformer cores with various volume fraction using multi-nanoparticles

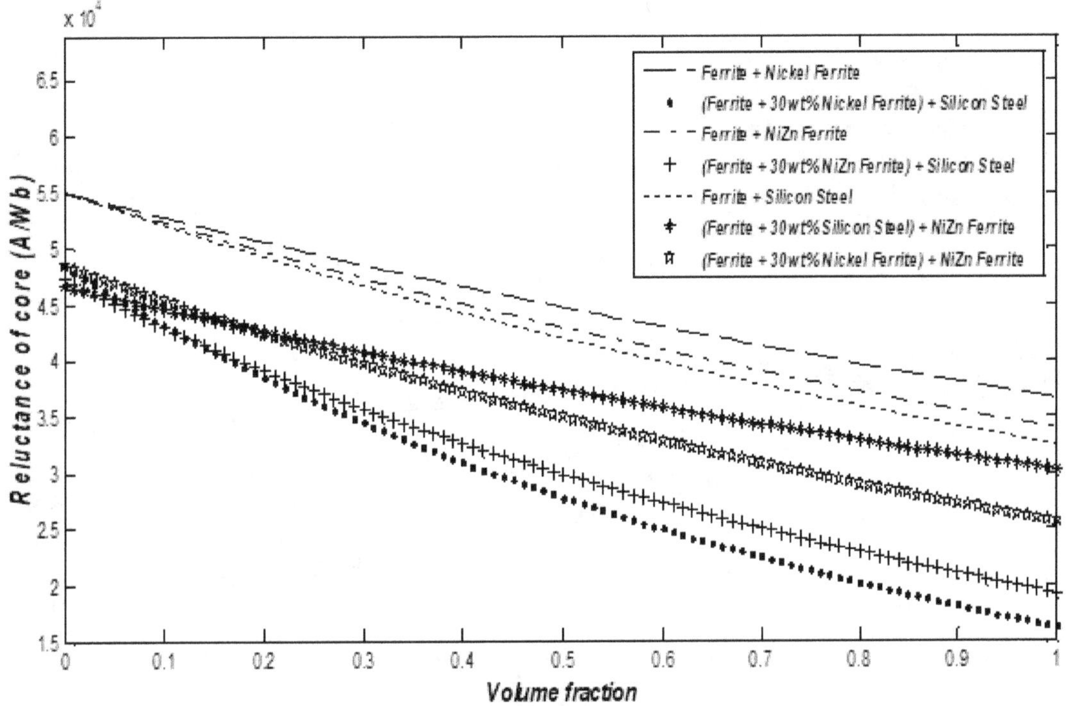

Figure 10. Total reluctance of Nickel Ferrite nanocomposites for three-phase transformer cores with various volume fractions using multi-nanoparticles

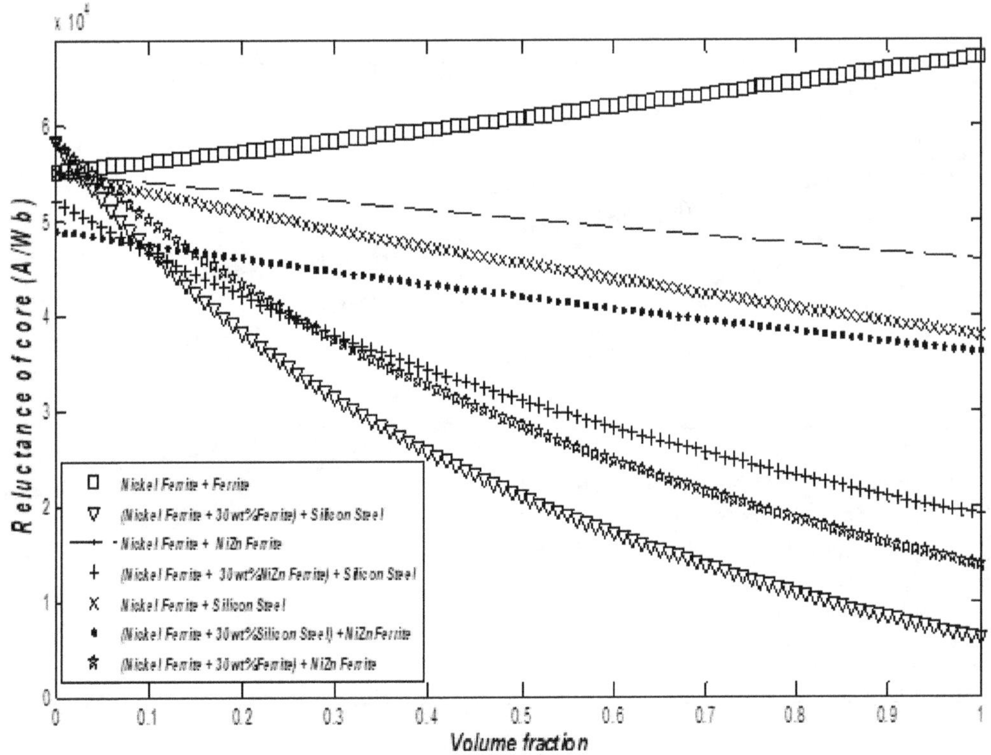

Figure 11. Total reluctance of NiZn_Ferrite nanocomposites for three-phase transformer cores with various volume fractions using multi-nanoparticles

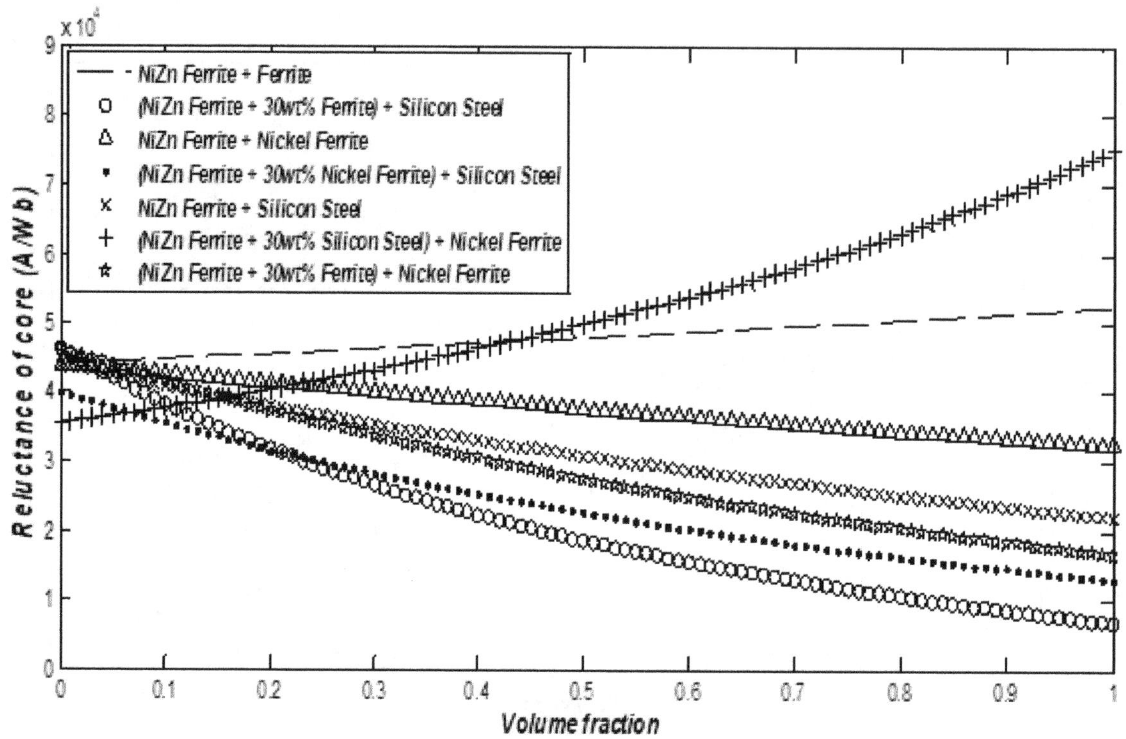

Figure 12. Total reluctance of Silicon Steel nanocomposites for three-phase transformer cores with various volume fractions of multi-nanoparticles

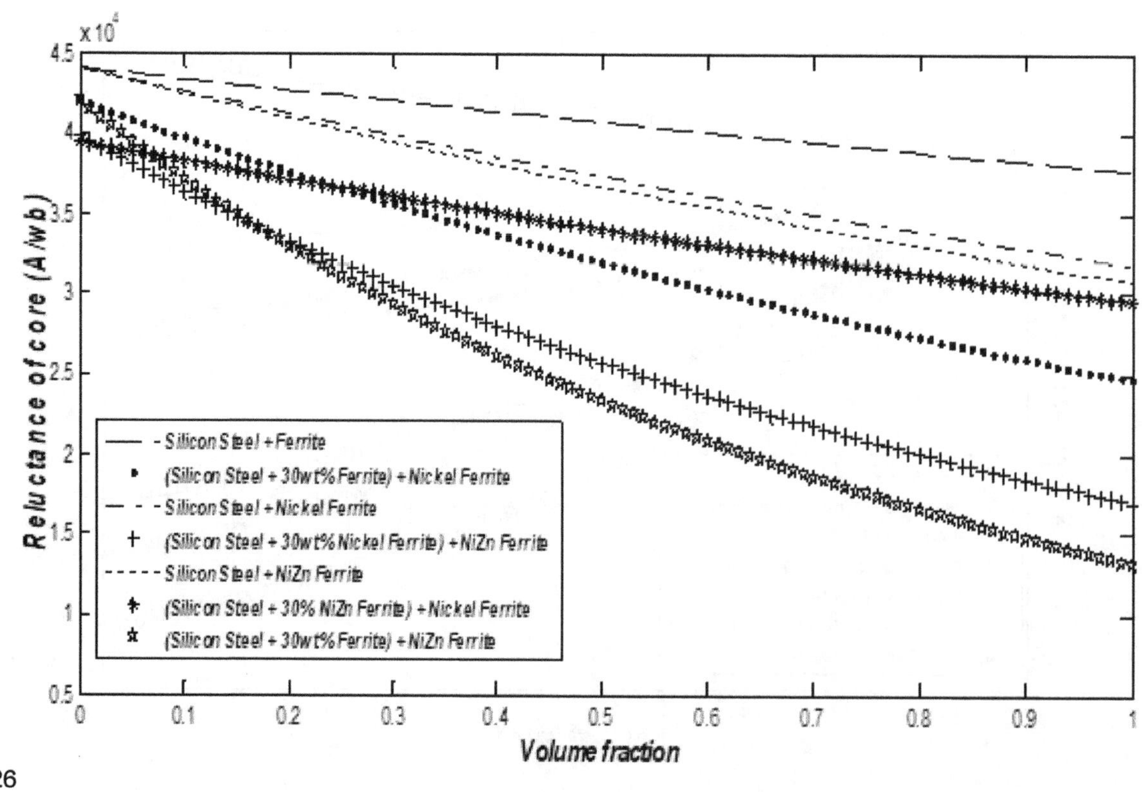

reluctance silicon steel nanocomposites for three-phase transformer cores has been comparable to the performance of the reluctance of the silicon steel nanocomposites of single-phase transformer cores and comparable to the performance of the reluctance of silicon steel nanocomposites to the central section of three-phase transformer cores.

The above results indicate the impact of distinct and multi-nanoparticles on the reluctance of new silicon steel core nanocomposites transformers. Utilizing Ferrite, nickel ferrite and NiZn ferrite distinct and multi-nanoparticles have been utilized for upgrading the magnetic characteristics of silicon steel nanocomposites to the transformer core by reducing the reluctance of core to single phase and three-phase transformers. Utilizing silicon Steel+ 30wt. % ferrite + 40wt. % NiZn_Ferrite multi-nanoparticles has been the main request for upgrading the magnetic characteristics of single phase and three-phase transformers by reducing the reluctance of transformer cores with the rate of 40.7%. Furthermore, the impact of individual and multi-nanoparticles on the reluctance of new ferrite core has been indicated. Utilizing ferrite + 30wt. % Ni _Ferrite + 40wt. % silicon steel nanoparticles recorded the best consideration utilized for decreasing the reluctance of ferrite nanocomposites core of single phase and three-phase transformers with rate 43.8%. Hence, the impact of individual and multi-nanoparticles on the reluctance of new nickel ferrite core has been indicated. Utilizing nickel ferrite + 30wt. %Ferrite + 40wt. % silicon steel as multi-nanoparticles recorded the best incorporation utilized for reducing the reluctance of ferrite nanocomposites core of single phase and three-phase transformers to the rate of 53.2%. Finally, the impact of distinct and multi-nanoparticles on the reluctance of new NiZn_ ferrite core has been exhibited. Utilizing NiZn ferrite + 30wt. % ferrite + 40wt. % silicon steel nanoparticles recorded the best consideration utilized for decreasing the reluctance of ferrite nanocomposites core of single phase and three-phase transformers with rate 49.3%.

6 MAGNETIZATION CURVES AND LOSSES

A standout amongst the majority of significant and exorbitant sections in the power transformer is the transformer core that serves as a low reluctance method to the magnetic flux linking grade and auxiliary windings. In doing so, heat from the core losses is produced (Tang et al., 2012). Because of the vitality of moved forward electrical core performance, analysts have been dynamic in the improvement of preferred transformer cores and core demonstrating strategies. Superior manufacturing strategies have been produced as an outcome of a superior observation of the elements that iñuence magnetic properties. Nowadays, the satisfaction of electrical steel has been significantly enhanced. Elements that sway core loss for electrical steel are accounted in: (a) nature of sheet insulation, (b) rate of silicon in the alloy, (c) concoction impurities, (d) grain size, (e) Precious stone introduction control, and (f) core lamination thickness (Córcoles et al., 2008; Fcf & Wsg, 2007; Juan et al., 2013; Kefalas et al., 2008; Moses, 1990; Schulz et al., 1995; Zirka, Moroz, Marketos, & Moses, 2008; Zirka, Moroz, Marketos, Moses et al, 2008). Under typical circumstances, control transformers are intended to do most of the work inside the straight part of the core's magnetization curve. The essential objective of this plan is to hold a sensible measure for linearity and to set tight the following flux density values that might have unreasonable losses. However, we are witnessing additional tests of immersion nowadays (Gottlieb, 1998). A standout amongst the practically vital magnetic properties of electrical steels is the nonlinear AC magnetization curve (B-H curve) which is a method of planning power transformers and motors to decrease measure and move forward performance. The B-H bend is additionally key for transformers and reactors demonstrated in control framework exchanging transient investigations, such as Ferro resonance. The nonlinear B-H curve expressed in an explicit equation form is, therefore, particularly useful and is preferred to yield accurate results while saving computer time. In the past, there have been a number endeavors on fit equations regarding test B-H data, and the examination falls under two fundamental categories. One arrangement uses hypothetical formulations in light of physical models, handling expressions which are mind boggling for utilization in building programming. Alternately, the magnetization characteristics are communicated in experimental scientific terms, which are suitable for quantitative estimation in spite of minimal hypothetical support (Charalambous et al., 2009; Fiorillo et al., 2002; Mészáros & Vértesy, 2010; Tang et al., 2013). The second classification stays well known at endeavoring the best delegate test for the B-H trademark (Jaafar et al., 2004). The losses of the transformer comprise no-load losses and load losses. No-load losses are steady and show up as long as a transformer is over service. At the same time, load losses change for the square of the load current conveyed by the transformer and are just critical under higher load states (Deren et al., 2012). The loss over transformer core is the electrical power lost as far as heat inside the transformer core is at a point when core is subjected to AC polarizing power.

The present ideas of physical beginning comprising losses of magnetic materials are reviewed under three customary categories: Hysteresis losses, eddy current or dielectric losses and lingering losses. This section focuses on initial and second categories. As a result, they are extensively answerable for losses of magnetic core materials. Core (iron) losses are basically identified with polarizing or stimulating the core and are displayed during the whole time the transformer is powered on, whether there's any load or not. Core losses are approximately consistent with no-load to full-load at nourishing straight loads. Although the core losses have little contrast with transformer yield power, as a rule, less than 1%, it reveals steady and huge working liability for a utility; particularly if its vitality costs are high (Hasan et al., 2013; John & Nough, 2002; Rejea & Ernst, 2005). It has been assessed that around 5% of the sum power produced is scattered as core losses; therefore, the requirement to decrease core loss progressively paramount's the outline and assembling attention of the measure of transformer (Galvan et al., 2010). Accordingly, this will decrease the core loss. Therefore, this section mainly and effectively focuses on examining the impact of nanocomposites core material on core loss and indicates the magnetization characteristics of the new transformer core nanocomposites.

6.1 Theory and Effective Parameters

The magnetization characteristics of the new transformer core nanocomposites can be ascertained by utilizing the custom MATLAB transformer (B-H Curve) reproduction, which indicates figuring and affirmation of a nonlinear transformer core charge trademark. Starting with key parameter values, the core trademark is inferred. This is utilized within a simscape™ model of a sample test circuit where camwood is used to plot the core charge trademark around an oscilloscope. Model outputs are at that point contrasted with the values indicated. Nonlinear magnetic flux density is decided by utilizing the following expression:

$$B = B_{sat} tanh\left(H.atanh\left(\frac{1}{2}\right)\frac{B_{sat}}{2\hat{\mu}_{clef}\mu_0}\right) \tag{4}$$

Where, H is the magnetic field strength

The magnetic field strength when the core saturation begins $H_{satbegin}$ is presented by the following expression:

$\hat{\mu}_{clef}$ is the effective relative permeability of core

B_{sat} is the maximum magnetic flux density in core that is determined from the following expression (Deren et al., 2012):

$$E_1 = 4.44NfB_{sat}A \tag{4.i}$$

Where,

E_1 is the rating of primary voltage

N is the number of primary winding

A is the cross section area of the core

f is the power frequency

Nonlinear magnetization curves characteristics of core transformers using individual nanoparticles have been described by the magnetic flux density B_i. In any case, the nonlinear magnetization curves characteristics of core transformer using multi-nanoparticles have been described by the magnetic flux density B_j. There are four routines that exist to unravel the transformer core loss under hysteresis and eddy-current losses, these strategies are: (1) structure component method, (2) 2-temperature method, (3) 2-frequency strategy and (4) dc hysteresis system (John & Nough, 2002). The magnetic flux density b way inside the magnetic core material doesn't benefit its individual esteem by the magnetic field strength. H applied on polarizing the magnetic material expands the et cetera and reduces the go to its individual worth. The charged magnetic field strength prompts magnetic flux density. This trademark has vitality loss, called hysteresis loss. The explanatory outflow to acquire hysteresis loss for every unit volume at power frequency excitation, as stated by (Barranger, n.d.; Juan et al., 2013) is:

$$W_h = \frac{2f\eta B_{sat}}{\mu_0\mu_r} \tag{5}$$

Where,

f is the power frequency

η is the shape factor

μr is the relative permeability of the material

$\mu 0$ is the free space permeability

The eddy current loss on transformer core is because of the eddy ebbs and flows framed in the body of magnetic core. At any point that the transformer core is subjected to an exchanging magnetic field, a prompted voltage is furnished in the constitution of magnetic core and prompts generating a limited circle present inside the core, known as eddy current. This can be minimized by making the transformer core lamination of slim layer and great insulation in the middle of one another (Moses, 1990). The explanatory statement to obtain eddy current losses for every unit volume in power frequency excitation, as stated by (Charalambous et al., 2009; Galvan et al., 2010; Rejea & Ernst, 2005), is:

$$W_e = \frac{(\ddagger.\pi.f\,B_{sat})^2}{6\rho} = \frac{\sigma(\ddagger.\pi.f\,B_{sat})^2}{6} \quad (6)$$

Where,

$\ddagger$ is the lamination thickness

ρ is the electrical resistivity of the material

σ is the electrical conductivity of the material

The total core-loss is deðned by using the following expression (John & Nough, 2002):

$$W_{total} = W_h + W_e \quad (7)$$

Magnetic loss characteristics of new nanocomposites core of transformer using individual nanoparticles have been described by W_{itotal}. The hysteresis losses and eddy current losses of new nanocomposites transformer core per unit volume at power frequency excitation use individual nanoparticles which can be described by using the following expression:

$$W_{itotal} = \frac{2f\eta B_{sat}}{\mu_0\mu_{reffi}} + \frac{\sigma_{effi}(\ddagger.\pi.f\,B_{sat})^2}{6} \quad (8)$$

Where,

μ_{reffi} is the effective relative permeability of magnetic nanocomposites materials using individual nanoparticles.

σ_{effi} is the effective electrical conductivity of magnetic nanocomposites materials using individual nanoparticles.

Similarly, the magnetic loss characteristics of new nanocomposites core of transformer using multi-nanoparticles have been described by W_{jtotal}. The hysteresis losses and eddy current losses of new nanocomposites transformer core per unit volume at power frequency excitation use multi-nanoparticles which are described by using the following expression:

$$W_{jtotal} = \frac{2f\eta B_{sat}}{\mu_0\mu_{reffj}} + \frac{\sigma_{effj}(\ddagger.\pi.f\,B_{sat})^2}{6} \quad (9)$$

Where,

μ_{reffj} is the effective relative permeability of magnetic nanocomposite materials using multi- nanoparticles.

σ_{effj} is the effective electrical conductivity of magnetic nanocomposites materials using multi-nanoparticles.

6.2 Effects of Nanoparticles on Magnetization Loss of Transformer Core

The following effects illustrate the nonlinear magnetization characteristics (B-H Curve) and magnetic loss characteristics of new nanocomposites transformer core regarding the different sorts and concentrations of chosen nanoparticles. The characteristics of magnetic materials utilized for new nanocomposites transformer core, such as the neighborhood relative magnetic permeability μr and the electrical conductivity of the core materials σ, are controlled by the same specifications of nanoparticles and core materials that are listed above. Nowadays, the nonlinear magnetization characteristics of ferrite nanocomposites to single- phase transformer core utilizing distinctive nanoparticles illustrate the increment of magnetic field strength with expanding the magnetic flux density of ferrite nanocomposites single phase transformer cores. Utilizing the silicon steel, NiZn_Ferrite, or nickel ferrite individual nanoparticles in the negative area of the magnetic field strength reduced the magnetic flux density of ferrite single phase transformer cores. However, in the sure zone of magnetic field strength, the expansion in the magnetic field strength-initiated expansion in the magnetic flux density of ferrite nanocomposites single phase transformer cores. Consequently, utilizing silicon Steel, NiZn_Ferrite, or nickel ferrite individual nanoparticles in the certain zone of the magnetic field strength improved the magnetic flux density of ferrite single phase transformer cores. Therefore, silicon steel nanoparticles have recorded the best incorporation utilized for expanding the magnetic flux density of ferrite nanocomposites transformers, followed by NiZn_Ferrite which has recorded the second request for expanding the magnetic flux density of ferrite nanocomposites transformer core in the sure range of the magnetic field strength. Moreover, the magnetic flux density of nickel ferrite single phase transformer core has been expanded by expanding the positive values of the magnetic field strength. NiZn_Ferrite or silicon steel nanoparticles have been utilized for expanding the magnetic flux density of nickel ferrite core of single-phase transformers for expanding the positive values magnetic field. However, ferrite nanoparticles reduced the magnetic flux density of nickel ferrite core for single phase transformers with expanding the certain characteristics of magnetic field strength.

For utilizing silicon Steel, NiZn_Ferrite singular nanoparticles in the negative range of the magnetic field strength reduced the magnetic flux density of nickel ferrite single phase transformer core. However, including fixation 30wt. % or 70wt. % of ferrite nanoparticles under nickel ferrite single phase transformer core improved the magnetic flux density with expanding the magnetic field strength in the negative area. Figure (13) depicts the nonlinear magnetization characteristics of NiZn_Ferrite nanocomposites core of single-phase transformer utilizing individual nanoparticles toward incorporation diameter 5nm and different rates of nanoparticles 30wt. % or 70wt. %. Utilizing 30wt. % or 70wt. % for ferrite nanoparticles reduced the magnetic flux density of NiZn_Ferrite transformer core with expanding the positive values of the magnetic field strength. The magnetic flux density of NiZn_Ferrite nanocomposites transformer core has been improved with expanding the magnetic field strength under most extreme esteem of magnetic flux density of single-phase transformer (1. 1T) toward utilizing 30wt. % or 70wt. % of nickel ferrite or silicon steel individual nanoparticles in the positive area of the magnetic field strength. Silicon steel has been the best incorporation utilized for upgrading the magnetic flux density of NiZn_Ferrite nanocomposites single phase transformer core in the certain region of the magnetic field strength. In the negative area of the magnetic field strength, the magnetic flux density of NiZn_Ferrite single phase transformer core expanded with expanding the magnetic field strength. The magnetic flux density of NiZn_Ferrite transformer core has been reduced with expanding magnetic field strength by utilizing 30 wt. % or 70 wt. % for nickel ferrite or silicon steel individual nanoparticles. Furthermore, including concentration 30 wt. % or 70 wt. % of ferrite nanoparticles under NiZn_Ferrite single phase transformer core improved the magnetic flux density of NiZn_Ferrite nanocomposites single phase transformer core for expanding the magnetic field strength in the negative zone. Figure (14) portrays the nonlinear magnetization characteristics of silicon steel nanocomposites to single phase transformer core utilizing individual nanoparticles toward different rates of nanoparticles 30wt. % or 70wt. %. The increment in

Figure 13. Nonlinear magnetization characteristics of NiZn_Ferrite nanocomposites for single phase transformer core using individual nanoparticles

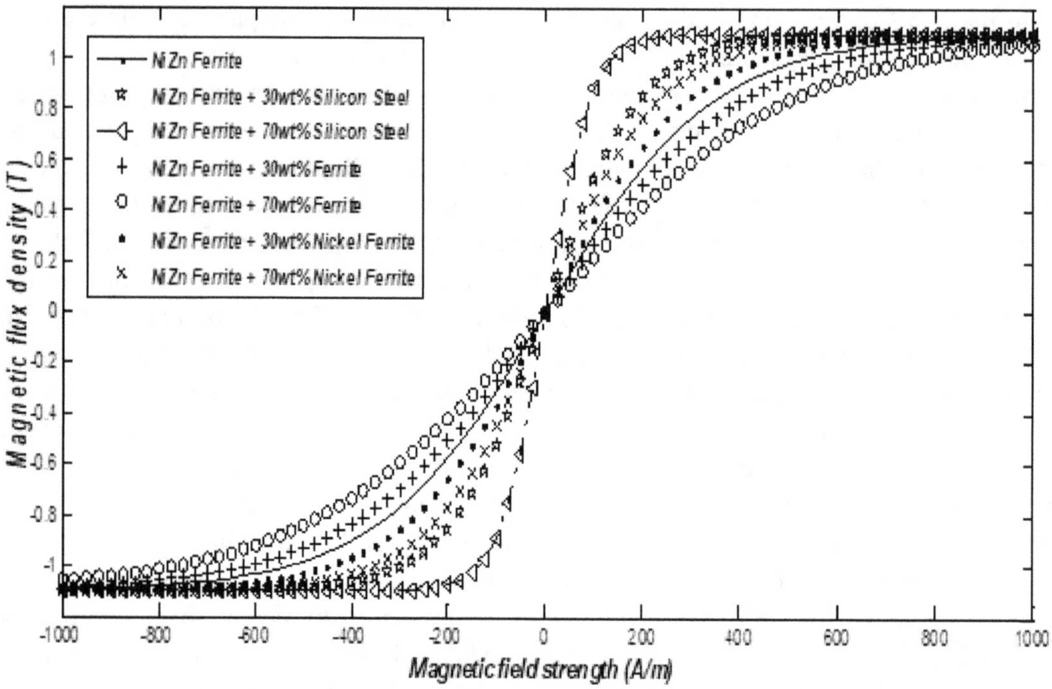

Figure 14. Nonlinear magnetization characteristics of Silicon Steel nanocomposites for single phase transformer core using individual nanoparticles

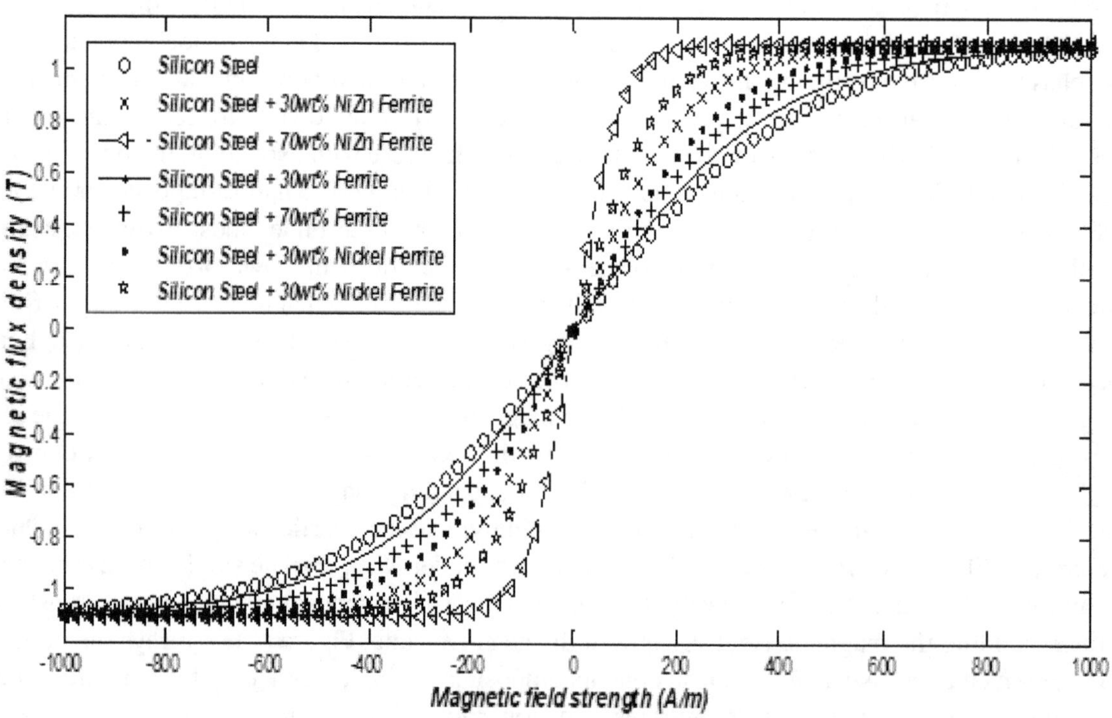

Figure 15. Effect of volume fraction on magnetic flux density of Ferrite nanocomposites for single phase transformer core using individual nanoparticles

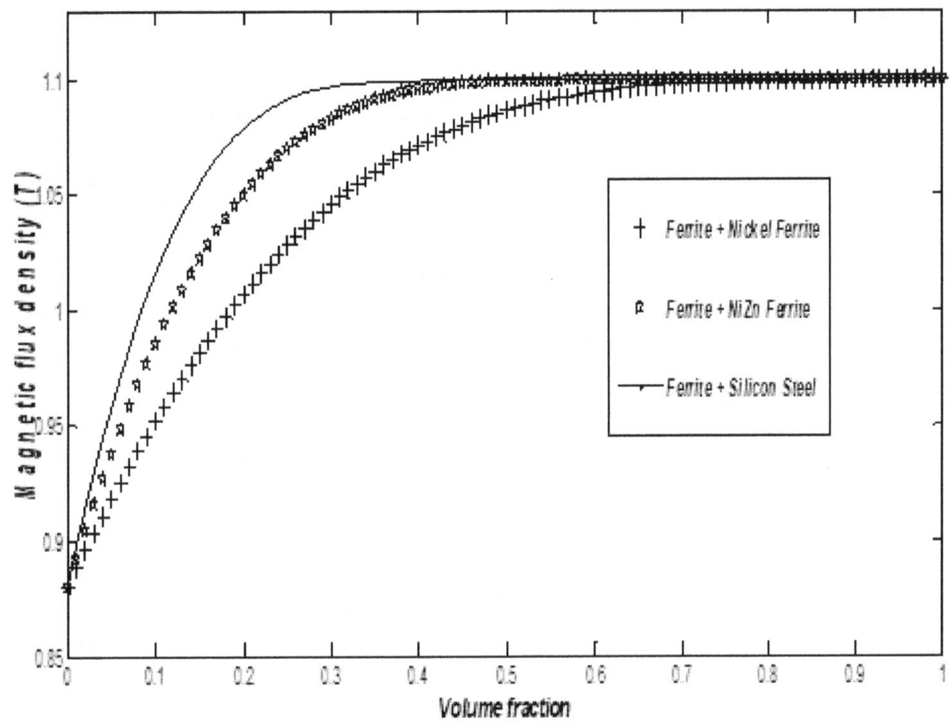

the positive values of the magnetic field strength results from expansion in the magnetic flux density of silicon steel single phase transformer core. The flux density expands in extent to the field strength until it achieves a sure strength, where it can't expand after that. Therefore, it turns into level and steady as the field strength proceeds to increment. This is because there is a breaking point of the measure of flux density that can be produced by the core. Thus, utilizing 30wt. % or 70wt. % of ferrite, NiZn_Ferrite, or nickel ferrite nanoparticles improved the magnetic flux density of silicon steel nanocomposites for single phase transformer core. Utilizing NiZn_Ferrite individual nanoparticles has been the best Incorporation for upgrading the magnetic flux density of silicon steel nanocomposites for the single-phase transformer core with expanding the certain values of magnetic field strength. The magnetic flux density of silicon steel single phase transformer core expanded with expanding the magnetic field strength. Utilizing Ferrite, NiZn_Ferrite, or nickel ferrite individual nanoparticles in the negative zone of the magnetic field strength reduced the magnetic flux density of silicon steel single phase transformer core.

Figure (15) The impact of volume fraction on the magnetic flux density of ferrite nanocomposites for single phase transformer core utilizing singular nanoparticles in most extreme magnetic field strength of ferrite core of single-phase transformer is self-evident. The magnetic flux density of ferrite nanocomposites transformer core has been expanded by expanding the volume fraction of NiZn_Ferrite, silicon steel or nickel ferrite nanoparticles in the ferrite transformer core under the greatest esteem of magnetic flux density of single-phase transformers. Silicon steel has recorded the best Incorporation to upgrading the magnetic flux density of ferrite nanocomposites transformer core. NiZn_Ferrite has been in the second request for upgrading the magnetic flux density of silicon steel nanocomposites transformer core. Nickel

Figure 16. Effect of volume fraction on magnetic flux density of NiZn Ferrite nanocomposites for single phase transformer core using individual nanoparticles

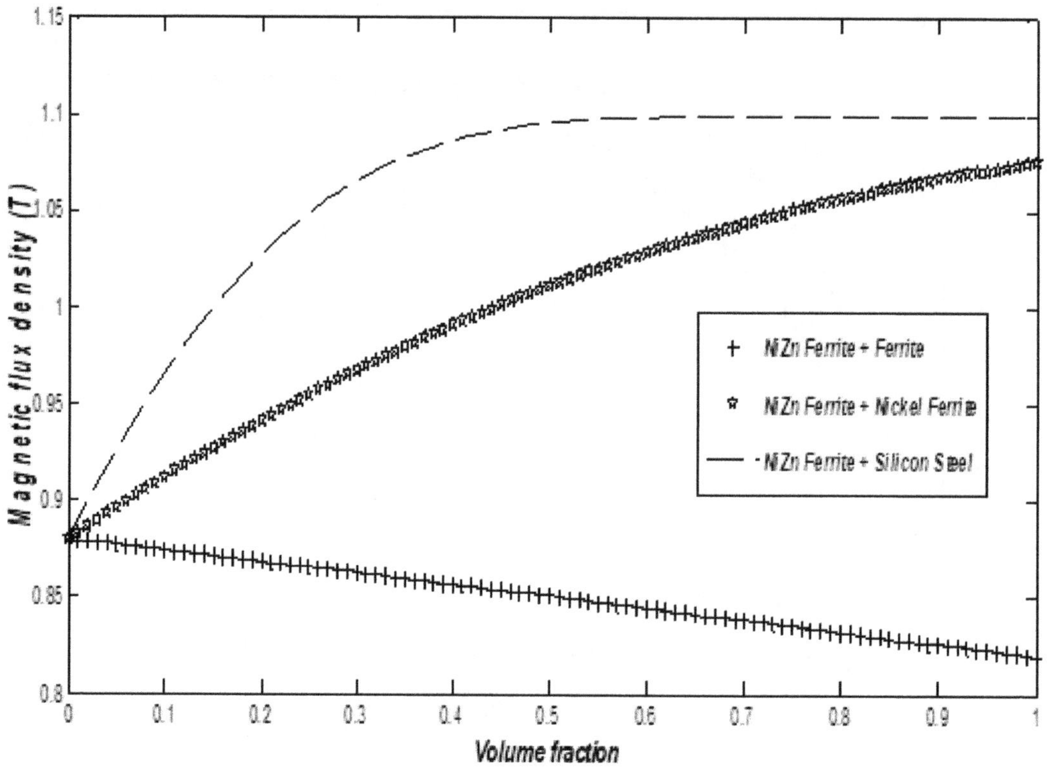

Figure 17. Effect of volume fraction on magnetic flux density of Silicon Steel nanocomposites for single phase transformer using individual nanoparticles

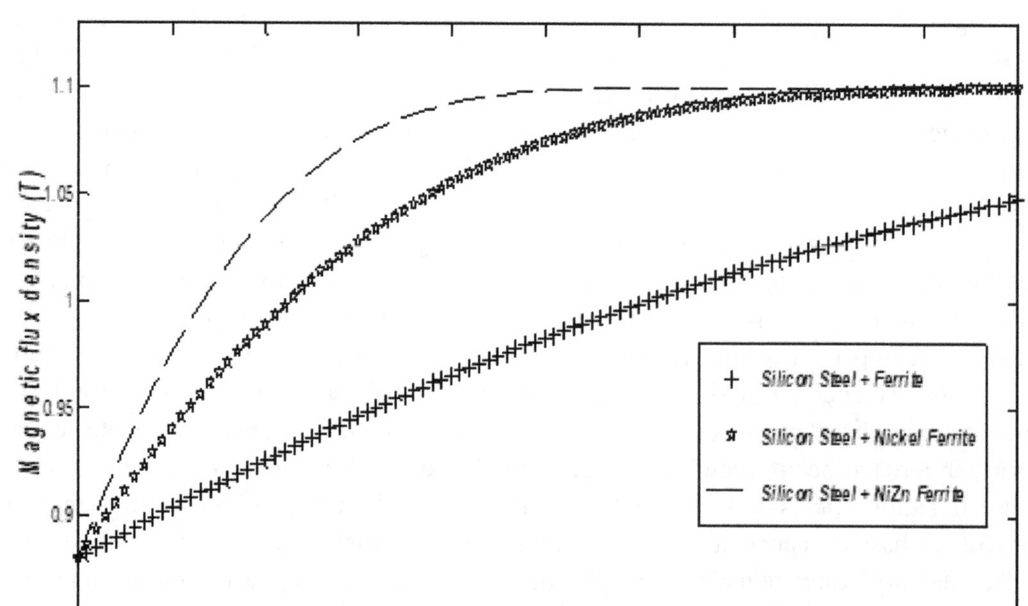

ferrite has been the slightest one in terms of expanding the magnetic flux density of ferrite nanocomposites transformer core with expanding the volume fraction. Furthermore, volume fraction is an important component in expanding the magnetic flux density of nickel ferrite single phase transformer core under immersion worth of single-phase transformer toward utilizing silicon steel or NiZn_ ferrite nanoparticles. The magnetic flux density of nickel ferrite nanocomposites transformer core has been reduced by expanding the volume fraction of ferrite nanoparticles in the nickel ferrite single phase transformer core. Figure (16) portrays the impact of volume fraction on the magnetic flux density of NiZn_Ferrite nanocomposites single phase transformer core utilizing singular nanoparticles during the greatest magnetic field strength of NiZn_Ferrite core for single phase transformers. The magnetic flux density of NiZn_Ferrite nanocomposites to single phase transformer core has been improved toward expanding the concentration of silicon steel or nickel ferrite nanoparticles in the NiZn_Ferrite nanocomposites for single phase transformer cores under immersion worth of magnetic flux density of single-phase transformers. However, it has been reduced toward expanding the fixation of ferrite nanoparticles in the NiZn_Ferrite nanocomposites for single phase transformer cores. Fig (17) portrays the impact of volume fraction around the magnetic flux density of silicon steel nanocomposites for single phase transformers utilizing distinctive nanoparticles in the greatest magnetic field strength of silicon steel single phase transformer cores. The magnetic flux density of silicon steel nanocomposites for single phase transformer cores has been expanded toward expanding the rate of NiZn_Ferrite, ferrite or nickel ferrite nanoparticles in the silicon steel transformer core. The vast majority enhancements have been struck by expanding the rate of NiZn_Ferrite nanoparticles in the silicon steel single phase transformer core for expanding the volume fraction. Nickel ferrite has been the second request for upgrading the magnetic flux density of silicon steel nanocomposites c single phase transformer core.

Fig. (18) shows the magnetic loss characteristics of Ferrite, nickel Ferrite, NiZn_Ferrite and silicon steel nanocomposites for three-phase transformer core utilizing singular nanoparticles. The total loss performance of Ferrite, nickel Ferrite, NiZn_Ferrite or silicon steel nanocomposites for three-phase transformer cores with expanding the magnetic flux density utilizing distinctive nanoparticles of chosen magnetic materials has been the same loss performance of Ferrite, nickel Ferrite, NiZn_Ferrite or silicon steel nanocomposites of single phase transformer for expanding the magnetic flux density utilizing distinct nanoparticles of chosen magnetic materials by utilizing the same nanocomposites magnetic materials. Fig. (19) shows the impact of volume fraction on the magnetic loss characteristics of Ferrite, nickel Ferrite, NiZn_Ferrite or silicon steel nanocomposites for three-phase transformers utilizing distinct nanoparticles in the greatest magnetic flux density 1. 2T of three-phase transformers. The impact of expanding the volume fraction of chosen distinct nanoparticles on the total loss of Ferrite, nickel Ferrite, NiZn_Ferrite or silicon steel nanocomposites for three-phase transformer cores has been comparative impact for expanding the volume fraction of chosen singular nanoparticles on the total loss of Ferrite, nickel Ferrite, NiZn_Ferrite or silicon steel nanocomposites magnetic materials to single phase transformers. The total loss values of the Ferrite, nickel Ferrite, NiZn_Ferrite or silicon steel nanocomposites of the three-phase transformer for expanding the volume fraction has been more than the total loss values of Ferrite, nickel Ferrite, NiZn_Ferrite or silicon steel nanocomposites to single phase transformers for expanding the volume fraction.

Fig. (20) portrays the nonlinear magnetization characteristics of Ferrite, nickel Ferrite, NiZn_Ferrite or silicon steel nanocomposites to three-phase transformers utilizing multi-nanoparticles toward diverse specified volume fractions. The impact of chosen rates of multi-nanoparticles on the of magnetic flux density of Ferrite, nickel Ferrite, NiZn_Ferrite or silicon steel nanocomposites of the three-phase trans-

Figure 18. Magnetic loss characteristics of Silicon Steel nanocomposites for the three-phase transformer core using individual nanoparticles

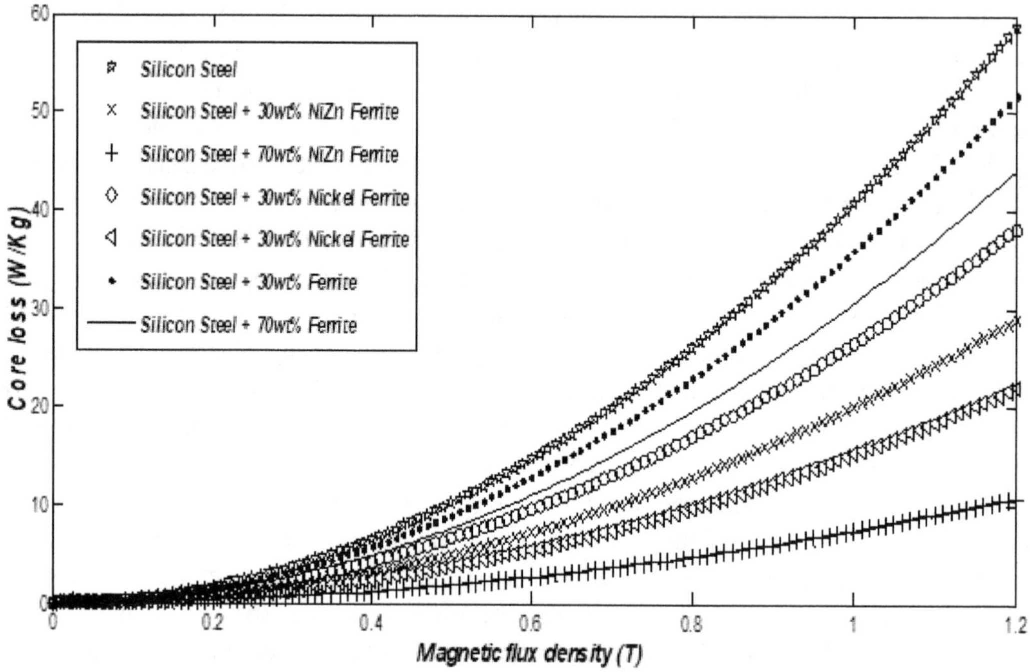

Figure 19. Effect of volume fraction on the magnetic loss characteristics of Silicon Steel nanocomposites for the three-phase transformer core using individual nanoparticles

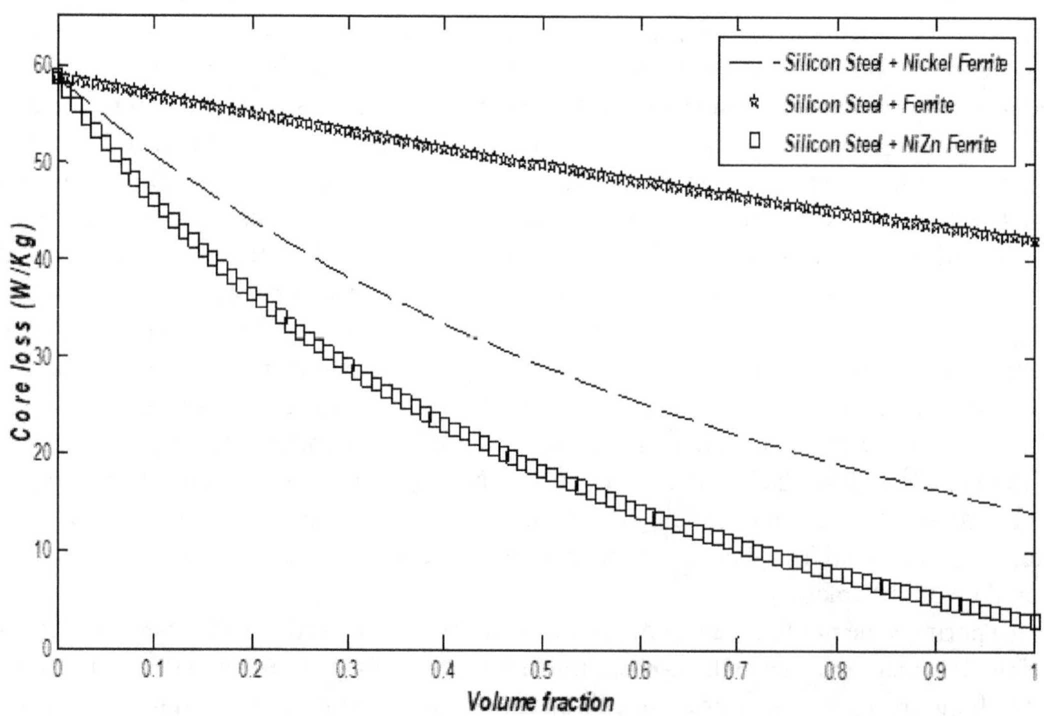

Figure 20. Nonlinear magnetization characteristics of Silicon Steel nanocomposites core of the three-phase transformer using multi-nanoparticles

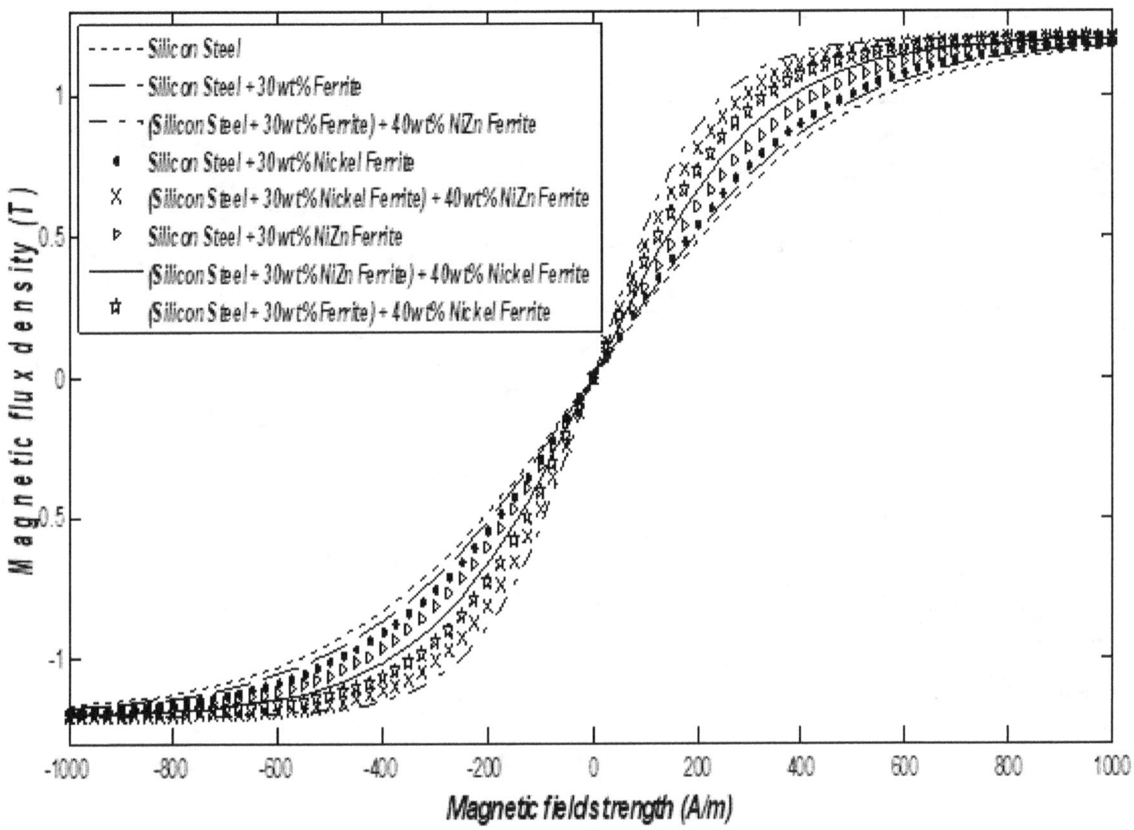

former core for expanding the magnetic field strength under the greatest strength of magnetic flux density for the three-phase transformer (1. 2T) has been the same impact of chosen rates of multi-nanoparticles on the of magnetic flux density of Ferrite, nickel Ferrite, NiZn_Ferrite, or silicon steel nanocomposites core of single-phase transformers for expanding the magnetic field strength under the greatest magnetic flux density of single-phase transformers (1.1T). The most extreme strength of magnetic flux density of three phase transformers (1.2T) has been higher than the magnetic flux density of single-phase transformers (1.1T). In this way, the characteristics of the magnetic flux density of NiZn_Ferrite, Ferrite, nickel ferrite or silicon steel nanocomposites for the three-phase transformer core have been more than the values of the magnetic flux density of NiZn_Ferrite, Ferrite, nickel ferrite or silicon steel nanocomposites to the single-phase transformer with expanding the magnetic field strength utilizing multi-nanoparticles. Figure (21) shows the impact of volume fraction on the magnetic flux density of Ferrite, nickel Ferrite, NiZn_Ferrite or silicon steel nanocomposites to the three-phase transformer utilizing multi-nanoparticles during the most extreme magnetic field strength of Ferrite, nickel Ferrite, NiZn_Ferrite, or silicon steel to the three-phase transformer. The impact of expanding the volume fraction for chosen multi-nanoparticles in the Ferrite, nickel Ferrite, NiZn_Ferrite, or silicon steel nanocomposites for the three-phase transformer core has been the same impact of expanding the volume fraction of chosen multi-nanoparticles in the Ferrite, nickel Ferrite, NiZn_Ferrite, or silicon steel nanocomposites for the single-phase transformer

Figure 21. Effect of volume fraction on the magnetic flux density of the Silicon Steel nanocomposites for the three-phase transformer core using multi-nanoparticles

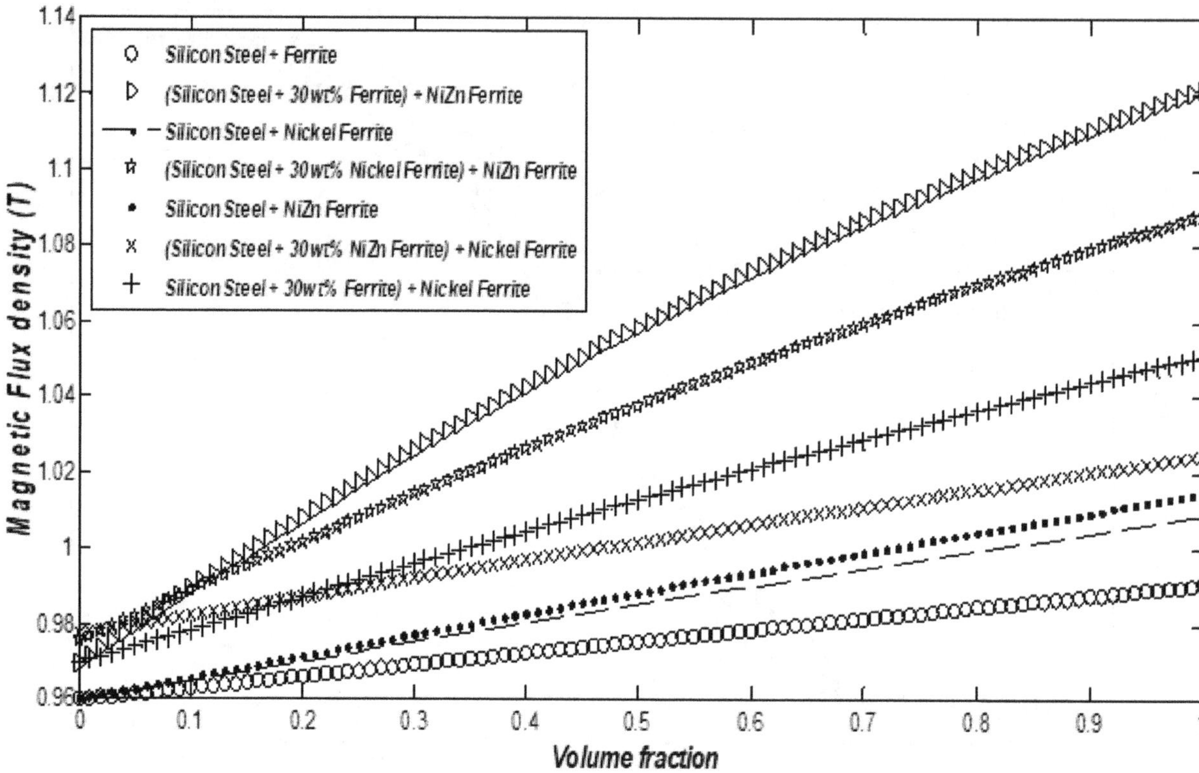

core. The values of the magnetic flux density of Ferrite, nickel Ferrite, NiZn_Ferrite, or silicon steel nanocomposites for the three-phase transformer core utilizing multi-nanoparticles for expanding the volume fraction have been more than the values of the magnetic flux density of Ferrite, nickel Ferrite, NiZn_Ferrite, or silicon steel nanocomposites for the single-phase transformer core. This is because the greatest magnetic flux density of the three-phase transformer (1.2T) is higher than the greatest magnetic flux density of the single-phase transformer (1. 1T).

7 EFFECT OF USING INDIVIDUAL AND MULTI-NANOPARTICLES IN MAGNETIC NANOCOMPOSITES TRANSFORMER CORE

The stated results indicate the impact of singular and multi-nanoparticles on the total core loss of new silicon steel nanocomposites core of single phase and three-phase transformers. Utilizing Ferrite, nickel ferrite and NiZn ferrite individual and multi-nanoparticles improved the magnetic characteristics of silicon steel nanocomposites transformer core by reducing the total loss for single phase and three-phase transformer cores. Including concentrations of 30 wt.% ferrite +40 wt.% NiZn_Ferrite under the silicon steel transformer core as multi-nanoparticles has been first request for reducing the total core loss of single phase and three-phase transformer cores to the of rate 52. 90%. In addition, the listed results indicate the impact of individual and multi-nanoparticles on the total loss of new ferrite nanocomposites for single phase and three-phase transformer cores. Utilizing ferrite + 30wt. %Nickel ferrite + 40wt. % silicon steel as multi nanoparticles has been the first request for reducing the total loss of ferrite nanocomposites core to the rate of 48.2%. The impact of singular and multi-nanoparticles on the total loss of new nickel ferrite nanocomposites for single phase and three-phase transformer cores has been depicted. Utilizing nickel ferrite + 30wt. % ferrite + 40wt. % silicon steel as multi-nanoparticles recorded the best Incorporation for reducing the total loss of nickel ferrite nanocomposites transformer core to the rates of 59. 6%. Furthermore, all results portray the impact of distinctive and multi-nanoparticles on the total loss of new NiZn_Ferrite nanocomposites transformer cores for single phase and three-phase transformers. Utilizing NiZn_Ferrite + 30wt.%Ferrite + 40wt.% of silicon steel nanocomposites transformer core recorded the best consideration utilized for reducing the total NiZn_Ferrite nanocomposites core loss for single phase and three-phase transformers to the rate of 63.6%. The impact of individual and multi-nanoparticles on the magnetic flux density of new silicon steel nanocomposites core for single phase and three-phase transformers has been indicated. Utilizing silicon steel + 30wt. % ferrite + 40wt. %NiZn_Ferrite as multi-nanoparticles recorded the best consideration for upgrading the magnetic flux density of silicon steel nanocomposites single phase and three-phase transformer cores to the rate of 9.1%. The evolving of host magnetic materials as demonstrated by the results in the addendum illustrates the impact of distinctive and multi-nanoparticles on the magnetic flux density of new ferrite nanocomposites core for single phase and three-phase transformers. Ferrite +30wt. % nickel ferrite + 40wt. % silicon steel have been the best multi-nanoparticles for upgrading the magnetic flux density of ferrite build matrix to the rates of 15.4% and 15.8% of new ferrite nanocomposites single phase and three-phase transformer cores. Utilizing 20wt. % or 30wt. % or 40wt. % Ferrite distinctive nanoparticles reduced the magnetic flux density of nickel ferrite nanocomposites core for single phase transformers to the rates of 1. 1%, 1.7% and 2.4%, since it reduced the magnetic flux density of nickel ferrite nanocomposites core of three phase transformers to the rates of 0.9%,1.5% and 2.2%. However, the magnetic flux density of nickel ferrite + 30 wt. % ferrite nanocomposites core of single-phase transformers and three-phase transformers has been expanded toward including 20wt.%, 30wt.%, or 40wt. % of silicon steel nanoparticles with rates 3.3%, 5.6%, 7.8% for single phase transformers and with rates 3.5%, 5.8%, 8%. Moreover, the impact of individual and multi-nanoparticles on the magnetic flux density of new NiZn_Ferrite nanocomposites core of single phase and three-phase transformers has been indicated. The vitality of the plan of multi-nanoparticles inside the NiZn_Ferrite base matrix is shown by utilizing nickel ferrite and silicon steel as multi-nanoparticles. Utilizing NiZn_Ferrite + 30wt. % nickel ferrite +40wt. %Silicon steel as multi-nanoparticles expanded the magnetic flux density of single phase and three-phase NiZn_ ferrite nanocomposites to the transformer core for rates of 8.1%, 9.2%. However, utilizing NiZn_Ferrite + 30wt. %Silicon steel + 40wt. %Nickel ferrite as multi nanoparticles reduced the magnetic flux density of single phase and three-phase NiZn_ ferrite nanocomposites transformer core with rates of 1.7% and 0.67%.

8 TRENDS OF NANOPARTICLES IN TRANSFORMER CORES

The importance of improving electrical core performance is to upgrade the power transformer conduct. Preferred manufacturing strategies have been produced as an outcome of a finer understanding of the factors that iñuence magnetic properties. The caliber for electrical steel, that has sway with respect to core loss for electrical steel, is described by: (a) satisfaction of sheet insulation, (b) rate of silicon in the alloy, (c) concoction impurities, (d) grain size, (e) Precious stone introduction control, and (f) core lamination thickness (Albach & Brockmeyer, 1996; Buswell, 2001; Li et al., 2016; Reinert et al., 1999; Sievert, 2000; Zhang et al., 2011). Under typical circumstances, control transformers are intended to work basically inside the straight and the core's magnetization curve only. Therefore, a sensible sum of linearity is held to stay below the flux density values that might have over-excessive losses. However, it is facing more challenges of saturation. One of the major critical magnetic properties of electrical steels, the nonlinear AC magnetization curve (B-H curve) is a way for the configuration of power transformers and motors for decreased extent and progressed performance. The B-H bend is like a key for transformers and reactors displaying power framework for exchanging transient investigations, such as Ferro resonance. The nonlinear B-H bend communicated in an express explicit equation form is particularly useful and preferred to yield accurate results while saving computer time (Agheb & Høidalen, 2013; Dermawan et al., 2016; Ionel et al., 2006; Mclyman & Colonel, 2004). The present ideas of the physical beginning and component of losses of magnetic materials are reviewed under three conventional categories: Hysteresis losses, eddy current, or dielectric losses and lingering losses (Chang et al., 2010; Kefalas & Kladas, 2012; Mu & Lee, 2014; Shin et al., 2010). The pattern in nanotechnology science prompts the improvement of electric and magnetic materials that will upgrade their possibility requisitions in future vitality storage/transmission units. Nanotechnology techniques have connections for the interfacial self-destructive considerations and conduct between the nanoparticles and the polymer matrix in such nanocomposites (Chang et al., 2010; Hernández et al., 2010; Liu et al., 2013; Shin et al., 2010).

8.1 Magnetic Characterization of Three Phase Transformer Cores

The magnetic characterization of nanocomposites and multi-nanocomposites for three-phase transformer cores follows the same conduct with respect to single phase transformer cores. In any case, the greatest strength of magnetic flux density of three-phase transformers (1. 2T) is higher than the magnetic flux density of single-phase transformers (1. 1T). Similarly, as demonstrated in Table 2, the magnetic flux density of Ferrite, or Fe-Si steel nanocomposites to the three-phase transformer core is higher than the magnetic flux density of the same nanocomposites to single phase transformers. An increment of the magnetic field strength happens in the event of utilizing multi-nanoparticles. Moreover, Table 2 indicates the total loss performance of three-phase transformer cores and single-phase transformers for expanding the magnetic flux density.

8.2 Multi-Nanoparticles Technique in Magnetic Transformer Cores

Nanoparticles of Fe-Si Steel, Ni-Ferrite and NiZn-Ferrite are used to improve the magnetic characteristics of nanocomposites and multi-nanocomposites. Table 2 demonstrates total core loss for new single phase and three-phase transformer cores that are planned starting with Fe-Si steel nanocomposites and multi-nanocomposites, individually. The plan of nanoparticles inside the base magnetic matrix is a fundamental parameter for specifying the magnetic characteristics of the transformer core, such as reducing the total core loss. New multi-nanocomposite to a transformer core (Fe-Si steel + 30 wt. % ferrite +40 wt. % NiZn-Ferrite) is regarded as the best choice for reducing the total core loss of the single-phase transformer core during 52.90%. Hence, the multi-nanocomposites (Fe-Si steel + 30wt. % ferrite + 40wt. %NiZn-Ferrite) are recorded as the best inclusions for upgrading the magnetic flux density of Fe-Si steel nanocomposites single phase and three-phase transformer cores to the rates of 9. 16% and 9. 19% separately.

9 FORECAST AND RECOMMENDATIONS

Multi-nanoparticles strategy provides high magnetic characterization, low core loss and low magnetic reluctance of transformer core lamination against distinctive nanoparticles and universal magnetic materials. Th new plan enhances the successful relative permeability by controlling the kind and volume fraction of each nanoparticle inside multi-nanocomposites. Multi-nanoparticles system is the best pattern for controlling the reluctance of magnetic nanocomposites core laminations regarding the utilization

Table 2. Total Core Loss of Fe-Si Steel Transformers Cores

Parameter	Silver	Gold
High frequency dielectric constant, ε_{∞}	2.48	7.0
Plasma frequency, WP	1.35×10^{16} rad/s	1.40×10^{16} rad/s
Damping frequency, Wd (bulk)	7.62×10^{13} rad/s	3.78×10^{13} rad/s
Fermi velocity, Vf	1.40×10^{6} m/s	1.40×10^{6} m/s

of singular nanoparticles; therefore, the recently planned multi-nanocomposites have the best magnetic characteristics of power transformer cores. The type, concentration and plan of magnetic nanoparticles have the key parameters in multi-nanoparticles system for reducing the total loss of nanocomposites core and building the magnetic flux density higher than toward utilizing distinctive nanoparticles system. Moreover, Ferrite, Ni-Ferrite and NiZn-Ferrite nanoparticles are intriguing for upgrading the performance of Fe-Si steel nanocomposites transformer core. To fabricate process, Fe-Si steel nanoparticles are recorded to make the best individual nanoparticles for upgrading the effective relative permeability and reducing the reluctance of ferrite nanocomposites core lamination of single phase and three-phase transformers. In any case, the NiZn-Ferrite is recorded to make the best individual nanoparticles for upgrading the effective relative permeability and reducing the reluctance of Fe-Si steel nanocomposites core lamination for single phase and three-phase transformers. Incorporation radii are the vital elements for upgrading the permeability of magnetic nanocomposites materials; moreover, the diminish in the incorporation radii expands the effective relative permeability of nanocomposites magnetic materials. In addition, the relative permeability of nanoparticles should have further bolstering to have higher polarizing properties over the group base matrix for acquiring the high change of magnetic characterization. Using multi-nanoparticles has investigated the best new patterns for upgrading the effective relative permeability regarding the utilization of distinct nanoparticles; therefore, the new magnetic nanocomposites should host multi-nanoparticles for upgrading the permeability of base matrix magnetic materials. The plan of multi-nanoparticles inside the group build matrix is a critical figure for upgrading the effective permeability of nanocomposites. Distinctive multi-nanoparticles of chosen magnetic materials will upgrade the magnetic performance of new nanocomposites core lamination of single phase and three-phase transformers. This will be through expanding the effective relative permeability of core lamination and reducing the magnetic reluctance with the expanding volume fraction at frequency 50 Hz toward reducing the total loss of the nanocomposites core and expanding the magnetic flux density of the nanocomposites core under the greatest worth of the magnetic flux density.

The plan design of nanoparticles in the base matrix is answerable for upgrading the effective relative permeability of nanocomposites core lamination and reducing the magnetic reluctance of nanocomposites transformer core lamination, expanding the magnetic flux density and reducing the total nanocomposites core loss more than what has been upheld toward including interesting nanoparticles. Silicon steel nanoparticles have recorded the best Incorporation for upgrading the performance of single phase and three-phase transformer cores toward expanding the effective relative permeability of NiZn_Ferrite, ferrite, or nickel ferrite nanocomposites core lamination of single phase and three-phase transformers and by expanding the magnetic flux density of the transformer core. On the other side, the best Incorporation for reducing the reluctance of NiZn_Ferrite, ferrite, or nickel ferrite nanocomposites core lamination for single phase and three-phase transformers and reducing the total nanocomposites core loss. Ferrite, nickel ferrite and NiZn_ ferrite exceptional nanoparticles have been utilized for upgrading the magnetic trademark for silicon steel nanocomposites core lamination of single phase and three-phase transformers by expanding the viable relative permeability, reducing the magnetic reluctance of silicon steel nanocomposites core lamination, reducing the total silicon steel nanocomposites core loss and upgrading the nonlinear magnetization characteristics of silicon steel nanocomposites core for single phase and three-phase transformers by expanding the magnetic flux density of core below the maximum value of magnetic flux density for single phase and three-phase transformers by expanding the volume fraction at most extreme magnetic field strength of silicon steel core or expanding the magnetic strength in chosen characteristics of volume fraction.

REFERENCES

Abeywickrama, N., Podoltsev, A., Serdyuk, Y., & Gubanski, S. (2007). Computation of parameters of power transfromer windings for use in frequency response analysis. *IEEE Transactions on Magnetics*, *43*(5), 1983–1990. doi:10.1109/TMAG.2007.891672

Abeywickrama, N., Serdyuk, Y., & Gubanski, S. (2008). High-frequency modeling of power transformers for use in frequency response analysis (FRA). *IEEE Transactions on Power Delivery*, *23*(4), 2042–2049. doi:10.1109/TPWRD.2008.917896

Agheb, E., & Høidalen, H. (2013). Modification of empirical core loss calculation methods including flux distribution. *IET Electr. Power*, *7*(5), 381–390. doi:10.1049/iet-epa.2012.0368

Albach, M. D., & Brockmeyer, T. (1996). Calculating core losses in transformers for arbitrary magnetizing currents a comparison of different approaches. *Power Electronics Specialists Conference*. 10.1109/PESC.1996.548774

Barranger. (n.d.). Hysteresis and eddy-current losses of a transformer lamination viewed as an application of the poynting theorem. *NASA Technical Note*, 1-18.

Bohren, C. F., & Huffman, D. R. (2004). *Absorption and Scattering of Light by Small Particles*. Wiley.

Buswell. (2001). *Wire core for induction coils*. U.S. Patent 6,239,681.

Chang, Y., Hsu, C., & Tseng, C. (2010). Magnetic properties improvement of amorphous cores using newly developed step-lap joints. *IEEE Transactions on Magnetics*, *46*(6), 1791–1794. doi:10.1109/TMAG.2010.2045738

Charalambous, C., Wang, Z., Jarman, P., & Osborne, M. (2009). 2-D Finite-element electromagnetic analysis of an autotransformer experiencing ferroresonance. *IEEE Trans*, *24*(3), 1275–1283. doi:10.1109/TPWRD.2009.2016629

Chen, Z., Bai, B., Chen, D., & Chi, W. (2016). *Design of distribution devices for smart matrix based on nanocomposite magnetic material* (Vol. 8). IEEE.

Collett, W., & Buswell, H. (2011). *Calculation of core loss in a novel transformer design* (Vol. 5). IEEE. doi:10.1109/SSST.2011.5753788

Córcoles, F., Sáinz, L., Pedra, J., Sánchez, J., & Salichs, M. (2008). *Three-phase transformer modelling for unbalanced conditions. part1: core modeling and introductory examples.* IET Electr Power.

Davoudi, A., Chapman, P. L., Jatskevich, J., & Khaligh, A. (2009). Reduced-order modeling of high-fidelity magnetic equivalent circuits. *IEEE Transactions on Power Electronics*, *24*(12), 2847–2855. doi:10.1109/TPEL.2009.2031436

Deren, L., Zhang, L., Guangmin, L., Zhichao, L., & Shoaxiong, Z. (2012). Reducing the core loss of amorphous cores for distribution transformer. Progress in natural science: Materials international, 7, 244-249.

Dermawan, Almanda, Zahro, Ilmar, & Diniardi. (2016). Analysis magnetization current of harmonic phenomena and power factor as indicators of core saturation at transformer 3-phase. *Journal of Power and Power Engineering*, *6*, 1–6.

Easwarlal, C., & Palanisamy, V. (2013). Two variable approach for optimum design of a distribution transformer. *Anna University, Faculty of Electrical and Electronic Engineering*, *6*, 31–36.

Fcf, G., & Wsg, M. (2007). Magnetic core model. *IET Sci.Meas.Tech*, (l), 145–151.

Fiorillo, F., Dupre, L., Appino, C., & Rietto, A. (2002). Comprehensive model of magnetization curve, hysteresis loops and losses in any direction in grain-oriented fe-si. *Magn. IEEE Trans.*, *38*(3), 1467–1476. doi:10.1109/20.999119

Galvan, O., Escarel, R., Littlewood, C., Alviles, E., & Cruz, C. (2010). Separation of core losses in distribution transformers using experimental methods. *Can. Elect. Comput. Engineering*, *35*(1), 33–39. doi:10.1109/CJECE.2010.5783382

Gottlieb, I. (1998). *Practical transformer handbook.* Newnes.

Hasan, S., Taib, S., Hardi, S., Rahim, A., & Shukri, A. (2013). Core loss characteristics analysis of power transformer under different frequencies excitation. *IEEE 7th International Power Engineering and Optimization Conference, 5.*

Hernández, I., Olivares, J., Georgilakis, P., & Cañedo, J. (2010). A novel octagonal wound core for distribution transformers validated by electromagnetic field analysis and comparison with conventional wound core. *IEEE Transactions on Magnetics*, *46*(5), 1251–1258. doi:10.1109/TMAG.2010.2040623

Ionel, D., Popescu, M., Dellinger, S., Miller, T., Heideman, R., & Gilp, M. (2006). The variation with flux and frequency of core loss coefðcitens in electrical machine. *IEEE Trans*, *42*(3), 658–666.

Jaafar, M., Markovski, V., & Elleuch, M. (2004). Modelling of the differential permeability and the initial magnetization curve for ferromagnetic materials. *IEEE International Conference*, 460–465. 10.1109/ICIT.2004.1490334

Jieli, L., Abdallah, T., & Sullivan, R. (2001). Improved calculation of core loss with nonsinusoidal waveforms. *IEEE Industry Applications Society Annual Meeting*, 2203–2210. 10.1109/IAS.2001.955931

John, B., & Nough, G. (2002). Classics in magnetics; summary of losses in magnetic materials. *IEEE Transactions on Magnetics*, *38*(5), 3398–3408. doi:10.1109/TMAG.2002.802741

Juan, C., Galvan, O., Georgilakis, P., Campero, E., & Escarela, R. (2013). Core lamination selection for distribution transformers based on sensitivity analysis. *Electrical Engineering*, *7*, 33–42.

Kefalas, D., Georgilakis, P., Kladas, A., Souflaris, A., & Paparigas, D. G. (2008). Multiple grade lamination wound core: A novel technique for transformer iron loss minimization using simulated annealing with restarts and an anisotropy model. *IEEE Transactions on Magnetics*, *44*(6), 1082–1085. doi:10.1109/TMAG.2007.916019

Kefalas, D., & Kladas, G. (2012). Mixed si-fe wound cores five legged transformer losses and flux distribution analysis. IEEE.

Kulkarni, S., & Khaparde, S. (2004). *Transformer Engineering Design and Practice*. Marcel Dekker. doi:10.1201/9780367800598

Li, Y., Cao, L., Zhang, C., Yang, Q., & Li, E. (2016). Rotational core loss of silicon steel laminations based on three-dimensional magnetic properties measurement. *IEEE Transactions on Applied Superconductivity*, *26*(4), 1–5. doi:10.1109/TASC.2016.2532039

Liu, Y., Han, Y., Liu, S., & Lin, F. (2013). *Pulse magnetic properties measurement and characterization of Fe-based nanocrystalline cores for high voltage pulse magnetics applications* (Vol. 14). IEEE.

Martin, L. (2008). Theoretical study of multi-ferroic thin films based on amicroscopic model. *Journal of Physics Condensed Matter*.

Mclyman, & Colonel. (2004). *Transformer and inductor design handbook*. CRC.

Mészáros, I., & Vértesy, G. (2010). Modelling of normal magnetization curves of soft magnetic alloys. *Materials Science Forum*, *659*, 429–434. doi:10.4028/www.scientific.net/MSF.659.429

Moses, A. (1990). Electrical steels: past, present and future developments. *IEE Proc A*, *137*, 233–245. 10.1049/ip-a-2.1990.0039

Mu, M., & Lee, C. (2014). *Anew core loss model for rectangular ac voltages* (Vol. 7). IEEE.

Muhlethaler, J., Biela, J., Kolar, J. W., & Ecklebe, A. (2012). Improved core loss calc- ulation for components employed in power electronic systems. *IEEE Transactions on Power Electronics*, *27*(2), 964–973. doi:10.1109/TPEL.2011.2162252

Pleite, J., Gonzáles, C., Vázquez, J., & Lázaro, A. (2006). Power transformer core fault diagnosis using frequency response analysis. *IEEE Melecom Conf.*

Radun, A. (2014). Development of dynamic magnetic circuit models including iron saturation and losses. *IEEE Transactions on Magnetics*, *50*(5), 7027010. doi:10.1109/TMAG.2013.2293638

Reinert, J., Brockmeyer, A., & Dencker, R. (1999). Calculation of losses in ferro- and ferrimagnetic materials based on the modified steinmetz equation. *Industry Applications Conference, Thirty-Fourth Annual Meeting*, 3, 2087-2092. 10.1109/IAS.1999.806023

Rejea, A., & Ernst, H. (2005). Measurements and correction of no-load losses of power transformers. *IEEE Transactions on Instrumentation and Measurement*, *54*(2), 503–506. doi:10.1109/TIM.2004.843407

Schulz, R., Alexandrov, N., Tdtreault, J., Simoneau, R., & Roberge, R. (1995). Development and application of amorphous core-distribution transformers. *Journal of Materials Engineering and Performance*, *4*(4), 430–434. doi:10.1007/BF02649303

Shin, S., Schäfer, R., & Decoman, B. (2010). Grain boundary penetration by lancet domains in Fe-3%Si grain-oriented steel. *IEEE Transactions on Magnetics*, *46*(9), 3574–3581. doi:10.1109/TMAG.2010.2045508

Shintemirov, A., Tang, W., & Wu, Q. H. (2010). Transformer core parameter identiðcation using frequency response analysis. *IEEE Transactions on Magnetics*, *46*(1), 141–149. doi:10.1109/TMAG.2009.2026423

Sievert, J. (2000). The measurement of magnetic properties of electrical sheets steel. *J. Magn. Mater*, *5*, 647–651. doi:10.1016/S0304-8853(00)00251-1

Sweetser, C. (2002). Experience with sweep frequency response analysis (SFRA) measurements. *Proc. IEEE/PES Transformers Committee*.

Tang, Q., Wang, Z., & Jarman, P. (2012). Electrical steels and power transformer cores in deep saturation. *IEEE International Conference*, 4. 10.1109/CMD.2012.6416333

Tang, Q., Wang, Z., Philip, I., Anderson, I., Jarman, P., & Moses, J. (2013). *Approximation and prediction of ac magnetization curves for power transformer core analysis* (Vol. 8). IEEE Transactions on Magnetics.

Thabet, A., Abdel-Moamen, M. A., & Abdelhady, S. (2016). Effective magnetic characterization for new nanocomposites industrial materials using multi-nanoparticles technique. In *International Middle East Power System Conference*. Cairo: IEEE. 10.1109/MEPCON.2016.7836871

Thabet, A., Abdelhady, S., & Abdel-Moamen, M. A. (2018, June). Design of multi-nanoparticles technique for enhancing magnetic characterization of power transformers cores. *Advances in Electrical and Electronic Engineering Journal*, *16*(2), 167–177. doi:10.15598/aeee.v16i2.2354

Zhang, Y., Cheng, M., & Pillay, P. (2011). A novel hysteresis core loss model for magnetic laminations. IEEE Transactions on Power Conversion, 26(4). doi:10.1109/TEC.2011.2160866

Zirka, S., Moroz, Y., Marketos, P., & Moses, A. J. (2008). Comparison of engineering methods of loss prediction in thin ferromagnetic laminations. *J. Magn. Mater*, *320*(20), 2504–2508. doi:10.1016/j.jmmm.2008.04.083

Zirka, S., Moroz, Y., Marketos, P., Moses, A., Jiles, D., & Matsuo, T. (2008). Generalization of the classical method for calculating dynamic hysteresis loops in grain-oriented electrical steels. *IEEE Transactions on Magnetics*, *44*(9), 2113–2126. doi:10.1109/TMAG.2008.2000662

Chapter 9
Electronic NanoDielectrics

ABSTRACT

This chapter describes properties of electronic materials (electrical, mechanical, magnetic properties, optical transparency, luminescence). As a progress in technology, it has been detailed the properties of nanodielectrics electronic materials and the fabrication procedures of conducting nanodielectrics. Furthermore, this chapter presents the effects of nanoparticles in optical light nanodielectrics as an effect of nanotechnology on industrial applications. Finally, this chapter draws attention to the suggested investment procedures for industrial electronic materials in the future.

1 DESIGN AND INVESTMENT

Molecular electronics is perceived as a guaranteeing nomination to the nano-electronics of the future. It can offer solutions for the ultimate scaling down issue in nano-electronics. Molecular electronics can be anticipated as a workable and sensible route to amass extensive numbers of nanoscale questions (molecules, nanoparticles, nanotubes and nanowires) to structure new devices and circuit architecture. Its semiconductor nanowires NW represent a perfect framework for investigating low dimensional material science and are expected to play an important role as both interconnects and functional device elements in nanoscale electronic and optoelectronic devices. This chapter presents the plan keys for developing molecular polymers properties and its electronic applications. It mainly presents a general approach for the synthesis of conducting polymer nanomaterials with precisely controlled chemical composition, physical dimension, as well as electronic and optical properties. Subsequently, adscription of normal methodologies to the progressive gathering of NW building blocks into functional devices and complex architectures based on electric field or micro-fluidic flow. Next, discussions of a variety of new nanoscale electronic device ideas including crossed NW p-n diode and crossed NW field effect transistors FETs likewise offers an intriguing methodology of molecular electronics as resistive memory and molecular level arbitrary access memory RAM concerning illustration in this field with an extensive range of investigations. Sub-molecular electronics, i.e., data transforming toward the molecular-scale, are increasingly investigated as guaranteeing benefits in the nano-electronics of the future. Person meaning is the data transforming utilizing photo-, electro-, ionic magneto-, thermo-, mechanic- or chemo-active

DOI: 10.4018/978-1-7998-8536-8.ch009

impacts structurally. This definition excludes devices dependent upon thicker organic polymer materials, alluded as organic electronics. In this chapter, a short diagram of the nanofabrication of leading polymer nanomaterials is offered, then the electronic properties of few fundamental devices are audited starting with basic particles, such as molecular wires. The fast scaling down of electronics of the submicron scale has prompted momentous expansions to registering power. As a rule, coordinated circuit nanoelectronics will oblige conceptually new device fabricating blocks, versatile circlet architectures and, in a far-reaching way, the distinctive creation methodologies. Characterized nanoscale building-blocks have the possibility of try significantly past the breaking points of top-flight engineering organization by characterizing magic nanometer-scale measurements through concoction amalgamation (Bumm & Arnold, 1996). Past investigations have prompted an expansive extent of proof-of-concept nanoscale devices, including diodes and transistors dependent upon distinctive organic polymer molecules, quantum spots or carbon nanotubes (NTs) (Patrone et al., 2003). However, these investigations about distinctive nano-devices represent just an introductory step to nano-electronic circuits. It remains an incredible challenge to development from a single device level to the functional circuit level because of the following reasons: (1) the insufficient control of the properties of individual building blocks, (2) low device-to-device reproducibility, (3) the absence of dependable techniques for proficiently amassing and coordinating fabricating pieces under device arrays and circuits.

Next, the discussion of new nanoscale electronic device ideas includes polymer transistors, crossed NW p-n diode and crossed NW field effect transistors (FETs). Proliferation gathering of these versatile crossed NW device components empowers an inventory of coordinated structures. Organic transistor backplanes are ideally suitable to electronic paper applications and different presentation schemes. Low-cost and processing advantages, in addition to moving forward execution have prompted organic-based radio frequency identification tag advancement. The chemical interaction between various organic and polymer semiconductors can be exploited in chemical and biological sensors based upon organic transistors. Recently, a fascinating methodology of molecular electronics as resistive memory and sub-molecular level RAM has emerged in this field with an extensive range of investigations. The point portrayal of an extensive variety of photonic and optoelectronic devices, including nanoscale light-emitting diodes LEDs and incorporated nano-LEDs, concerning illustration organic vast scale coordinated circuit circuits has been executed for this methodology (Fiore et al., 2002).

2 PROPERTIES OF ELECTRONIC MATERIALS

Various possible applications exist for polymer nanocomposites. In particular, the enormous possibilities of distinctive combinations of nanocomposites between inorganic and organic species permit accomplishment of properties that can't be obtained with universal composite materials. This chapter concentrates principally on the consideration of the chemical viewpoints for nanoparticles and nanocomposites that direct their provisions. A couple of cases are viewed as points of interest. The samples address particular properties of the nanocomposites.

2.1 Thermal Properties

Consolidation of inorganic nanoparticles to a polymer matrix typically enhances its thermal stability, as described toward Thermal Gravimetric Analysis TGA (Liu & Su, 2005). The impact of thermal stability change is higher as the nanoparticles are functionalized for a coupling agent, which builds the interaction with the polymer matrix. These regularities are obviously exhibited by the case of PS/silica nanocomposites prepared by radical polymerization of styrene for silica nanoparticles or copolymerization for styrene or for a macro monomer, methacryloxypropyl silica nanoparticles. It has been officially stated that the thermal strength of polyether ether ketone look will be progressed to 40 C by the creation of nanocomposites for silica or alumina nanoparticles (Kuoa et al., 2005). The thermal strength of polymers for joined metal nanoparticles frequently exhibits exceptionally specific characteristics: following the relied upon change of the thermal stability by consolidation of 2 wt.%. of Cu nanoparticles under low-thickness polyethylene and further expanding of the measure of Cu nanoparticles rate beyond 2 wt.%. leading to deterioration in the thermal stability. This impact will be likely identified with the capacity of Cu to impact reactant oxidation of LDPE. This decision can also be understood in the perception that micro measured Cu–LDPE composites breakdown thermally toward 6wt.% metal content.

2.2 Mechanical Properties

Single and multi-walled carbon nanotube SWCNT and MWCNT have remarkable mechanical properties (tensile strength on the request of 100 GPa, firmness ca. 1000 GPa and disappointment strain up to 0. 4). These Lab-test parameters are near the hypothetically predicted strengths and they exhibit the incredible possibility to move forward the mechanical properties, particularly the crack toughness of polymer nanocomposites. However, a significant dissipating in the information and the distributed physical–mechanical properties of the nanocomposites tried in distinctive laboratories are regularly opposing (Wagner & Vaia, 2004). The workable makes for such inconsistencies are basically linked to the methodological issues. These incorporate varieties in nanotube nature and purity and scattering of the perspective proportion and degree of arrangement of the nanotubes examined. Therefore, unseemly systems utilized for dispersing the nanotubes in the polymer matrix, variable example preparation methods, and contrasts over nanotube–polymer interfacial chemistry are extremely vital. Often such data may be absent in publications. These issues which likewise assume a paramount role in universal composites processing can be great in additional discrimination of nanocomposites when requests of extent in the interfacial territory are made for every volume of filler included. Thus, near-term endeavors to decrease the variability of strength, thereby enhancing information unwavering strength and comparability, are incredulous and irreplaceable for the experimental group. Noteworthy advancement has been attained as of late in comprehension of the key regularities of deformity forms for absolute macromolecules, solitary nanotubes, and nanocomposites (Buchachenko, 2006), and particularly in fathoming the most troublesome issue of stress exchange through the interface between nanotubes and polymers (Wagner, 2002).

2.3 Optical Transparency

Nanocomposites can be prepared with tailored transparency using appropriate transparent polymer matrices and uniform nanoparticles below 100 nm 100 nm. Nanocomposites have been prepared by means of sol–gel syntheses that can be arranged optically transparent. Optical transparency is a property in light of the little period scale through which period division might exist. Hence, all the optoelectronic alternately photonic properties for devices have to be actualized, materials must satisfy high optical nature requests and must be stable again for a long time. Furthermore, the control of the refractive file and additionally the thickness and, therefore, the processibility of the materials are paramount. Profoundly transparent rutile (a crystalline polymorphous adjustment of titanium dioxide) nanoparticles with Normal diameter 2 nm are covered in situ by dodecylbenzenesulfonic corrosive (or polystyrenesulfonic acid) and separated starting with the response framework concerning the illustration of powder. Dispersions for this powder on toluene are essentially transparent at visible wavelengths but absorb UV radiation over a broad wavelength range. In the premise of these core–shell nanoparticles, profoundly transparent nanocomposites have become ready for PS and polycarbonate concerning illustration of matrix. These films go about as outwardly transparent UV blocking filters (Xia et al., 2006). Polymer-mediated creation for transparent alumina novel into a film and fiber for transmittance over 85% during 300–1000 nm is a new advancement (Fujimura et al., 2004). Films are prepared from inorganic/organic polymer composites and are tempered in 600–1000 c's to air. The composites are made of essential aluminum chloride and methylcellulose. Films are prepared by casting and fibers by spinning from aqueous solutions. Colored transparent alumina films can be prepared effortlessly starting with the composite by the acquaintance for move metal ions such as Au^{3+} and co^{2+}. Highly transparent films holding nano-sized silica domains are synthesized from acrylic polymer and watery monodisperse colloidal silica for a coupling agenize 3-(trimethoxysilyl) propyl methacrylate. The silica substance in the mixture dainty films is differed from 0 to 50wt.%, and the silica particles are measured starting with 20 to 40 nm and are controlled by the mole ratio of the coupling agent to silica. The thermal solidness of the ready mixture films is expanded for expanding silica substance (Zhuravlev, 1993). The refractive file decreased linearly with expanding the silica portion in the nanocomposite. The synthesized dainty films are prescribed for applications as latent films for optical devices.

2.4 Luminescence

Luminescence is critical principally in medical, biological, or pharmaceutical provisions. In general, Luminescence is a property of fragrant organic atoms or single separated semiconducting nanoparticles regarding nanoparticles, quantum dots in light of sulfides, selenides, or tellurides of zinc and cadmium that show the most noteworthy Luminescence effectiveness (Nussbaumer et al., 2002). These materials are harmful and cancer-causing and have a set of thermodynamic solidness against oxidation. These extreme hindrances revolve around troubles in provision. There is a solid requirement for non-toxic oxide luminescent nanocomposites that are stable like water. Luminescence of oxide nanoparticles is subject to fast aging brought about toward framing of hydroxides in the surface of the particles, prompting quenching of the Luminescence. Covering of the surface of the in situ produced nanoparticles for a polymer protecting of the surface of the oxide nanoparticles against the encompassing air, and likewise considerably against water. Moreover, polymer-coated oxide nanoparticles are suspended in water, above all to living applications. Recently, the possibility for this methodology has been exhibited, and it originated from the conclusion that polymer-coated oxide nanoparticles have the possibility to displace harmful luminescent materials. It has been indicated that nanocomposites, comprising an electrically insulating oxide center and PMMA covering, show solid Luminescence. This Luminescence relies on the interface, the place PMMA will be associated to by means of carboxylate holding of the surface of the ceramic nanoparticles. Luminescence probably begins at the carbonyl assembly of the covering polymer. Nanocomposites of this kind are synthesized utilizing a microwave plasma transform. Likewise, the Luminescence of these oxide/polymer nanocomposites is proportional to the surface, it will be workable to tailor the wavelength and the power of the Luminescence toward fluctuating the molecule extent.

On account of semiconducting oxides, such as ZnO, a mixture of the quantum restriction instrument with the surface cooperation system may be observed. Oxide–PMMA core–shell nanoparticles are ideally suitable to nanophotonic waveguides for the transport of electromagnetic vitality. It may be essential likewise to specify that they are not harmful or carcinogenic, rearranging taking care of applications. Mixes holding lanthanide ions have long been utilized concerning illustration of phosphors and laser materials due to their sharp, intensely luminescent electronic transitions, extensive Stokes shifts, long rot times and limited emanation groups. These emissions are generally an aftereffect of the absorption of vitality in the wide absorption band of the organic ligand emulated by proficient exchange of this vitality on energizing the national lanthanide ion and after that emulated by light emanation from particular cases or a greater amount of the eager states of the lanthanide particle. The lanthanides generally have utilized cathodoluminescent presentation phosphor screens, fluid lasers, and electroluminescent devices, and as probe ions for following the biochemical procedures, the sol–gel technique has been as of late utilized for the preparation of lanthanide complex-dispersed composites. The thermal and mechanical stability of the complexes may have been moved forward. The lanthanide complex-dispersed composites have handy benefits for phosphors, dynamic waveguides, optical sensors and markers of living atoms (De Stefano, 2004). A couple of recent illustrations of polymer-assisted creation for luminescent materials are examined. Phosphors produced of powders of Eu^{2+}, Dy^{3+} doped by strontium aluminate $SrAL_2O_4$ with high brilliance and long following gleam are obtained toward heating the forerunner gel (resulting from the sol–gel method) in 900 c under a reductive carbon air of dynamic carbon. The impact of changing focuses of $EuCI_3$ co-doped with 2-pyridinecarboxylic corrosive complexes determined through sol–gel union are accounted for by fan et al. (Fan et al., 2003). They found that the luminescence force level and the energy of its rot depend determinedly on the proportion of segments. Intriguing light-emitting polymer nanocomposite diodes have been created with exfoliated MoS_2 and SnS_2 distributed in polyfluorene matrix (Saraidarov et al., 2000). The intercalation of polyfluorene under the galleries of layered MoS_2 and SnS_2 are indicated to authorize a planar conformity on the conjugated polymer and forestall undesirable amass because of p–p interchain collaborations. This gives dramatically moved forward photograph physical strength and subsequently enlarged lifetime of the brilliant blue emanation. Europium-substituted hetero-polytungstates are joined under citric acid/polyethylene glycol matrices by the sol–gel strategy so as to move forward mechanical strengths. Esterification of citric corrosive happens promptly in the vicinity for pegging at the direct temperature (100–150 C). Europium-substituted heteropolytungstates are trapped in the gel polyester (Vollath et al., 2004). The films show the trademark emanation groups of the uncommon earth ion, which may not be discovered in the strong. The outcomes demonstrated that the creation of europium-substituted heteropolytungstates/ polyester slim novel into a film has a solid impact on the Luminescence of europium-substituted heteropolytungstates. $BaMgAl_{10}O_{17}$ phosphors doped for europium ions arranged through the sol–gel course by utilizing citric corrosive concerning the illustration of a chelating agenize and ethylene glycol as a polymerizing agenize turn out to hold great hopes for the blue-light segments of plasma show panels (Aharon & Albo, 2006). The sol–gel course prompts the amalgamation of pure-phase BaMgAl_10 O_17 phosphors regarding calcinations toward 1400C. The brilliance of the sol–gel determined phosphors can be higher over that of the phosphors obtained through the traditional strong-state course. The emanation intensities of the sol–gel-derived phosphors and then the afterward thermal corruption is more excellent than the individuals of the phosphors obtained by means of the solid-state rout.

2.5 Magnetic Properties

Characterization and Comprehension of the attractive properties of nanoparticles has been a challenge for decades and has provoked compelling exploration action motivated; particularly by provisions for attractive recording innovations. Some fascinating discoveries have as of late been new persons. Another manufactured course has been formed to get novel composites ready from the SWCNT joined poly (amic acid) holding a bis-thiazole ring and the comparing complexes framed by the chelation for poly (amic acid) and the bis-thiazole ring (Wang et al., 2004). These complexes show muddled attractive collaborations. Consolidation of uncommon earth ions in SWCNT– polymer frameworks through usage of polymer coordination science will be accounted for toward the same creators (Lu et al., 2005). Facilitated metal locales are used to expand request in the polymer and will regulate its attractive properties. Estimations demonstrate that this material possesses uncommon ferromagnetic properties. For expanding the temperature, charge progresses from certain to negative, and the attractive state proselytes starts with ferromagnetic to diamagnetic. These fascinating properties are unique in relation to the individuals of regular organic complexes and infer possible convenience of this nanocomposite material for attractive devices. Nickel nanoparticles with d =3–6 nm and limited measure conveyance are prepared toward particle trade and recuperation routines for every fluorinated sulfo-cation film polymer matrix (He, Suna, Wang, Liu et al, 2004). The ni nanoparticles in the polymer novel into a film are accepted to have a single attractive Web-domain and show super attractive properties. Regarding the greater part of the methods referred to above, the polymeric ion-exchange procedure is a standout amongst them for guaranteeing the straightforwardness of amalgamation and size control.

3 ELECTRICAL PROPERTIES

The lion's share for single person single walled carbon nanotube SWCNTs synthesized by CVD for an upheld impetus are semiconducting. This sort of nanotube exhibits field impact transistor FET conduct technique in room temperature and have been intensively investigated recently for a considerable length of time in nanoelectronics devices, including transistors and logic devices (Yoon et al., 2005). The semiconducting SWCNTs are developed as aid hole-doped p-type FETs for gap exhaustion and reduced conductance. It has been secured that sub-molecular oxygen adsorbed on the nanotubes is answerable for gap doping of SWCNTs (Abes et al., 2003). Evacuation of O2 could prompt an almost inalienable semiconducting conduct technique. Regardless of customary proclamations that CNTs have profoundly strong and inactive structures, their electrical properties are greatly delicate to charge exchange and concoction doping impacts by different atoms. Fictionalization of SWCNTs toward an amine-rich polymer, such as polyethyleneimine, haves about n-type FETs which are stable to air. A p-type semiconducting SWCNT as-made promptly reverts to a n-type FET, then afterward fictionalization by pei (due to be adsorbed O_2). Examination reveals that the adsorbed PEI donates roughly 1 electron for each 1000 carbon iotas on the nanotube (Liu et al., 2001). These realities unambiguously show that the vicinity of the polymer center of the SWCNT assumes a significant part of its requisition as a semiconductor over propelled nanoelectronics devices. The inter capitulars vein impact mostly crystalline POE copolymer for EO with oxymethylene that has been embedded in the exhibition space of MoSS_22 by utilizing the exfoliation/re-stacking properties of LiMoS_2. The coming about intercapitular vein intensify demonstrates improved electronic conductivity because of a structural conversion of the MoS_2. The polymer encapsulated can act as a robust electrolyte that can assist clinched alongside upgrading the dispersion of lithium starting with the anode of the cathode.

4 NANODIELECTRICS MATERIALS

Until 30 years ago, all carbon-based polymers were unbendingly viewed as insulators. The idea that plastics are settled on to direct power might have been viewed as absurd. Indeed, plastics have been extensively utilized by the electronics industry due to this real property. They were used as dormant bundling and protecting tape. This thick limited point of view is quickly evolving toward illustration of another population of polymers known as inalienably conductive polymer or electro dynamic polymers. Although this population will be clinched alongside its infancy, greatly similar to plastic business in the 30's and 50's, the possibility of employing these is very huge. In 1958, polyactelyene have started to be synthesized by Natta et al as a dark powder. This is found to make a semi-conductor with conductivity in the middle of 7×10^{-11} to 7×10^{-3} Sm^{-1}, depending on how the polymer is transformed and manipulated. It has been discovered that 1000 times more of the required Ziegler-Natta catalyst, ti (O-n-But)4 - Et3Al have been utilized. This film has been investigated and has been found to make semi conducting, with a comparable level of conductivity of the best of the leading bootleg powders. Further investigations, at first, meant to prepare slim films of graphite and demonstrated that the introduction of this type of polyactelyene on halogens expanded its conductivity billion-fold. Undoped, the polymer may have been silvery, insoluble and intractable, with conductivity comparative to that of semiconductors. When it is weakly oxidized toward mixes, such as iodine, it transformed toward a brilliant color and its conductivity expanded with around 104 Sm^{-1}. In the 1980's, polyheterocycles were initially created. Polyheterocycles were discovered to play a significant role in making air additionally stable over polyactelyene, although their conductivities are not high along these lines, normally regarding 10^3 Sm^{-1}.

Toward including different side bunches of the polymer backbone, subsidiaries which were dissolvable for different solvents were arranged. Other side bunches influenced properties, such as their color and their reactivity to oxidizing and decreasing operators. The metals regularly utilized for copper conductors have a high conductivity. Silver is all the more conductive, but due to its cost, it will not be useful by and large. However, it may be utilized within specific equipment, such as satellites and as a dainty plating to relieve skin impact misfortunes at high frequencies. Due to its simplicity of interaction by binding or clamping, copper may be even now for the greater part the regular decision of the majority of light-gauge wires. Aluminum has been utilized concerning the illustration of a conductor in lodging applications for expense reasons. It is more conductive than copper when compared in terms of unit weight. However, it has specialized foul issues identified with heat and its coefficient for thermal expansion, which has a tendency to slacken interactions. It is the basic metal utilized within high-voltage transmission lines, in consolidation of steel. Electrically conductive adhesives ECAs have been investigated for utilization on microelectronics bundling and inter connections without lead lead-in patch substitute because of such preferences as low curing temperature. They are normally made about conductive metal fillers and polymer tar. The metal fillers and polymer resins give electrical and mechanical interconnections between surfaces mount device segments and substrates.

To polymer resins, thermosetting resins, such as epoxy, polyimide and polyurethane are regularly utilized. For metal fillers, gold Au, silver Ag, copper Cu and nickel Ni in different sizes and shapes are utilized. Among them, silver is the most regularly utilized because of its high conductivity and Stability. However, the expense of conductive adhesives with silver filler is much higher than the traditional without lead lead-in solders. Therefore, ebbs and flows ECAs at present have a few limits as far as cost and additionally electrical and mechanical properties, such as displacements to without lead lead-in patch. A large portion of research endeavors has concentrated on the change of electrical conductivity and the unwavering strength of ECAs joints (Buchachenko, 2006; Bumm & Arnold, 1996; Fiore et al., 2002; Patrone et al., 2003; Wagner & Vaia, 2004). Additionally, the substitution cost of unreasonable Ag fillers by different metal fillers or new materials has been inspected for wider applications of ECAs (Kuoa et al., 2005; Liu & Su, 2005; Wagner, 2002) like Copper. Leading polymers—plastics that direct electricity—continue to reach business niches, such as customer electronics and antistatic textiles which can have military provisions. The majority of energizing applications is the utilization of leading polymers clinched alongside light-emitting devices LEDs, replant silicon as the accepted substrate material for clock radios, sound equipment, televisions, cell division telephones, car dashboard displays, and flying machine cockpit shows. Directing polymers furnish profits to commercial enterprises, like electronics by protecting against electromagnetic impedance EMI. Conductive polymers are likewise recently utilized within devices that recognize organically dangerous chemicals, plant emissions, and flavors or aromas in sustenance results. Currently, their conductivity is undoubtedly investigated in electrostatic materials, leading adhesives, electromagnetic shielding, simulated nerves, airplane structures, diodes, and transistors. In this study, ECAs utilizing surface-modified by different sorts of fillers and the impact of these distinctive sorts of fillers on the electrical safety of ECAs after curing and then after stability Lab-tests have been investigated.

4.1 Conducting Materials Properties

Conductor is a material which holds versatile electric charges. In metallic conductors, for example, copper or aluminum, the versatile accused particles, are electrons. Positive charges can be portable in the structure of iotas in a matrix that has absent electrons (known as illustration holes), or in the manifestation of ions, such as in the electrolyte of a battery. Constantly, conductors hold numerous electric charges which will move when an electric potential contrast is connected crosswise over separate focuses on the material. This stream of accuse will be implied by electric current. For the vast majority of materials, the immediate current may be proportional to the voltage, given that the temperature remains consistent and the material stays in the same shape and state. Most of the commonplace conductors are metallic. Copper is the usually normal material utilized for electrical wiring. Silver is the best conductor; however, it can be unreasonable for use. Gold is utilized for high-strength surface-to-surface contacts. However, there is likewise a large number of non-metallic conductors, including graphite, results of salts, and the sum plasmas. Moreover, there are a couple of basic cases of conductors: Silver, Copper, Gold, Aluminum, Iron, Steel, Brass, Bronze, Mercury, Graphite, Dirty water and Concrete. The contrasts in electrical conductivity of different materials utilized within marine electrical items are not always greatly comprehended. Settling on presumptions like the electrical conductivity of a material in view that it has a gander comparative to another conductive material of referred ampacity can prompt terrible effects. Maybe the vast widely recognized structure of this lapse is the substitution of metal or bronze to copper for electrical provisions. Metal is main material of 28% illustration conductive similar to copper. Bronzes as a conductive are 7%, as low as copper. Copper is the standard by which electrical materials are rated and conductivity appraisals are communicated regarding illustration of a relative estimation to copper. These appraisals are habitually communicated regarding illustration "28 IACS".

IACS is the abbreviation of universal tempered copper standard, and the number first "IACS" is the rate of conductivity a material has relative to copper, which may be viewed as a chance to be 100% conductive. This does not, of course, mean that copper has no resistance, but that instead it is the standard by which other materials are measured. The higher the % IACS, the more conductive the material becomes. This standard is dependent upon a tempered copper wire Hosting a thickness of 8. 89 g/cm3, 1-meter-long, weighing 1 gram, with a safety of 0. 15328 ohms. This standard is allocated the worth of 100 to 200°C (680°F). It is intriguing to inspect the IACS conductivity strengths of a portion as a relatable point material. A person may effortlessly accept that alloys, such as brasses and bronzes, since they are basically copper, are almost as conductive regarding illustration of copper. This is not the case. The little rates of tin, aluminum, nickel, zinc and phosphorus that make up these alloys corrupt the electrical execution of the coming about chemical to a significantly higher rate than their compositional rate in the chemical. A particular case ought not be inferred from this. However, it is concluded that metal needs to be further bolstered and can never be utilized within electrical provisions. There are instances in which the predominant pliable and machining strengths of metal make it a superior decision over copper concerning illustration in length, as the sectional zones are expanded proportionately to accomplish the conductivity of copper. However, some pieces might be in the provision. Regarding size, however, copper may be surpassed mainly by silver "around the materials regularly utilized for electrical provisions". Conductive polymers are organic polymers that direct power. Such exacerbates can make accurate metallic conductors or semiconductors. It is by and large acknowledged that metals direct power greatly and that organic exacerbates are insulating, but this class of materials combines the properties of both.

The great Lab-test focal point of conductive polymers will be their process ability. Conductive polymers are likewise plastics (which are organic polymers), and subsequently can consolidate the mechanical properties (flexibility, toughness, malleability, elasticity, and so forth throughout this way, observing and stock arranging of all instrumentation may be enhanced) for plastics with high electrical conductivities. Their properties can be fine-tuned utilizing the techniques of organic union. It has been found that around a dozen of different polymers and polymer subsidiaries experience this move when doped for a feeble oxidation agenize or lessening operator. They are in the greater part different conjugated polymers. This initial worth of effort has prompted an understanding of the components of charge stockpiling and accuse exchange in this framework. They all have an exceedingly conjugated electronic state. This likewise causes the primary issues with the utilization of these systems for process ability and stability. Most promptly conjugated polymers are flimsy in air and are not skilled for constant transformation. The vast majority of later scrutinization has been the advancement of exceedingly directing polymers with useful stability and worthy preparing strengths (Buchachenko, 2006; De Stefano, 2004; Fujimura et al., 2004; Kuoa et al., 2005; Liu & Su, 2005; Nussbaumer et al., 2002; Wagner, 2002; Xia et al., 2006; Zhuravlev, 1993). It is greatly indicated that graphite is a great conductor, formerly it might have been thought that polymers which substitute a carbon for another molecule cannot conduct. However, our more excellent information of conjugated frameworks has enabled the revelation of directing polymers. Similarly, as clinched alongside a conjugated framework, the electrons are just inexactly bound; electron stream can be reasonably expected. In any case, the polymers have covalently fortified the material necessities so that doped for electron stream will happen.

Doping is possible in the electrons or in the evacuation of electrons starting with the polymer. Once doping has occurred, the electrons in the pi-bonds have the ability to "jump" around the polymer chain. Likewise, the electrons will be moving along the atom, so an electric current happens. Thus, the conductivity of the material will be set as the electrons must "jump" over particles with the goal of exceptional conductivity. The particles must be well requested and nearly stuffed to the breaking point where separation "jumps" toward the electrons. This occurs better in Tran's un-doped polyactelyene. By doping, the conductivity of the polymer increases from 10^{-3} S m^{-1} to 3000 S m^{-1}. An oxidation doping (removal of electrons) can be done using iodine. The iodine attracts an electron starting with the polymer with a standout amongst the phytotoxin bonds. Hence, the remaining electrons can move along the chain (Fujimura et al., 2004; Xia et al., 2006; Zhuravlev, 1993). Conjugated twofold groups to an atom imply that the single and twofold groups exchange. These empower the electrons with a chance to be delocalized through the entirety framework to be imparted toward huge numbers of iotas. This implies that the delocalized electrons can move around the entirety framework (De Stefano, 2004; Nussbaumer et al., 2002). The developed p-systems of conjugated polymer is profoundly defenseless with chemical or electrochemical oxidation alternately decreasing. These change the electrical and optical properties of the polymer. Furthermore, toward regulating this oxidation and reduction, it is allowed to unequivocally control these properties. Since these responses are regularly reversible, it is allowed to deliberately control the electrical and optical properties for an incredible bargain of precision. It may be actually time permits with switch starting from a directing state to an insulating state. There are two principal aggregations regarding the applications of these polymers. The main aggregation uses their conductivity as its primary property. The second bunch uses this electro action (De Stefano, 2004; Fujimura et al., 2004; Nussbaumer et al., 2002; Vollath et al., 2004; Xia et al., 2006; Zhuravlev, 1993), as demonstrated in Table (1).

Table 1. Two Main Groups of Conducting Polymers Applications

Group 1	Group 2
1. Electrostatic materials 2. Conducting adhesives 3. Electromagnetic shielding 4. Printed circuit boards 5. Artificial nerves 6. Antistatic clothing 7. Piezoceramics 8. Active electronics (diodes transistors) 9. Aircraft structures	1. Molecular electronics 2. Electrical displays 3. Chemical, biochemical and thermal sensors 4. Rechargeable batteries and solid electrolytes 5. Drug release systems 6. Optical computers 7. Ion exchange membranes 8. Electromechanical actuators 9. 'Smart' structures 10. Switches

5 CONDUCTING NANODIELECTRICS FABRICATION

An assortment for creation techniques have been created for leading polymer nanomaterials. Regarding the different manufactured strategies, format strategy is the guaranteeing and capable apparatus to manufacture directing polymer nanomaterials. Format strategy includes the Incorporation of visitors, such as inorganic or organic constituents inside the void spaces of a group material. These voids are like the template, deforming the shape, size, and introduction of the prepared intensify. In general, format system is ordered by delicate and diligent templates. In as much as anodic aluminum oxide (AAO) membrane, track-etched polycarbonate (PC) and zeolite can be utilized as tricky templates; delicate templates incorporating surfactant, cyclodextrin, fluid crystal, and so forth. Compared for delicate and hard templates, the template-free system speaks to the creation strategy of leading polymer nanomaterials without the format.

5.1 Soft Template Method

Recently, a delicate format system has been utilized for the creation of different morphologies for polymer nanomaterials. There are a few delicate templates, such as surfactant, fluid crystalline polymer, cyclodextrin and functionalized polymer (Collins et al., 2000; Dai, 2002). Among them, surfactants, which suggest cationic, anionic and non-ionic amphiphiles are basically utilized for the arrangement of micelle as a nanoreactor. Microemulsions are macroscopically homogeneous mixtures of oil, water and surfactant which, on the minute level, comprise unique domains of oil and water differentiated toward a monolayer of amphiphilic. Microemulsions cannot be viewed as emulsions with thick, since the little droplet size, micro-and macroemulsions are in a far-reaching way separate. Macroemulsions mean traditional emulsions. In as much as emulsions are organically flimsy frameworks in which the droplets in the end will experience coalescence, microemulsions are thermodynamically stable with a high degree of progress with respect to the interior structure. In emulsion, stage detachment may be fast unless the framework is greatly blended. Droplets ceaselessly impact and coalesce and are broken by the shear exertion on the framework. The droplet measure may be subject to the framework segments (oil, stabilizer, period ratio) and the blending strengths. On the other hand, micro emulsions are thermodynamically stable (i.e., uncertainly stable) for droplet sizes changing from 10 will 100 nm. Generally, vast amounts of blended emulsifiers commonly comprising an ionic surfactant (e.g., sodium dodecyl sulfide SDS) and a short chain liquor are normally used to prepare these emulsions. (Raez, 2000). Throughout the polymerization, in an accepted emulsion polymerization, the monomer can be found in the Emulating four locations: (1) monomer droplets; (2) dormant monomer swollen micelles; (3) animated micelles that turn into monomer-swollen polymer particles when the polymerization occurs; (4) solute monomer in a watery period. Two strengths of oil-in-water microemulsion polymerization are unique in relation to the individuals of routine emulsion polymerization: (1) No monomer droplets and no dormant micelles exist; (2) the framework is optically transparent. Microemulsions go about alluring networking to polymerization responses. Microemulsion polymerization is a novel creation system which permits the preparation of ultrafine latex particles inside the span extent of 10 nm to100 nm and with limited extent circulation. However, the plan of microemulsions is liable to extreme imperatives because of high emulsifier level (over 10 wt.%.) necessary for accomplishing their thermodynamic stability. Mini emulsion systems are somewhere between Macro- and Microsystems. They hold numerous micelles and monomer droplets, but the monomer droplets are more modest in Microsystems.

For both micro- and mini-reaction frameworks in which the initiator is dissolvable in the nonstop phase, the component of polymerization may be resolved by the relative surface ranges of micelles versus monomer droplets. Compared with the mini emulsion (5– 10 wt.% of surfactants used), high fixation (15–30 wt.%.) of surfactant manifestations of hearty and conservative micelle, and the inward space of micelle can be utilized as a nano-reactor. Furthermore, regarding circle and layer morphologies, an extensive variety of morphologic spectra can be obtained by precisely regulating the manufactured states. The surfactant templates of sphere, rod, and layer nanomaterials are schematically spoken to clinched alongside. In general, micro emulsion polymerization has been broadly acknowledged for synthesizing directing polymer nanoparticle, empty nanosphere, center shell nanostructures and nanofiber. Polypyrrole (PPy) is an organic polymer obtained by oxidative polymerization of pyrrole. The different morphologies of PPy nanomaterials, such as ellipse, hexagon, tetrahedron, rod, and brush shapes were observed in a particular condition. The main impetus to figuring out the morphine may not be clear; it is shown that the delicate templates assume a paramount part in the structural advancement of PPy Nanomaterials. However, it is challenging to control the micelle structuring throughout microemulsion polymerization. In general, the polymerization procedure may be kinetically and thermodynamically flimsy due to ostwald ripening; the development by impacting the middle of monomer droplets and monomer utilization during polymerization. It is foremost that exact control of the micelle is crucial for processing monodisperse and nano-sized directing polymer nanomaterials.

5.2 Hard Template Method

Tough format system has been utilized in the 1-D nanostructures, such as nanotubes, nanorods and nanofibers for directing polymers. The regularly utilized templates are AAO membrane and track-etched pc membrane, whose pore size ranges from 10 nm to 100 µm. Hard format routines to synthesizing and directing polymer nanomaterials have been extensively reviewed ever since a long time ago. In general, the polymerization of a leading monomer has been performed toward nano-channel, since nano reactors and diligent templates are uprooted following polymerization in order to manufacture 1-D directing polymer nanomaterials. When the pore is loaded for proper material, it generates a self-amassed nanofiber. The membranes are economically accessible in settled sizes for particular pore diameters, and likewise arranged in the research center utilizing electrochemical implies. Fancied pore period and diameter can be controlled toward manufactured parameters. Martian et al. have utilized the tough format technique

Figure 1. VDP of PPy nanotubes using AAO membrane (Lewis, 2000; Lovell & Al-Aasser, 1997; Toshima, 2000)

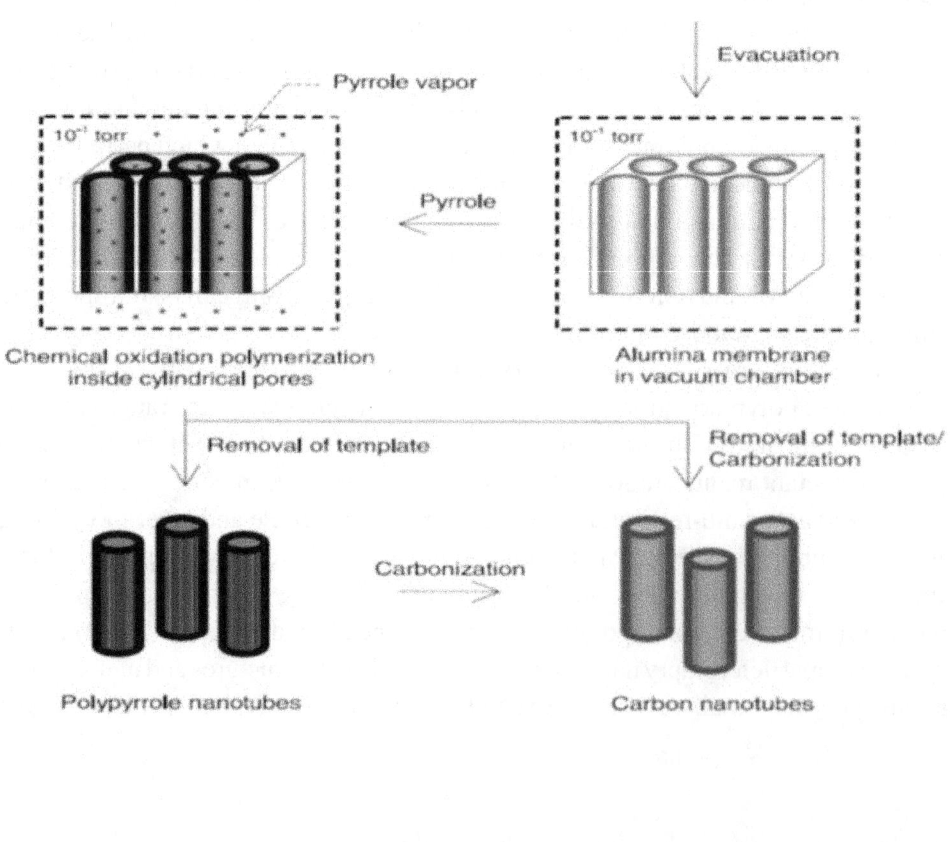

to prepare polymer nanomaterials (Lovell & Al-Aasser, 1997). In particular, nanotubes and nanofibers made of directing polymers were created inside the pores of nanoporous membranes. They synthesized the nanotubes of PPy, PANI, and poly (3-methylthiophene) with diligent templates utilizing concoction oxidation and electrochemical polymerization. Throughout the polymerization process, the directing polymer preferentially nucleates and grows on the pore dividers of membranes. Resultant polymer tubular structures are tuned by polymerization in the long run. While short polymerization chance gives the slim divider of the leading polymer nanotube, in length polymerization time handles thick dividers. In addition, PPy nanotubes and nanofibers are specifically created contingent upon the polymerization duration. Figure 1 shows few analysts as centered with respect to tricky format system so as to orchestrate different leading polymer nanomaterials (Lewis, 2000; Toshima, 2000).

Most films have a vital role in the semiconductor and coating industries. In addition, vapor affidavit transform has been produced to acquire prominent metal, semiconductor, or ceramic films. The affidavit procedure can be comprehensively sorted under Possibly physical vapor affidavit (PVD) or concoction vapor affidavit (CVD). Concoction and physical vapor period syntheses are well-established advances to extensive scale handling of metal, metal oxides, and inorganic nanoparticles. At the same time, a substance response happens clinched alongside CVD, PVD incorporates vanishing and sputtering. On VDP, a monomer or pre-polymer may be 1st deposited, et cetera polymerization is conveyed crazy on the surface by means of thermal, photochemical, or different start methodology. VDP has unique favorable circumstances through basic affidavit of polymer to uniform process, format strategies can be finished by essentially immersing the diligent format under a monomer/oxidant result. Recently, Jang et al. Prepared PPy nanotube and carbon nanotube (CNT) utilizing vapor deposition polymerization (VDP) interceded AAO film strategy. A Lab-test plan of this technique can be quelled, as shown in Fig. (7) Vapor affidavit of inorganic slim defect-free films, and it has the focal point again CVD in that different beginning monomers to be utilized. Since the thickness of monomer films can be decisively and reproducibly controlled, polymer films starting with a couple of nanometers with sub micrometers are structured. This methodology dispenses with the utilization of organic solvents, so that the undesired pinhole defects and darkness never present any issues again. There is likewise a developing enthusiasm toward eliminating dissolvable waste starting with techniques because of organic worries. Since every last bit of transforming has led to a well-controlled vacuum environment, any remaining monomer and dissolvable can be unequivocally figured out and recouped. Although the VDP procedure is tentatively convenient, the nature of polymerization has not been completely investigated. Additionally, in this configuration, it may be challenging to recognize the polymers that have been framed in the gas period for the resulting affidavit onto the surface, from the individuals polymerized by monomers adsorbed on the surface.

5.3 Template-Free Method

Template-free strategies have been extensively under the focus for directing polymer nanomaterials creation. Compared with difficult and delicate format methods, these methodologies give an easy and useful course to process pure, uniform, and prominent nanofiber. Format allows routines to envelop different systems, such as electrochemical synthesis, chemical polymerization, aqueous/organic interfacial polymerization, electrochemical synthesis, and scattering polymerization. The principle of all is that the forerunning worth of effort can be performed for electrochemical polymerization of PANI micro- and nanorod creation. Iridium and platinum needles are utilized as electrodes, and a PANI micro-Pole can be developed on the cathode surface. In turn, approaches can be identified with the fullerene subsidiaries doping for electrochemical polymerization so as to manufacture PANI microfibers and nanofibers. In addition, PPy micro tubes have been synthesized in the vicinity of β-naphthalene sulfur (β-NSA) toward electrochemical polymerization. This microtube has the diameter of 0. 8–2. 0 μm and the period of 15–30 μm. Electrochemical affidavit of PANI has been likewise conveyed crazy inside multi-walled CNT. Clinched alongside addition, PPy microtube (a couple of mm long and hundreds μm in diameter) will be framed in the vicinity of Na-toluene sulfur along platinum wire. PANI micro- and nanotubes have been ready toward concoction polymerization utilizing fullerene, nsa derivative and azobenzene sulfur. PPy microfiber and microtube are likewise created in the vicinity of β-NSA, and the morphological tenet of directing polymer can be tuned by regulating the fixation of β-NSA.

PANI nanofibers are created under encompassing nature's domain by utilizing aqueous/ organic interfacial polymerization. Notwithstanding vast scale production, the diameter of the resultant PANI nanofibers went from 30 to 50 nm, and fiber length differed between the middle of 500 nm and a few micrometers. Two results were prepared preceding the interfacial polymerization of aniline. Aniline monomer can be broken down in an organic dissolvable (e. g., CCl4, benzene, toluene, CS2). The opposite result comprises ammonitic peroxydisulfate and camphorsulfonic corrosive (CSA) to water. Then after blending two solutions, green PANI is framed during the interface between two layers and consecutively diffused under watery stage. Dull green PANI nanofiber can be gathered for water period. Polymerization yields start with 6 to 10 wt.% and give or take 95 vol% of the example of PANI nanofiber for the diameter of 30–50 nm. Bertino and coworkers created PANI nanofibers utilizing electrochemical amalgamation as another template-free pathway. PANI nanofibers were structured over watery results of aniline, ammonitic peroxydisulfate, and hydrochloric corrosive for gamma beam illumination. Commonplace PANI nanofibers have the diameters of 50–100 nm and the period of 1–3 μm. Enormous exploration to meet expectations has been performed on the amalgamation of directing polymer nanomaterials utilizing the scattering polymerization strategy (Martin et al., 1998). There are two classifications of scattering polymerization in order to manufacture the leading polymer colloids. The primary approach structures polymer stabilizer covered by leading polymer nanoparticles. In this case, the monomer and oxidant are broken-down for a settled fluid medium, and the structuring of insoluble directing polymer nanoparticles happens regarding the illustration of the polymerization returns. An alternate methodology includes inorganic oxide leading polymer center shell nanostructures. In-situ polymerization of monomer can be performed in the vicinity of silica nanoparticles, and the directing polymer will be adsorbed onto the silica nanoparticles as a seed. The resultant silica-conducting polymer core-shell nano material will be acquired in the non-attendance of extra polymers or surfactants. Thc molecule size and morphological tenet of silica-conducting polymer can be promptly shifted by changing the extent of silica, focus of silica and monomer, and kind of oxidant and monomer. In order to orchestrate directing polymer nanoparticle concerning illustration a core, different water solvent polymers, such as poly (vinyl alcohol) (PVA), poly (vinyl methyl ether), cationic and anionic polyelectrolyte's have been utilized as polymeric stabilizers. This water solvent polymer can be physisorbed onto the surface of leading polymer center because of hydrogen holding by and large. Large portions Examine Assemblies primarily create PPy and PANI covered silica nanoparticle utilizing scattering polymerization. Separate inorganic oxides (hematite, ceria and titania) and polymer latex particles, such as polyurethane, poly (vinyl acetate), and alkyl resins are used as the center a major aspect. In recent years, a scattering polymerization system has been produced for the creation of leading polymer nanorods (Cochrane et al., 2007).

Contrasted with accepted polymerization utilizing hydrochloric corrosive (HCl) as a dopant, 35wt.% of hydrochloric corrosive can be used in order to structure amphiphilic monomer ions. PANI nanorods with Normal diameter of 21–53 nm and length of 0. 5–1. 0 μm have been transformed contingent upon the sort of salt, mixing speed and response temperature. Scattering polymerization with APS oxidant transforms shorter PANI nanorods compared with FeCl3 oxidant. In addition, attractive blending decreases the period of PANI nanorods and the normal diameter of nanorods decreases with expanding the polymerization temperature. Currently, electrospinning system has been distinguished as a simple, proficient and versant method for the creation of leading polymer nanofibers. This electrostatic procedure includes the requisition of a high voltage electrostatic field on accuse of the surface of a directing polymer result droplet and the launch of a fluid plane through a spinneret. It has been shown that the diameter and state of leading polymer nanofibers can be influenced by Lab-test parameters, such as result viscosity and conductivity, surface tension, the sort of polymer and solvent, connected electrical possibility and separation between the slim and accumulation screen. Scientists have contemplated the impact of result viscosity on the shaping of PANI-CSA fibers [They discovered that a stable polymer droplet can be framed including 2wt.% poly (ethylene oxide) (PEO) under solution, in as much as no fiber creation of PANI/CSA happens for chloroform without PEO expansion]. It can likewise be shown that polystyrene (PS) and polyacrylonitrile fibers have been covered with PPy alternately au from watery result by electrostatic affidavit (Cochrane et al., 2007; Martin et al., 1998).

6 ELECTRONIC NANODIELECTRIC CHARACTERIZATION

6.1 Dielectric Constant Effect

A charge q placed in water (dielectric constant ε_w=80), in the vicinity of an external medium of low dielectric constant εe induces polarisation charges of similar sign in this medium, therefore repelling it. This effect leads to an increase in the surface energy of an electrolyte solution (investigated by Onsager and Saramas). The calculation of the electric field of an ion placed in the center of an infinitely long cylindrical channel passing through a membrane of dielectric constant εe $_a$nd filled with water can be found in standard textbooks (Armes et al., 1998). It was extended by A. Parsegian and others to the case of a finite length pore (Patrone et al., 2002). The original idea was attributed to P.-G. De Gennes. The external dielectric medium is less permeable than water to the electric field and channels the flux of the electrostatic field created by the charge q along the pore. Due to polarization charges, the normal component of the electric field at the surface is practically zero when the ratio $\frac{\varepsilon_w}{\varepsilon_e}$ tends to infinity.

6.2 Nanometric Size Effect

The surface Plasmon resonance doesn't straightforwardly include quantum restriction of the electronic states, but all the things considered are impacted by it. Indeed, the dielectric steady ε happening here may be that of the metal in the nanoparticle, and it will be changed regarding the greater part of metal. Clinched alongside a metal, two electronic components underlie the optical properties and must be made under account with respect to light–matter collaborations. This initially will be the interbond component. This is due to the quasi-free electrons of thickness ne in the conduction band, and it may be connected with optical moves without change of band commitment of the dielectric steady (Parsegian, 1969). The absorption, provided for by the fanciful part of the dielectric constant, is proportional to the normal optical impact rate of the electrons. This expression corresponds to three-body forms within which an impact happens with concurrent absorption of a photon frequency. This reflects the way that immediate absorption of a photon toward an electron can't happen, since the wave vector of the photon may be thick, as little compared to that of the electrons and the extent of the Brillouin zone. This will correspond to an almost vertical transition with constant electron wave vector. To achieve the goal of absorption, third particle's help is needed, i.e., in turn electron or a phonon, so as to ensure protection of vitality and energy (Phonons are ordinary vibration modes of the occasional matrix of iotas dependent upon a crystal). It is noteworthy that Electron–phonon interactions rule at room temperature.

The Drude Model was introduced by P. Drude at the beginning of the twentieth century to describe electrical conduction in metals. It is based on a kinetic approach in which conduction electrons are treated as a free electron gas subject to an external electric field E. The motion driven by the electric field is damped by collisions with other particles, i.e., electrons (electron–electron interactions) or ions in the crystal lattice (electron– lattice interactions), or with impurities and defects. This effect which leads to the resistivity of metals is introduced phenomenologically by a fluid friction force with damping rate. This approach has been generalized to the highest frequencies in the case of a field oscillating at frequency ω in the infrared or visible region of the spectrum. The damping rate depends on the frequency. The static term *s*, i.e., at zero frequency ω » 0, which governs the properties of the metal in non-alternating conditions, e.g., conductivity, must then be replaced by the optical rate 0, with 0 ω. Although it is based on a classical approach that appears at first sight to be poorly suited to the description of electrons in a solid, the Drude model has been successful in describing many phenomena and providing simple physical interpretations for them. In optics, the quantum microscopic description of light absorption does justify it and can be used to calculate the relaxation rate 0 explicitly, showing that it exhibits little dispersion in the visible. In a perfect crystal, both the electron–phonon and electron–electron interactions contribute; the latter only slightly at room temperature, i.e., about 10% of the total optical collision rate. The second mechanism is the interbond mechanism, related to optical transitions between two electronic bands, from filled bands to states in the conduction band or from the conduction band to empty bands of higher energy. In the alkali metals, these transitions occur at high frequencies and produce only a small correction to the real part of ε in the optical region.

In the noble metals, this contribution is much greater and is essentially due to transitions between the d bands and the s–p conduction band. When a photon is absorbed, an electron changes band and can only be excited into an unoccupied state of the conduction band (Pauli exclusion principle), i.e., above the Fermi energy E_f. Although quantum confinement is dominant for semiconductors (confinement of excitons with which the Bohr radius becomes comparable to the particle size and confinement of electronic states to low quantum number, close to the center of the Brillouin zone in the bulk), it is much less important for metal particles. The properties of the latter are related to the response of electrons in the vicinity of the Fermi level. They involve states of high quantum number for which a continuum of states can be used in conjunction with an approach similar to the one for the bulk medium. Size effects appear as corrections, and in the optical properties, they are revealed mainly through an increase in the optical collision rate of electrons in the Drude model (Parsegian, 1969).

6.3 Alloying Effect

Electronic adjustments are prompted by cooperation between the two metals, e.g., creation of a concoction bond between two partners, accuse transfer, polarization, and so on., or by the stress exerted on the catalytically pertinent metal toward its accomplice. Once again, these impacts will change the valence orbitals of the surface destinations of the catalyst, and subsequently the collaborations in the middle of the sub-molecular orbitals of the reactants and the response results. Most key investigations regarding electronic impacts are conveyed in chemical monocrystals or on metal monolayers kept for a metallic substrate. In this case, the electronic properties of surface atoms, the principal performing artists in reactant reactions can be probed and weakening impacts dismissed. Some Lab-test (Nussbaumer et al., 2002) and hypothetical fill in infer that significantly without the ligand effect, stress alone has an impact on the adsorption of co. varieties because tensional anxiety is shown to be connected with a narrowing of the band width when the separation between iotas decreases, taken after toward a progress in the middle of gravity of the band, and henceforth progressing to a cooperation for the reactants. The inverse happens for a layering anxiety. Every last one of the above considerations is concerned with the relative varieties in the vitality position of the middle of gravity of the valence band regarding the fermi level, so as to anticipate or clarify the contrasts in chemical reactivity in comparison to immaculate metals. However, on outright terms, the fermi level of the chemical varies starting with that of the immaculate metals, and this vitality parameter must be additionally taken under account, and of the relative varieties of the electronic band structure. This is a challenging issue and frequently dismissed. It is not a difficult matter to recognize the different alloying impacts (dilution impacts and electronic adjustments prompted by ligand impacts and stresses) happening on model monocrystalline surfaces. The regulate provision with genuine underpinned impetuses can be significantly convoluted by the multitude of sites and number of heterogenous cases for measure and arrangement in connection with the help that cannot be dismissed.

6.4 Quantum Size Effect

Physicists predicted that nanoparticles in the diameter extend 1-10 nm (intermediate between the middle of the measure of little atoms and that of heft metal) can reveal electronic structures, reflecting the electronic band structure of the nanoparticles, owing to quantum- mechanical guidelines (Brust & Kiely, 2002). The coming about physical properties are not the individuals of heft metal or the individuals of sub-molecular chemicals, since they determinedly rely on the molecule size, bury molecule distance, nature of the securing organic shell, and state of the nanoparticles. (Ashcroft & Mermin, 1981) The few "last metallic electrons" are utilized to tunneling forms in the middle of neighboring particles, an impact that can be distinguished by impedance estimations that recognize intra- and intermolecular techniques. The quantum size impact included in the de broglie wave length of the valence electrons will be of the same request regarding the illustration of the size of the molecule itself. Then, the particles act electronically as zero-dimensional quantum dots (or quantum boxes) important on quantum-mechanical standards (Varshni, 2001). Uninhibitedly versatile electrons are trapped over such metal boxes and show a trademark aggregate swaying frequency of the plasma resonance, providing for Ascent of the alleged Plasmon thunder band PRB watched close 530 nm in the 5-20 nm-diameter go. For nanoparticles, there will be a hole in the middle of the valence band and the conduction band, dissimilar to heft metals. The span prompted metal-insulator transition, depicted in 1988, will be observed if the metal molecule is little sufficient (about 20 nm) so that size-dependent quantization impacts happen. Then, the remaining electron waves with discrete vitality levels will be structured. Single electron moves happen between a tip and a nanoparticle, creating the perception of alleged coulomb blockades assuming that the electrostatic energy $E_{el}=e_2/2C$ is larger than the thermal energy E_T= KT. The capacitance C becomes smaller with smaller particles. This implies that single-electron moves can be observed toward a provided for temperature if c is very small, i.e., to nanoparticles since they are little sufficient (C < 10-18 F).

Huge varieties of electrical and optical properties will be observed when the vitality dividing line surpasses the temperature, and this adaptability will be for extraordinary useful benefits for provisions (transistors, switches, electrometers, oscillators, biosensors, catalysis) (Armes et al., 1998; Ashcroft & Mermin, 1981; Cochrane et al., 2007; Parsegian, 1969; Patrone et al., 2002; Smythe, 1968), such as single electron tunneling identified with the electrical safety of a solitary Pole molded atom giving a strength of 18±12 MΩ to self-amassed monolayers on gold substrate used to tie AuNPs stored from a group pillar. The move starting with metal-like capacitive charging to redox-like charging can be observed with alkanethiolate-gold nanoparticles of low dissimilarity to an electrochemical setup to coulomb staircase examinations. (Alivisatos, 1996; Brust & Kiely, 2002) Indeed, is shown at first that these AuNPs can suit 10 redox states (Brust & Kiely, 2002). It is indicated that easier temperatures upgrade the determination of quantized double-layer charging peaks on differential pulse voltammetry DPV perceptions. This prompts the determination of 13 peaks clinched alongside CH_2Cl_2 at 263 K for Au140 particles at the same time. However, a production by Quinn's aggregation uncovered strikingly well-resolved DPV practically equivalent to Au_140 particles, demonstrating 15 uniformly dispersed peaks at room temperature (295 K) compared to 15 oxidation states. The arranged solutions hold a remaining portion of Au that smears out the charging reaction over e locales in QDL peaks cover. The cathode possibility has scanned negative to certain. It can be foreseen that the number of noticeable charge states are set toward the extent of the accessible possibility window, extra peaks ought to be observed for a controlled climate and decreased temperature states (Alivisatos, 1996).

Thus, AuNPs act as different delocalized redox molecules, uncovering redox falls that are greatly referred to for inorganic and organometallic electrochemistry regarding other move metal groups and bi-sandwich complexes. The pioneering work by Schmid and co-workers on well-defined phosphine-stabilized gold clusters has shown the properties of quantum-dot particles for the first time. The number of atoms in these gold clusters is based on the dense packing of atoms taken as spheres, each atom being surrounded by 12 nearest neighbors. Thus, the smallest cluster contains 13 atoms, and the following layers contain $10n^2 + 2$ atoms, n being the layer number. For instance, the second layer contains 42 atoms, which leads to a total of 55 atoms for a gold cluster, and the chemical $[Au_{55}(PPh_3)_{12}Cl_6]$ has been well characterized by Schmidt's group. Recently, spectroscopic data have revealed discrete energy level spacings of 170 m eV that can be attributed to the Au_{55} core. Larger clusters containing, respectively, 147, 309, 561, 923, 1415, or 2057 (n= 3-8) atoms have been isolated (Chen & Ingram, 1998). Notwithstanding the respectable mixture of contributions, we will focus on the synthesis, stabilization, and different sorts of assemblies, et cetera of physical properties, with respect to chemical, supramolecular, and sensor properties. Finally, the most recent applications of biochemistry, catalysis, and nonlinear optical properties will be tackled preceding the closing of perspectives about AuNPs clinched alongside nanosciences and nanotechnology. Accordingly, our order is arbitrary, but the onlooker will regularly reach the soul of every paper from its title provided for in the reference section.

6.5 Thermal Expansion Coefficient Effect

The high thermal development coefficients of flawless plastics cause dimensional transforms during the formation and the illustration of the encompassing temperature progressions that will lead to undesirable and unsuitable instances to certain provisions. The last is of specific concern to car parts when plastics is incorporated into metals which have a great deal of down coefficients of thermal expansion, CTE. Fillers habitually include plastics to decrease the CTE. For low perspective proportion filler particles, the diminishment of CTE follows, less than the straightforward added substance principle, and may not be exactly large. In these cases, the straight CTE transforms are comparable to the three coordinate directions. However, when high perspective proportion fillers in fibers or platelets are included and greatly oriented, the impacts are substantially larger. In these cases, the CTE in the three coordinate directions can be really separate. The fibers or platelets regularly have a higher modulus and an easier CTE over the matrix polymer. Concerning illustration, the temperature of the composite changes, the matrix tries to augment or contract in its typical manner. However, the fibers or platelets stand up to this progress making restricting focuses in the two periods. When the filler to matrix modulus is large, the limitation to the dimensional change does not cause any noteworthy noise inside the heading of arrangement. Platelets can be acceptable in their limitation of two directions. When suitably oriented, the fibers can do so over particular case headings. Due to their state differences, fibers can cause more stupendous diminishment in the bearing of their introduction over the platelets. The CTE in the ordinary course of the fibers or the platelet plane can really increment when such fillers are included. Hypotheses in light of the components depicted are accessible to quantitatively foreseeing CTE conduct (Braunstein et al., 1999; Schmid et al., 1993). Montmorillonite MMT platelets are especially successful for diminishing CTE of plastics for well-exfoliated nylon 6 nanocomposites (Dasari et al., 2006). This information was measured in the stream heading of infusion formed bars. At the semi crystalline nylon, 6 matrixes are over its glass move temperature, the CTE decrease may be more than when underneath the Tg. Of course, the slick nylon 6 has a higher CTE over Tg beneath. However, due to its down modulus over Tg, the MMT platelets are powerful in diminishing the CTE. The CTE conduct technique is also a significant thought to the TPO materials utilized within car provisions (Deng & Mitchell, 2009). MMT may be considerably more productive during decreasing CTE than talc, as previously mentioned. These materials in composite hypotheses catch these patterns in a great way.

7 OPTICAL LIGHT NANODIELECTRICS

The strength and state of the portability hole clinched alongside PVC rely on the extent of states, such as the substrate temperature, degree of impurity, and abandon of the material. Any variety in such parameters prompts a movement in the absorption edge towards higher or have down vitality (Yousif et al., 2013). Recently, investigations on the electrical and optical properties of polymers have attracted great attention in perspective regarding their applications on optical devices for momentous reflection, antireflection, impedance and polarization properties (Ahmed, 2010; Al-Ramadin, 2000; Reddy, 2006). This article studies and examines the impacts of inorganic nanoparticles (ZnO, Al_2O_3, SiO_2) on the morphology, concoction holding and optical properties of polyvinyl chloride nanocomposite films. The refractive index, the band hole energy, the optical conductivity, the dielectric constant and the vitality misfortune work were identified methodologically for the room-temperature reflectance and transmittance information. This single section displays the effects of the proportion of inorganic nanoparticles on the optical properties of immaculate and altered PVC films.

7.1 PVC Nanocomposites Preparation

Films of (thickness ~100 – 200 μm) polyvinyl chloride, (polyvinyl chloride + metal oxide) (99:1 w/w), (95:5 w/w) and (90:10 w/w) were arranged by the solution-cast strategy utilizing tetrahydrofuran (THF) concerning the illustration dissolvable. The mixture of these results can be blended for 12h in room temperature. The PVC/metal oxide nanocomposite films were ready for dissolving aniline together for the economically accessible polyvinylchloride clinched alongside THF and throws under films by vanishing of the dissolvable. These films framed after the fact by glass plates. Diminishment can be attained toward treating films for watery results of NH_4OH or NaOH. The PVC/metal oxide nanocomposite films were presented with low weight and delicate heating for fluctuating span and uprooting of any voids inside nanocomposite films. Chemical arrangement of the arranged PVC and polyvinyl chloride nanocomposite

Figure 2. SEM images for polyvinyl chloride nanocomposites

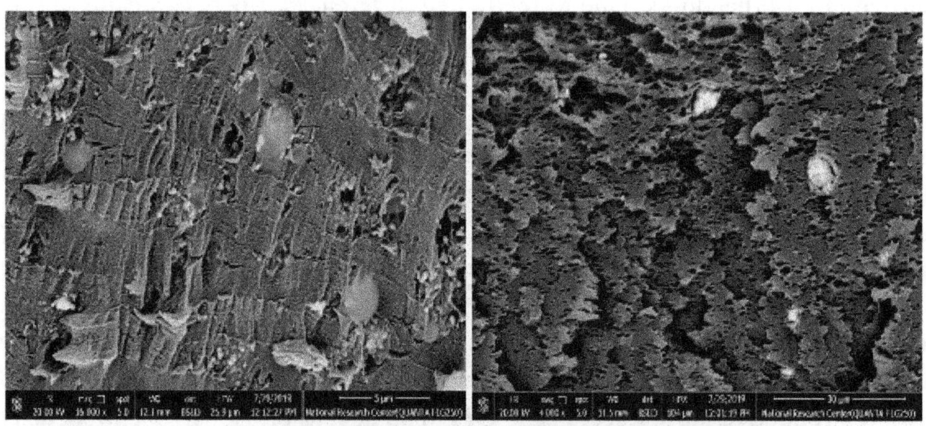

(a) ZnO /PVC nanocomposite (b) SiO_2 / PVC nanocomposite

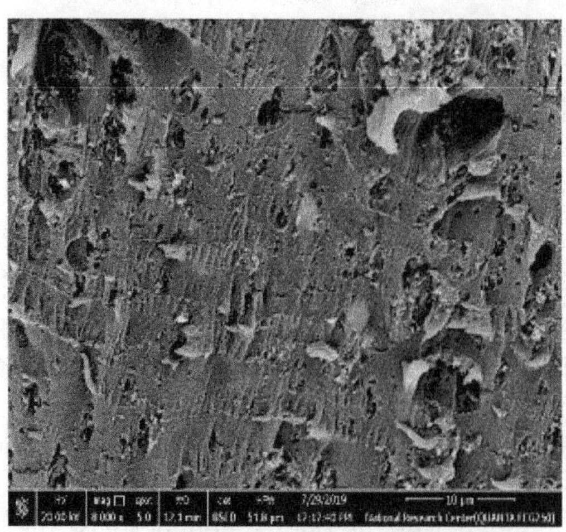

(c) Al_2O_3 / PVC nanocomposite

films have been tried by the FTIR ghastly examination. Every example utilized for infrared spectroscopic dissection has become ready as stated by standard. Every last one of the spectra was measured by utilizing (Jasco model 4100 – Japan) infrared spectrometer with a determination of 4. 00 cm^{-1} during space temperature, which blankets the wave number extent for 4000 – 400 cm^{-1} to assess the utilitarian aggregations in the ready Lab-tests. The morphological tenet of the arranged specimens was analyzed utilizing (JEOL SEM model JSM – 5500 – Japan) for accelerated voltage 10 kv. The optical properties investigations of polyvinyl chloride and polyvinyl chloride nanocomposite films have been performed by method for an electronic UV-2300 spectrophotometer with 5 nm step, toward typical frequency in room temperature in the wavelength go 200 – 1100 nm. The reproducible of the information has been checked by measuring few samples. SEM pictures of inorganic polyvinyl chloride nanocomposites slim films are demonstrated in Fig. (2) (a, b, c). It has flakes similar to the morphological tenet for high surface range. Furthermore, it illustrates the circulation of nanoparticles scattered in the polymer matrix.

7.2 Effects of Nanoparticles in Dielectric Characterization

The vibrational peaks of complex polymer nanocomposites were moved towards higher wavenumbers with the increment of nanoparticles proportion. Fourier transforms infrared dissection of PVC nanocomposites. Some other peaks were observed at around 1950, 1420 and toward 1772 cm^{-1}, these peaks were allocated to CH_3 deviated extending and bowing vibration of polyvinyl chloride (Reddy, 2006). The manifestation of new peaks alongside transforms in existing peaks to IR spectra demonstrates the muddling of polyvinyl chloride for metal oxide nanoparticles as shown in Fig. (3).

Figure 3. IR spectra for PVC and PVC – SiO_2 nanocomposites

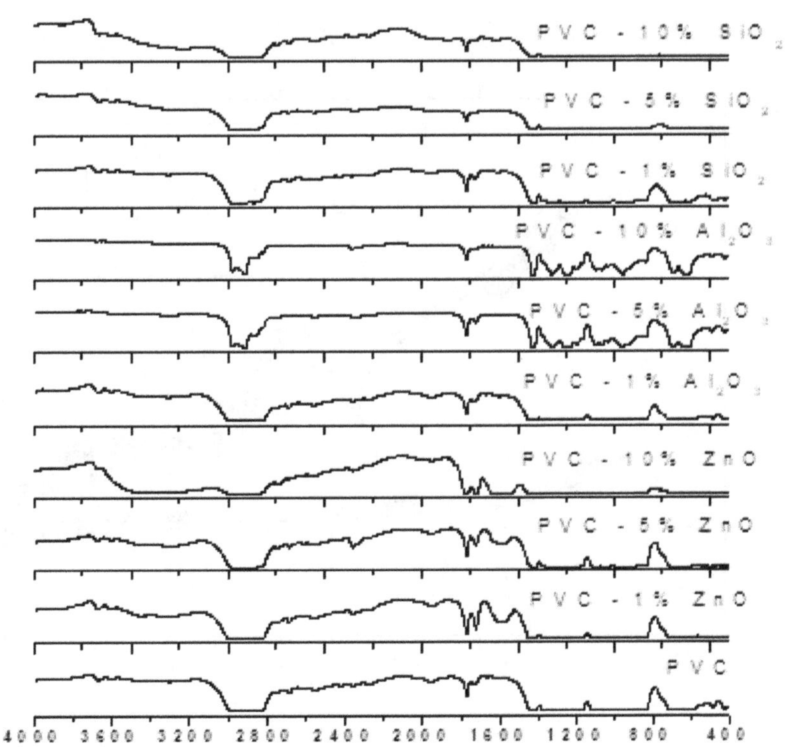

7.3 Optical Transmittance and Reflectance

The transmittance (T%) and reflectance (R%) of the arranged inorganic polyvinyl chloride nanocomposites slim films is preferable to be measured in the wavelength reach of 200 – 900 nm. Characterization and preparation of nanodielectrics have been detailed, furthermore, the examination of the range starting with easier and higher wavelength uncovers handy transmission of the arranged Lab-tests in the whole unmistakable area recommends its suitableness for second symphonious ERA devices (). The reliance on optical absorption coefficient with the photon vitality serves to ponder the band structure and the kind of move electrons (Denton et al., 1972; Ebnalwaled & Thabet, 2016). Therefore, Figs. (4-6) indicate the spectral distribution of the absorption at typical frequency of light to the analyzed compositions of inorganic polyvinyl chloride nanocomposites films. The expanding of metal oxides (SiO_2) nanoparticles proportion (1% wt., 5% wt. and 10% wt.) expands the absorption of polyvinyl chloride nanocomposite films. Otherwise, the expanding of the metal oxide proportion expands the absorption and declines the absorption of polyvinyl chloride nanocomposite films.

The optical constants (n and k) can be resolved from the transmittance and reflectance range in light of the known relations. It is clear that expanding the nanoparticles proportion builds the optical constants n and k. Figs. (7-9) show the variety of true and only optical conductivity for photon vitality. Likewise, (Ebnalwaled & Thabet, 2016), it is apparent from these figures that the worth of optical conductivity expands for expanding the proportion of dirt nanoparticles. It can be expected that the mud nanoparticles increment the excitation of electron toward photon vitality (El-Korashy et al., 2003), and likewise because of the increment of absorption coefficient by expanding dirt proportion (Abeles & Abeles, 1972; Sarkar et al., 2014; Usha et al., 2013). This demonstrates that the commitment of electron moves between the valence and conduction groups expands with expanding the proportion of the metal oxide

Figure 4. Spectral dependence of the absorption of pure polyvinyl chloride and ZnO /PVC nanocomposite

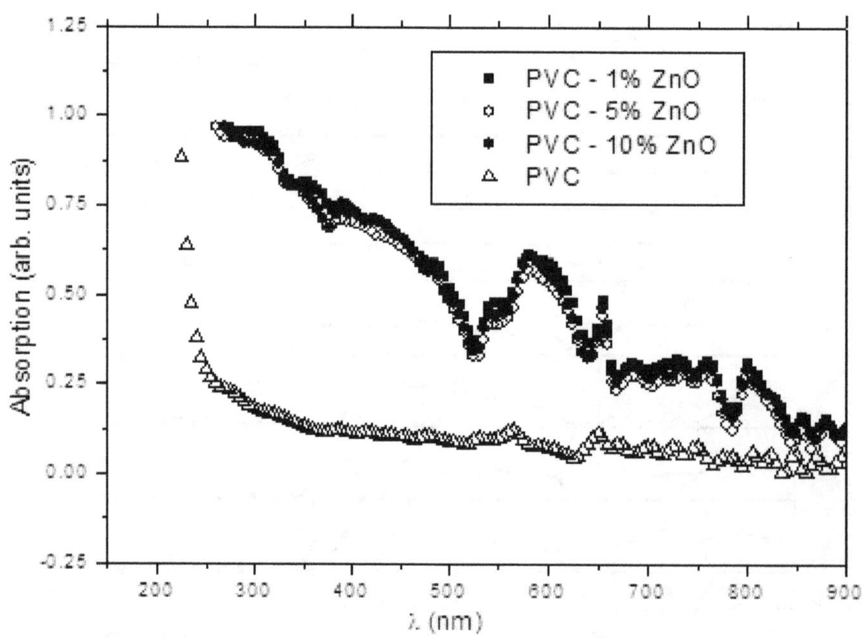

Figure 5. Spectral dependence of the absorption of pure polyvinyl chloride and SiO_2/PVC nanocomposite

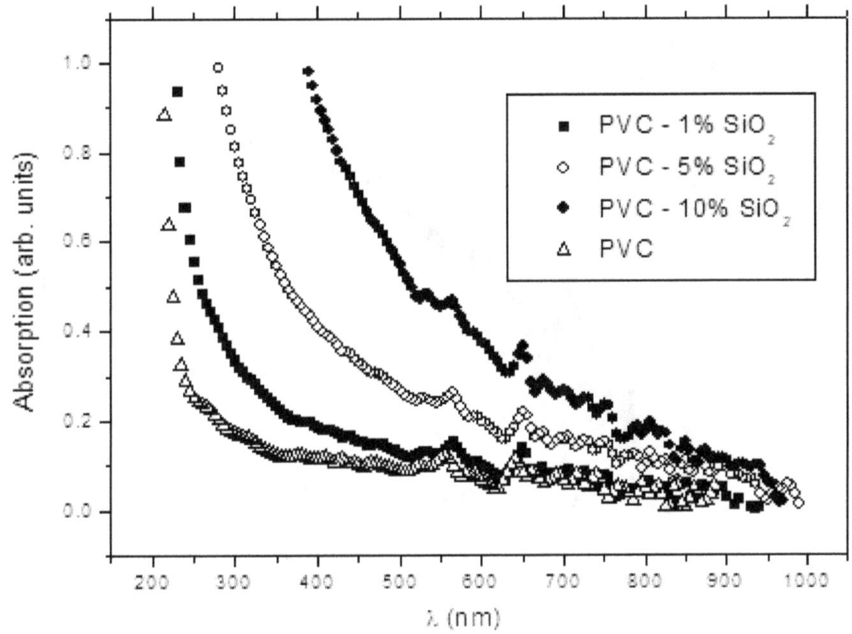

Figure 6. Spectral dependence of the absorption of pure polyvinyl chloride and Al_2O_3/PVC nanocomposite

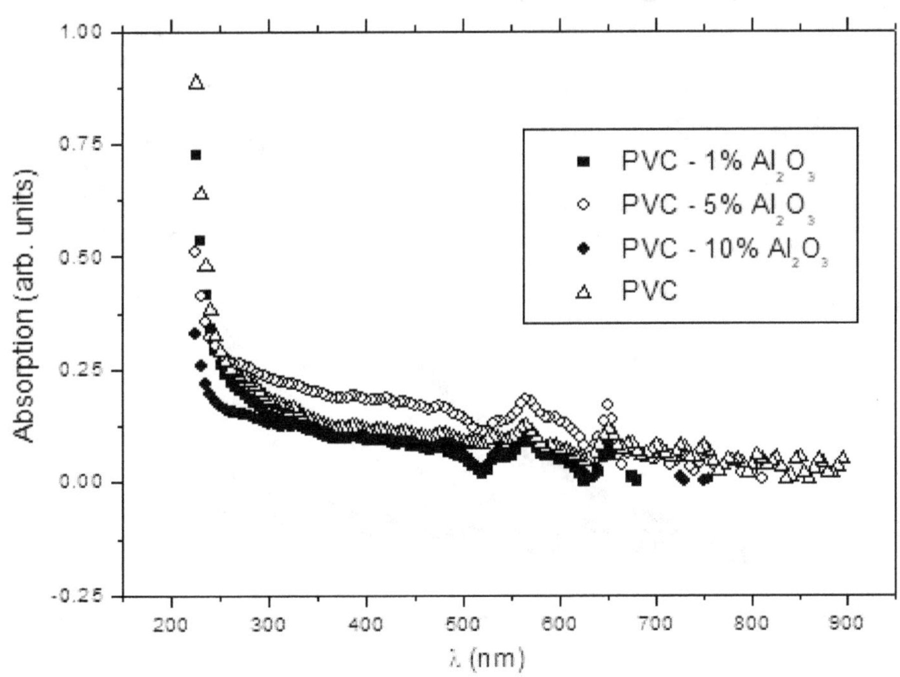

Figure 7. Dependence of real parts of optical conductivity on photon energy for pure polyvinyl chloride and ZnO / PVC nanocomposite

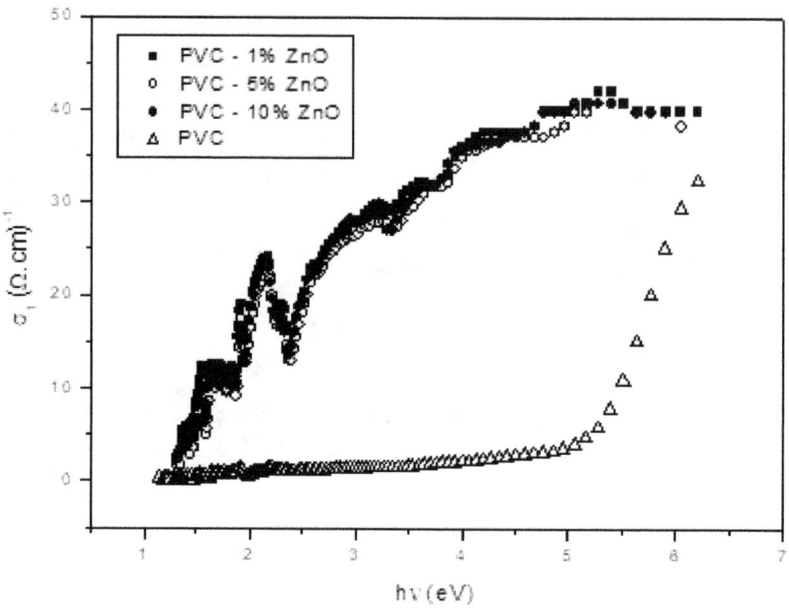

Figure 8. Dependence of real parts of optical conductivity on photon energy for pure polyvinyl chloride and SiO_2 / PVC nanocomposite

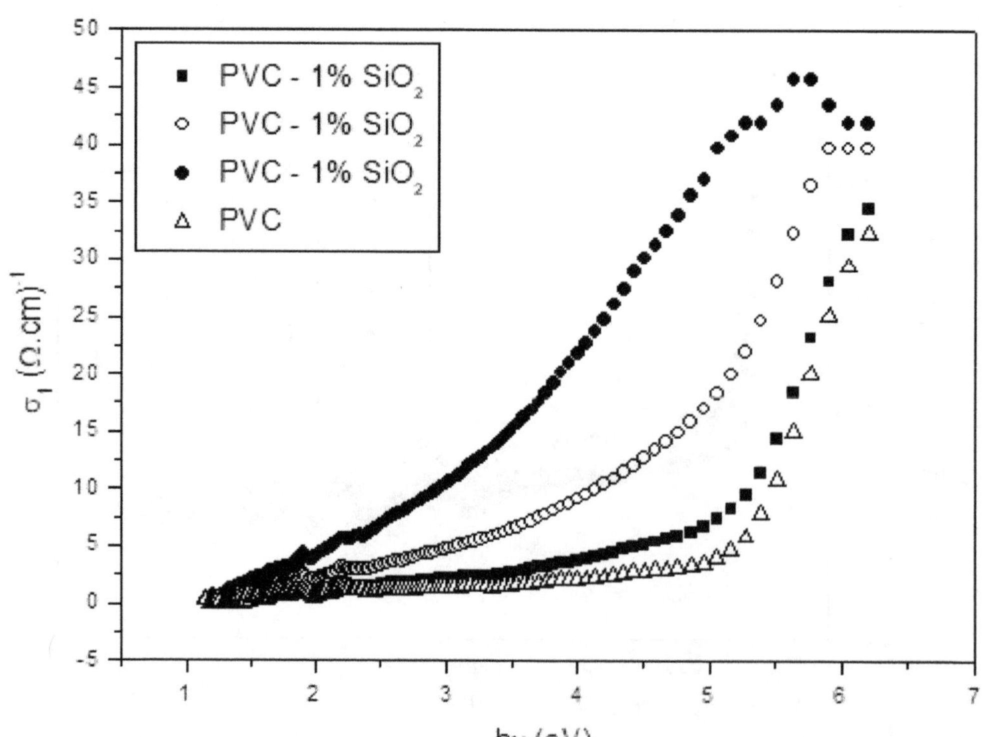

Figure 9. Dependence of real parts of optical conductivity on photon energy for pure polyvinyl chloride and Al_2O_3 / PVC nanocomposite

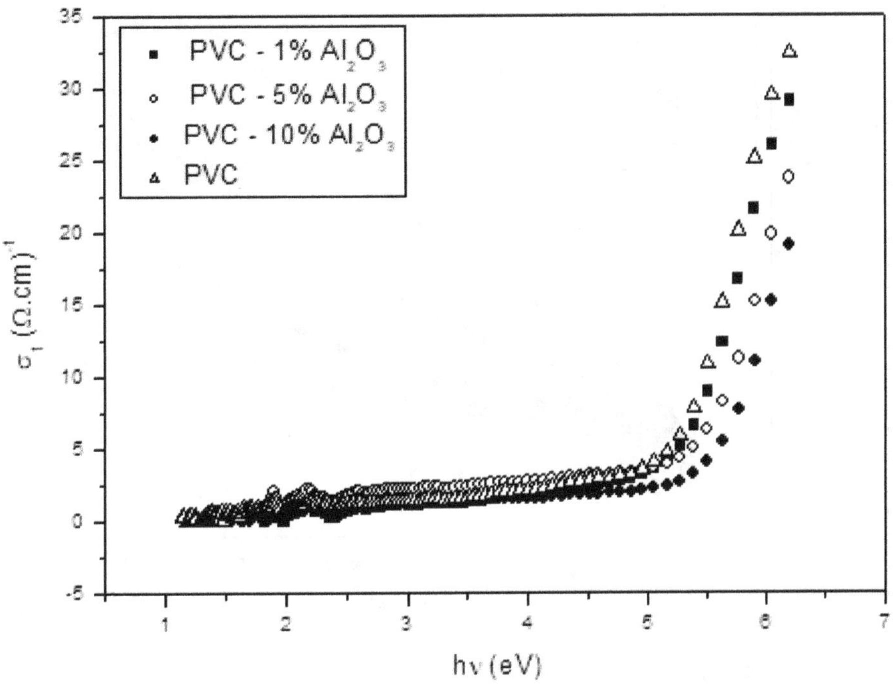

Figure 10. Dependence of ε_1 on photon energy for pure polyvinyl chloride and ZnO / PVC nanocomposite

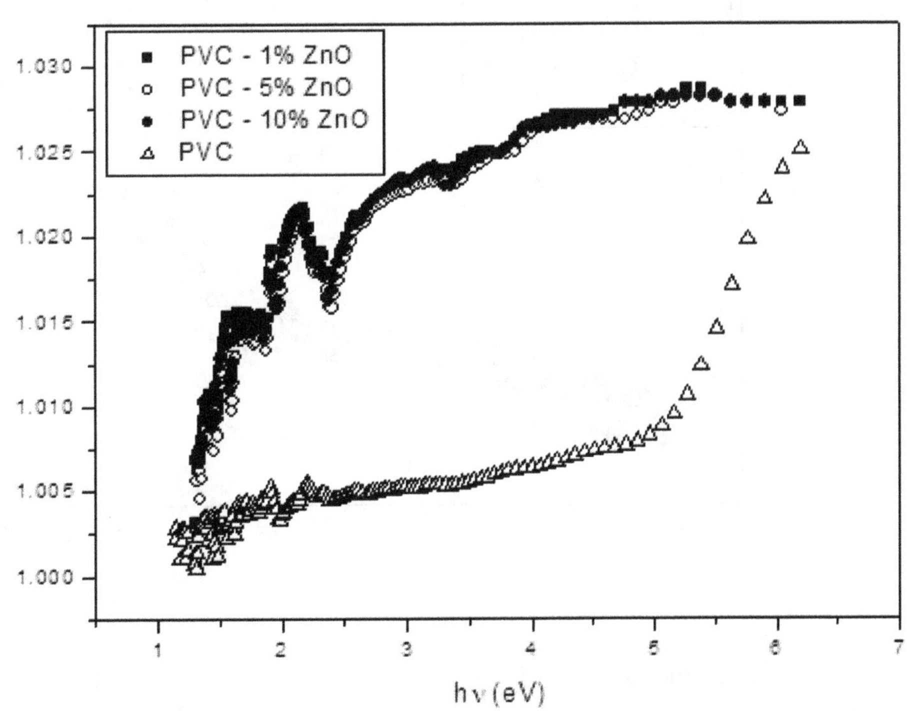

Figure 11. Dependence of ε_1 on photon energy for pure polyvinyl chloride and SiO_2/PVC nanocomposite

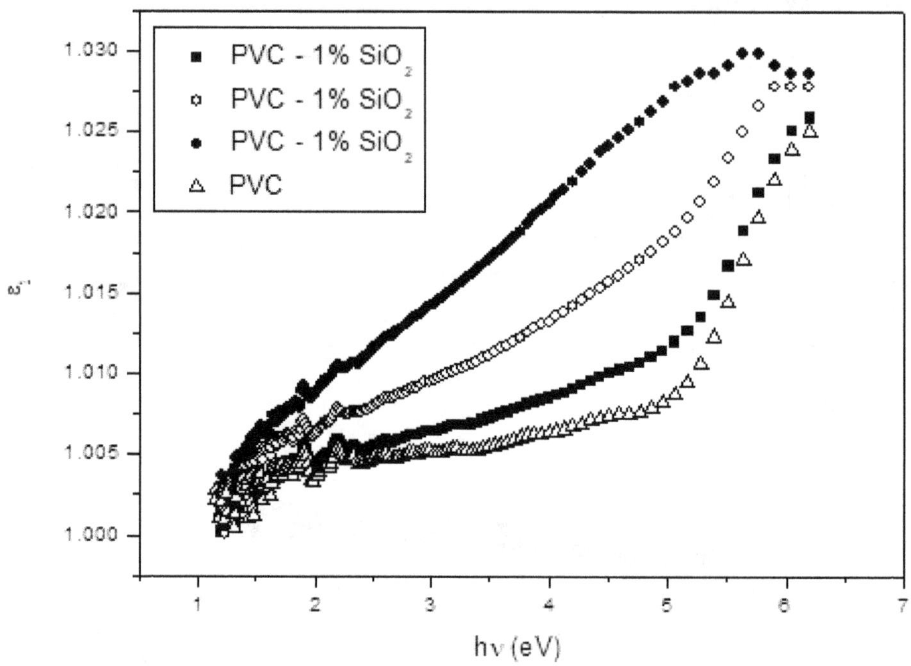

Figure 12. Dependence of ε_1 on photon energy for pure polyvinyl chloride and Al_2O_3/PVC nanocomposite

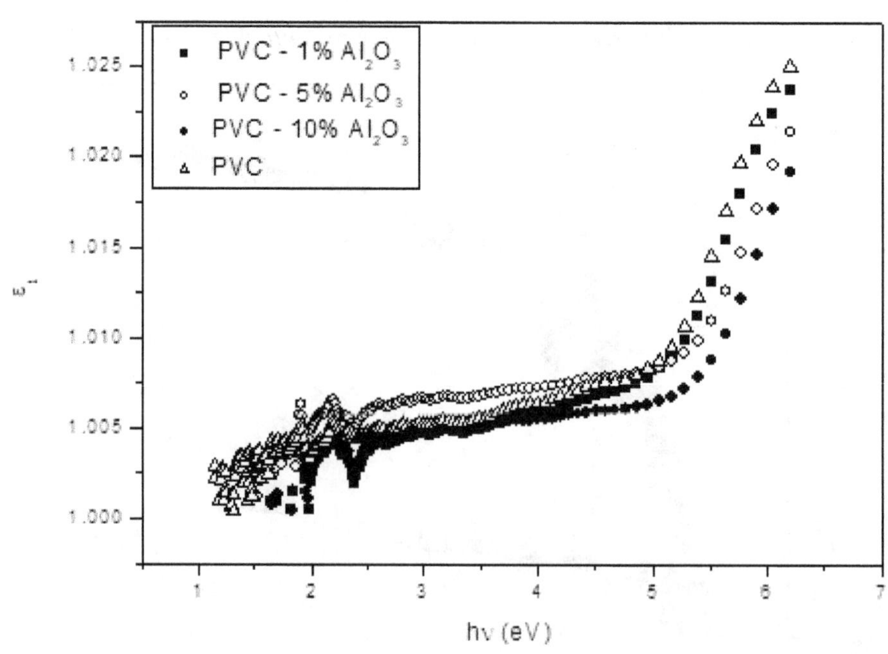

nanoparticles. The true and fanciful parts of electronic dielectric constants ε_1 and ε_2 are joined with n and k values (Abeles & Abeles, 1972). The reliance of ε_1 with respect to photon vitality can be indicated in Figures (10-12). It is self-evident that the true parts of dielectric constants expand with expanding the metal oxide nanoparticles and demonstrate the changeover optical reaction of polyvinyl chloride for including variant metal oxide proportions (Ebnalwaled & Thabet, 2016).

8 FORECAST AND RECOMMENDATIONS

In this chapter, a short study of the properties of molecular polymers and its electronic applications has been presented. This chapter started by presenting an approach for the union of directing Polymer Nanomaterials for unequivocally regulated concoction composition, physical dimension, and electronic and optical properties. Subsequently, adscription of the normal methodologies to the progressive gathering of NW building squares under utilitarian devices an arrangement of key progresses that should eventually empower nano-electronics creation and manufacturing starting with the base up been settled on. Afterwards, a methodology for the union of NW fabricating squares for well-controlled chemical compositions, physical dimensions, and electronic properties is devised. Then, the outline and operation of future electronic devices that utilize these nanostructures, such as field-effect transistors, concoction and living sensors, super capacitor and nanoscale optical devices are presented. Additionally, an extensive variety of nanoscale electronic and optoelectronic devices and circuits starting with NW building blocks, which hold the guarantee of powering future electronic devices has been investigated.

Nanotechnology is a revolutionizing electronic engineering, and nanoparticles continue to intrigue to researchers due to the compound and actual properties that come from incorporating such little particles. For instance, scientific experts and scholars build nanoparticles to take on wanted constructions with the goal that they will tie to harming particles in infected patients. Comparative thoughts are taken into developing semiconductors, since electrical attributes are adjusted when the size of semiconductors are significantly diminished. The quantum dab, for instance, utilizes semiconducting materials to frame a nanocrystal which can be utilized to create semiconductors, diodes, lasers, and sun-based cells. Beside its semi-directing properties, quantum dabs are especially intriguing due to their fluorescence. The shading that is delivered from the nanocrystal relies upon its size which changes the light that is discharged. One application that rose up out of quantum spots is the Sony TV that fuses innovation from QD Vision to make brighter Sony TVs. Venturing into bigger nanomaterials and nanowires, which are a couple of nanometers in measurement and a couple of micrometers long, is the reason for making little electrical gadgets. Since 1991, interest and exploration in nanowires has been developing dramatically. These nanowires are silicon-based but have electrical properties that are vastly different from bigger silicon parts and they bring about specific sorts of nano-transistors. One sort is the field impact semiconductor, which is utilized in basically every electrical gadget and clears the way for progressively little gadgets.

Novel approaches to electronic and optoelectronic materials and device assembly are some of the challenges in solar power generation, sensing, high gain materials and nanoscale lasers with particular interests in bio-assisted assembly, photoelectrochemical processes and optical cavity design. There has been a development of green and sustainable batteries and supercapacitors for power portable electronics and electric vehicles, as well as 2D-materials and graphene films for future electronic and spintronic devices. The researcher's challenges have been to conduct advanced research on integrated circuits and contribute, to the most possible extent, to the advancement of micro-nano-electronics. Nowadays, there are theoretical and experimental investigations of the properties of inorganic/organic/hybrid nanostructures and development of novel electronic/optical/thermoelectric devices and circuits based on these nanostructures. Also, there is a theory of quantum computing with superconducting qubits, continuous quantum measurement, and quantum feedback control which explores Silicon-based and Zinc Oxide-based thin films, nanowires and quantum dots and their nanoelectronics and optoelectronic devices.

REFERENCES

Abeles, F. (1972). *Optical Properties of solids* (F. Abeles, Ed.). North – Holland Publishing Company.

Abes, J. I., Cohen, R. E., & Ross, A. (2003). Selective growth of cobalt nanoclusters in domains of block copolymer films. Chemicals Material Journal, 15, 1125–1131.

Aharon, E., & Albo, A. (2006). Stable blue emission from a polyfluorene/layered-chemical guest/host nanocomposite. Advanced Functional Materials, 16, 980–986.

Ahmed. (2010). Optical properties modification of poly (vinyl chloride) using complexes of 2-amino acetate benzothiazole. *Journal of Engineering and Applied Sciences (Asian Research Publishing Network)*, *5*(12).

Al-Ramadin, Y. (2000). Optical properties of poly (vinyl chloride)/poly (ethylene oxide) blend. *Optical Materials*, *14*(4), 287–290. doi:10.1016/S0925-3467(00)00007-0

Alivisatos, A. P. (1996). Semiconductor Clusters, Nanocrystals, and Quantum Dots. *Science*, *271*(5251), 933–937. doi:10.1126cience.271.5251.933

Armes, S. P., Skotheim, T. A., Elsenbaumer, R. L., & Reynolds, J. R. (1998). *Handbook of conducting polymers*. Marcel Dekker.

Ashcroft, N. W., & Mermin, N. D. (1981). *Solid State Physics*. Holt-Saunders.

Ashour, A., El-Kadry, N., & Mahmoud, S. A. (1995). On the electrical and optical properties of CdS films thermally deposited by a modified source. *Thin Solid Films*, *269*(1-2), 117–120. doi:10.1016/0040-6090(95)06868-6

Brust, M., & Kiely, C. J. (2002). Some Recent Advances in Nanostructure Preparation from Gold and Silver. *Physic Chem. Eng. Asp.*, *202*, 175–186. doi:10.1016/S0927-7757(01)01087-1

Buchachenko, A. L. (2006). New horizons of chemistry single molecules. *The Russian Review*, *75*, 3–26.

Bumm, L. A., & Arnold, J. J. (1996). *Are single molecular wires conducting?* (Vol. 271). Science Journal.

Chen, S., & Ingram, R. S. (1998). Gold Nanoelectrodes of Varied Size: Transition to Molecule-Like Charging. *Science*, *280*(5372), 2098–2101. doi:10.1126cience.280.5372.2098 PMID:9641911

Cochrane, C., Koncar, V., Lewandowski, M., & Dufour, C. (2007). Design and Development of a Flexible Strain Sensor for Textile Structures Based on a Conductive polymer Composite. *Sensors (Basel)*, *7*(4), 473–492. doi:10.33907040473

Collins, P. G., Bradley, K., Ishigami, M., & Zettl, A. (2000). Extreme oxygen sensitivity of electronic properties of carbon nanotubes. *Science*, *287*(5459), 1801–1804. doi:10.1126cience.287.5459.1801 PMID:10710305

Dai, H. (2002). Carbon nanotubes: synthesis, integration, and properties. Account of Chemical Research, 35, 1035–1044.

Dasari, A., Yu, Z. Z., Yang, M. S., Zhang, Q. X., Xie, X. L., & Mai, Y.-W. (2006). Micro- and nano-scale deformation behavior of nylon 66-based binary and ternary nanocomposites. *Composites Science and Technology*, *16*(16), 3097–3114. doi:10.1016/j.compscitech.2005.03.020

De Stefano. (2004). *Optical Sensors for Vapors, Liquids, and Biological Molecules* (Vol. 3). IEEE Transactions on Nanotechnology.

Deng, K., & Mitchell, A. (2009). *Characterizing interfacial structure of TPO/CPO/TPO adhesive joints* (Vol. 50). Polymer.

Denton, Campble, & Tomlin. (1972). The determination of the optical constants of thin films from measurements of reflectance and transmittance at normal incidence. *J. Physics D: Applied Physics, 5.*

Ebnalwaled, A. A., & Thabet, A. (2016, July). Controlling the optical constants of PVC nanocomposite films for optoelectronic applications. *Synthetic Metals Journal*, *220*, 374–383. doi:10.1016/j.synthmet.2016.07.006

El-Korashy, A., El-Zahed, H., & Radwan, M. (2003). Optical studies of [N(CH3)4]2CoCl4, [N(CH3)4]2MnCl4 single crystals in the normal paraelectric phase. *Journal of Physics. B, Atomic and Molecular Physics*, *334*, 75–81.

Fan, X., Wang, F., Lu, X., Wang, M., Qiu, J., & Kawamoto, Y. (2003). The in situ synthesis process and luminescence behavior of 2- pyridinecarboxylic acid europium complexes in the sol–gel derived host materials. *Materials Chemistry and Physics*, *82*(1), 38–43. doi:10.1016/S0254-0584(03)00218-9

Fiore, A., Chen, J. X., & Ilegems, M. (2002). Scaling quantum-dot light-emitting diodes to sub micrometer sizes. *Applied Physics Letters*, *81*(10), 1756–1758. doi:10.1063/1.1504880

Fujimura, K., Ohkoshi, Y., Nagura, M., Akamatsu, K., Deki, S., & Gotoh, Y. (2004). Preparation of transparent alumina film and fiber from a composite of aluminum polynuclear complex methyl cellulose. Material Chemicals Phys, 83, 54–9.

He, B., Suna, W., Wang, M., Liu, S., & Shen, Z. (2004). Preparation and characterization of a series of novel complexes by single walled carbon nanotubes connected poly(amic acid) containing bithiazole ring. *Materials Chemistry and Physics*, *84*(1), 140–145. doi:10.1016/j.matchemphys.2003.11.018

He, B., Suna, W., Wang, M., & Shen, Z. (2004). Synthesis and magnetic property of a novel SWNT-poly(Schiff base)-Nd3+ complex. *Materials Chemistry and Physics*, *87*(1), 222–226. doi:10.1016/j.matchemphys.2004.05.042

John Xavier, K. V. (2005). Article. *Spectrochimica Acta. Part A: Molecular and Biomolecular Spectroscopy*, *60*, 709.

Krishnakumar, R. (2004). Article. *Spectrochimica Acta. Part A: Molecular and Biomolecular Spectroscopy*, *60*(3), 709. doi:10.1016/S1386-1425(03)00281-6

Krishnakumar, R. (2005). Article. *Spectrochimica Acta. Part A: Molecular and Biomolecular Spectroscopy, 63*(3), 499. doi:10.1016/j.saa.2004.04.014

Kuoa, M. C., Tsaia, C. M., Huanga, J. C., & Chena, M. (2005). PEEK composites reinforced by nano-sized SiO2 and Al2O3 particulates. *Materials Chemistry and Physics, 90*(1), 185–195. doi:10.1016/j.matchemphys.2004.10.009

Laguna, A. (1999). *In Metal Clusters in Chemistry* (P. Braunstein, L. Oro, & P. R. Raithby, Eds.). Wiley-VCH.

Lewis, T. W. (2000). *A study of the over oxidation of the conducting polymer polypyrrole* (Dissertation). University of Wollongong.

Liu, P., & Su, Z. (2005). Thermal stabilities of polystyrene silica hybrid nanocomposite via microwave-assisted in situ polymerization. *Materials Chemistry and Physics, 6*(2-3), 94–412. doi:10.1016/j.matchemphys.2005.05.023

Liu, X., Lee, C., Zhou, C., & Han, J. (2001). Carbon nanotube field-effect inverters. *Applied Physics Letters, 79*(20), 3329–3331. doi:10.1063/1.1417516

Lovell, P. A., & Al-Aasser, M. S. (1997). *Emulsion polymerization and emulsion polymers*. Wiley.

Lu, C.-H., Hsu, W.-T., Huang, C.-H., Godbole, S. V., & Cheng, B.-M. (2005). Luminescence characteristics of europium-ion doped phosphors prepared via a sol–gel route employing polymerizing agents. *Materials Chemistry and Physics, 90*(1), 62–108. doi:10.1016/j.matchemphys.2004.09.031

Martin, C. R., Skotheim, T. A., Elsenbaumer, R. L., & Reynolds, J. R. (1998). Handbook of Conducting Polymers (2nd ed.). Marcel Dekker.

Nussbaumer, R. J., Caseri, W., Tervoort, T., & Smith, P. (2002). Synthesis and characterization of surface-modified rutile nanoparticles and transparent polymer composites. *Journal of Nanoparticle Research, 4*(4), 319–323. doi:10.1023/A:1021141124257

Parsegian. (1969). Energy of an ion crossing a low dielectric membrane: Solutions to four relevant electrostatic problems. *Nature, 221*(5183), 844–846. doi:10.1038/221844a0 PMID:5765058

Patrone, L., Palacin, S., Bourgoin, J.-P., Lagoute, J., Zambelli, T., & Gauthier, S. (2002). Direct comparison of the electronic coupling efficiency of sulfur and selenium anchoring groups for molecules adsorbed onto gold electrodes. *Chemical Physics, 281*(2-3), 325–332. doi:10.1016/S0301-0104(02)00373-7

Patrone, L., Palacin, S., Charlier, J., Armand, F., Bourgoin, J. P., Tang, H., & Gauthier, S. (2003). Evidence of the key role of metal–molecule bonding in molecule- metal transport experiments. *Physical Review Letters, 91*(9), 096802. doi:10.1103/PhysRevLett.91.096802 PMID:14525200

Pei, Q., & Inganäs, O. (1992). Electrochemical applications of the bending beam method. *Journal of Physical Chemistry, 96*(25), 10507–10514. doi:10.1021/j100204a071

Quinn, B. M., Liljeroth, P., Ruiz, V., Laaksonen, T., & Kontturi, K. (2003). Electrochemical Resolution of Oxidation States for Monolayer Protected Gold Nanoparticles. *Journal of the American Chemical Society, 125*(22), 6644–6645. doi:10.1021/ja0349305 PMID:12769569

Raez. (2000). A Super molecularly Assisted Transformation of Block-Copolymer Micelles into Nanotubes. Angewandte Chemicals International Edition, 39, 3862.

Reddy, Q. (2006). Optical, electrical and discharge profiles for (PVC + NaIO4) polymer electrolytes. *Journal of Applied Electrochemistry*, *36*(9), 1051–1056. doi:10.100710800-006-9158-3

Roshan, S. (2001). Article. *Materials Letters*, *49*, 299.

Saraidarov, T., Reisfeld, R., & Pietraszkiewicz, M. (2000). Luminescent properties of silica and zirconia xerogels doped with europium (III) salts and europium (III) cryptate incorporating 3, 30-biisoquinoline-2,20-dioxide. *Chemical Physics Letters*, *330*(5-6), 515–520. doi:10.1016/S0009-2614(00)01059-9

Sarkar, S., Das, N. S., & Chattopadhyay, K. K. (2014). Optical constants, dispersion energy parameters and dielectric properties of ultra-smooth nanocrystalline BiVO4 thin films prepared by rf-magnetron sputtering. *Solid State Sciences*, *33*, 58–66. doi:10.1016/j.solidstatesciences.2014.04.008

Schmid, G., Harms, M., Malm, J. O., Bovin, J. O., Van Ruitenbeck, J., Zandbergen, H. W., & Fu, W. T. (1993). Ligand-stabilized Giant Palladium Clusters: Promising Candidates in Heterogeneous Catalysis. *Journal of the American Chemical Society*, *115*(5), 2046–2048. doi:10.1021/ja00058a066

Sergei, L. (2013). *Optoelectronics - Advanced Materials and Devices*. InTech Publisher.

Shankar, G., Joseph, P. S., Suvakin, M. Y., & Sebastiyan, A. (2010). Investigation of optical band gap in pyrrolidinomethylphthalimide crystal. *Physica B, Condensed Matter*, *405*(19), 4231–4234. doi:10.1016/j.physb.2010.07.016

Smythe, W. R. (1968). *Static and Dynamic Electricity* (Vol. 204). McGraw-Hill.

Thabet. (2020). *Design and Investment of High Voltage NanoDielectrics*. IGI Global. doi:10.4018/978-1-7998-3829-6

Toshima, N. (2000). *Direct synthesis of conducting polymers from simple monomers*. Department of Applied Chemistry.

Usha, K. S., Sivakumar, R., & Sanjeeviraja, C. (2013). Optical constants and dispersion energy parameters of NiO thin films prepared by radio frequency magnetron sputtering technique. *Journal of Applied Physics*, *114*(12), 123501. doi:10.1063/1.4821966

Varshni, Y. P. (2001). Binding energy of a hydrogenic impurity in a 2D circular quantum dot. *Superlattices and Microstructures*, *30*(5), 253–259. doi:10.1006pmi.2001.1012

Venkataramanan, Maheswaran, Sherwood, & Bhat. (1997). Crystal growth and physical characterization of the semiorganic bis(thiourea) cadmium chloride. *Journal of Crystal Growth, 179*, 605-610.

Vollath, D., Szabo, D. V., & Schlabach, S. (2004). Oxide/polymer nanocomposites as new luminescent materials. *Journal of Nanoparticle Research*, *6*(2/3), 181–191. doi:10.1023/B:NANO.0000034624.56366.ac

Wagner, H. D. (2002). Nanotube–polymer adhesion a mechanics approach. *Chemical Physics Letters*, *361*(1-2), 57–61. doi:10.1016/S0009-2614(02)00948-X

Wagner, H. D., & Vaia, R. (2004). Nanocomposites issue at the interface. *Materials Today, 11*(11), 38–42. doi:10.1016/S1369-7021(04)00507-3

Wang, Z., Wang, J., & Zhang, H. (2004). Luminescent sol–gel thin films based on europium-substituted heteropolytungstates. *Materials Chemistry and Physics, 87*(1), 44–48. doi:10.1016/j.matchemphys.2004.04.016

Xia, X., Cai, S., & Xie, C. (2006). Preparation structure and thermal stability of Cu/LDPE nanocomposites. *Material Chemicals Physics, 9*, 95-122.

Yakuphanoglu, F., Cukurovail, A., & Yilmaz, I. (2005). Refractive index and optical absorption properties of the complexes of a cyclobutane containing thiazolyl hydrazone ligand. *Optical Materials, 27*(8), 1363–1368. doi:10.1016/j.optmat.2004.09.021

Yoon, M., Kim, Y., Kim, Y. M., Volkov, V., Song, H. J., Park, Y. J., & Park, I.-W. (2005). Super paramagnetic properties of nickel nanoparticles in an ion-exchange polymer film. *Materials Chemistry and Physics, 91*(1), 104–107. doi:10.1016/j.matchemphys.2004.10.059

Yousif, E., Abdallh, M., Hashim, H., Salih, N., Salimon, J., Abdullah, B. M., & Win, Y.-F. (2013). Optical properties of pure and modified poly(vinyl chloride). *International Journal of Industrial Chemistry, 4*(1), 4. doi:10.1186/2228-5547-4-4

Zhuravlev, L. T. (1993). Characterization of amorphous silica surface. *Reaction Kinetics and Catalysis Letters, 50*(1-2), 15–25. doi:10.1007/BF02062184

Chapter 10
Nano-Tech Electronic Applications

ABSTRACT

This chapter contains the theories and effective parameters for developing the electronic applications by using nanotechnology techniques. This chapter sheds light on developing characterization of surface plasmon resonance (SPR) sensor, organic light emitting diode (OLED), thin-film transistors (TFTs), compact microstrip patch antenna, semiconductors, resistive memories, memory ram, chemical sensors, biosensors, and super capacitors. Furthermore, this chapter contains the detailed literature of the effects of different types and concentrations of nanoparticles for developing characterization of electronic applications. Finally, this chapter draws attention to the recommendations for investment in electronic applications by using nanotechnology techniques.

1 NANO-TECH VISION

Electronic devices are made up of active and passive elements and smaller integrated circuits (IC) memories. The ICs, diodes, and transistor are made of semiconductor materials and their working is dependent on current flow through them. Nowadays, the nanotechnology easy used for improving the sensitivity and Signal to Noise Ratio (SNR) of Surface Plasmon Resonance (SPR) sensor by utilizing the nanocomposite material Polyaniline/Silver. The novel nanocomposite material Polyaniline/Silver is included to the silver layer of SPR sensor. The gold layer is utilized for dependability and security of novel nanocomposite materials starting with oxidation. The sensitivity of the proposed model is higher than the traditional and nanoparticle compound SPR sensor. SNR in the recommended model is higher than the late nanoparticle compound sensor, but it is more level than accepted sensors. Also, this chapter investigates a hypothetical model for Organic Light Emitting Diode (OLED) characteristic for including an insulating buffer layer at cathode. The recommended support has easier dielectric constant ZnO and high potential drop. Furthermore, it can improve the conductivity of OLED by mixing polymeric emissive layers of nano-oxide particles TiO_2. Finally, the previously mentioned model in this chapter has shown the improved electron injection of OLED which came about from bringing down the compelling barrier

DOI: 10.4018/978-1-7998-8536-8.ch010

height to electron injection and expanding the electron-tunneling probability. Furthermore, an explanatory suggested model for upgrading the performance of Thin-Film Transistors (TFTs) by supplanting the customary silicon dioxide entryway dielectric with high-k entryway dielectric novel nanocomposite PVP/La_2O_3 (k_{ox}=25) is also presented. Our recommended model has a single nanocomposite material entryway SMG MOSFETs. Furthermore, the impacts of the gate-dielectric permittivity, spacer oxide permittivity, entryway length, and the width of TFTs structure.

In addition, this chapter is investigated to improve the dimensions of the microstrip antenna; namely the height of the microstrip antenna with and without slot substrate and the width of the patch microstrip antenna, by utilizing nanocomposite materials which would create RT Duriod 5880 for nanoparticles (Fumed Silica, aluminum oxide and Graphite). Novel nanocomposite materials have been designed as a substrate to the microstrip antenna toward including nanoparticles to RT Duriod. Furthermore, it has been shown that the height and width of the recommended model for the microstrip antenna are smaller than the conventional rectangular microstrip antenna. The suggested nanocomposite material has investigated productive and compact microstrip antenna, compact size, simple creation and expansion radiation, bandwidth and surface wave force of microstrip antenna with and without opening. Optical properties, such as direct and indirect optical energy gap and optical absorption edge, have been experimentally investigated on immaculate and polyvinyl chloride (PVC) nanocomposites dainty films. The impacts of variant sorts and concentrations of inorganic nanoparticles (ZnO, Al_2O_3, and SiO_2) looking into optical light characterization of polyvinyl chloride under variant wave lengths have been examined. Moreover, the successful sorts and concentrations of inorganic nanoparticles regarding absorption and transmission light performance have been specified in comparison with unfilled base grid polymers. The ideal sorts and concentrations of the recommended inorganic nanoparticles can be controlled clinched alongside optical light characterization of polyvinyl chloride and their electric and electronic industrial applications.

2 SURFACE PLASMON RESONANCE SENSOR

This new class of polymers is known as directing polymers. Directing polymers with polyaromatic spine including Polypyrene, polythiophene, polyaniline, and so on has accepted an incredible bargain for consideration in the final two decades (Jong Seol Yuk, 2006). These polymers have their interesting electrical, optical, attractive and compound properties prompting the extensive variety of innovative unrest applications for rechargeable batteries, electrodes, sensors, erosion insurance coatings, microwave absorption, light emitting diodes, and so forth. In this way, observing and stock arrangement of all instrumentation may be enhanced (Jing et al., 2007). Polymers have their exceptional electrical, optical, attractive and chemical properties prompting the extensive variety of innovative provisions for rechargeable batteries, electrodes, electromagnetic impedance EMI shielding, sensors, erosion security coatings, microwave absorption, light Emitting Diodes LED, and so on (Gupta et al., 2010). In the display worked, a legitimate correlation has been conveyed for the suggested model, customary sensor and late nanoparticle compound sensors. The changeover sensitivity and SNR have been shown by utilizing novel nanocomposite materials (Polyaniline/Silver) and again the previously mentioned SPR sensor. The electrical conductivity of Polyaniline–Silver nanocomposite increments for the increment of silver nanoparticle content over that of immaculate Polyaniline. This may be a simple approach toward which optical and electrical properties of Polyaniline can be improved by doping for suitableness nanoparticles and useful electrical, optical, attractive and compound properties.

2.1 Theory and Parameters of SPR Sensor

Attenuated total reflection (ATR) is a sampling technique used in conjunction with infrared spectroscopy. The weakened downright reflection ATR technique alongside $K_{retschmann}$ setup will be for the most part utilized within SPR estimations. In the available plan of a fiber optic, SPR sensor and the cladding around the center of a multimode step-index optical fiber are uprooted and covered with a metallayer silver of thickness d_1. This internal metal layer will be further secured with a metal layer Gold for thickness d_2. An additional model structure has a bimetallic compound layer comprising round silver and gold nanoparticle. In the hypothetical modeling, the optical fiber is viewed as a multimode that you quit offering on that one with silica center (Zeng, 2000). Fiber core, metal layer, reflection coefficient, and transmitted energy are the effective parameters of the SPR sensor. The wavelength dependence of the refractive index $n_1$of the fiber core is given by Sellmeier dispersion relation:

$$n_1 = \sqrt{1 + \frac{A_1\lambda^2}{\lambda^2 - B_1^2} + \frac{A_2\lambda^2}{\lambda^2 - B_2^2} + \frac{A_3\lambda^2}{\lambda^2 - B_3^2}} \tag{1}$$

Where λ denotes the wavelength (in μm), and the Sellmeier coefficients A_1, A_2, A_3, B_1, B_2, and B_3 have certain numeric values.

Consider N-layer model to obtain the expression of amplitude reflection coefficient for p-polarized incident beam. The layers are assumed to be stacked along the z-axis. The arbitrary medium layer is defined by thickness d_k, dielectric constant ε_k, permeability μ_k, and refractive index n_k. The tangential fields at the first barrier $Z=Z_1=0$ is related to the final barrier Z=ZN_1 given by (Zeng, 2000):

$$\begin{bmatrix} U_1 \\ V_1 \end{bmatrix} = M \begin{bmatrix} U_{N-1} \\ V_{N-1} \end{bmatrix} \tag{2}$$

Where U_1 and V_1 are the tangential components of electric and magnetic fields at the barrier of first layer, respectively, UN-1 and VN-1 are the corresponding fields at the barrier of Nth layer and M, known as characteristic matrix of the combined structure.

The light is propelled under the one end of the fiber through a broadband light source for correct optics. The opposite limit is associated with an identification system to record the yield sign. Think about the proliferation of every last one of the guided beams starting in the fiber utilizing a collimated sourball and a magnifying instrument destination. The goal is to keep tap of the shaft in the end-face of the fiber toward the pivotal side of the point. Acquiring λ_{res} wavelength of the light pillar will be varied, and the comparing force of the transmitted light will be P_{trans}. If the refractive index of the sensing layer is altered by δns, the resonance wavelength will shift by $\delta\lambda_{res}$. There are two important parameters of the SPR sensor. The sensitivity Sn of a SPR sensor with spectral interrogation is defined as (Kicklbick, 2008):

$$S_n = \frac{\delta\lambda_{res}}{\delta n_s} \tag{3}$$

$$\delta\lambda = \frac{n_1\varepsilon_{Ar}}{\frac{n_1}{\lambda}n_s^3\left(\frac{1}{\varepsilon_{Ar}}-1\right)+\frac{dn_1}{d\lambda}n_s\left(n_s^2+\varepsilon_{Ar}\right)} = S_n\delta n_s \quad (4)$$

As a tradition, the sensitivity λ_{res} is plotted as a function of ns, the slope of this curve gives the sensitivity of the sensor (Bernius & Inbasekaran, 2000). In the ghastly interrogation-based fiber optic SPR sensor, the wavelength of the light will be varied, and the related transmitted energy will be measured. In λ_{res}, a sharp dip to the transmitted control will be watched. The strength of λ_{res} relies on the refractive file of the sensing medium. The precision of the identification about SPR wavelength λ_{res} relies on the width of the reaction bend. The narrower width may be most astounding the identification correctness. Therefore, though $\delta\lambda_{0.5}$ may be the ghastly width of the reaction bend comparing on half of the transmitted power, the identification precision of the sensor can be expected to be conversely proportional with $\delta\lambda_{0.5}$. The SNR of the SPR sensor using ghastly session can be composed as the following illustration (Kicklbick, 2008):

$$SNR = \frac{\delta\lambda_{res}}{\delta\lambda_{0.5}} \quad (5)$$

2.2 Nano-Tech SPR sensor

Clinched alongside Later years, the enthusiasm toward the improvement of inorganic–organic composite has developed quickly owing to an extensive variety of potential utilization of these materials. Around these materials, you quit offering on that one paramount population in which the organic piece will be the leading polymers similar to Polyaniline, Polypyrene, and so forth. The optical properties of a metallic nanoparticle rely principally on its SPR and it is referred to the plasmon full peaks and line widths for a delicate span and state of the nanoparticles, the metallic species, and the encompassing medium. The optical properties of metal nanoparticles inserted in a dielectric medium bring have also been a subject of enormous enthusiasm for quite some time lately due to their novel aspects (Baraton, 1998). Polyaniline–Silver nanocomposite has been synthesized effectively by the chemical oxidative polymerization of aniline for ammonitic Peroxydisulphate as an initiator for vicinity of contrarily charged silver nanoparticles. Thermo-Gravimetric Analyzer (TGA) and different examining calorimeter DSC outcomes indicate that Polyaniline silver nanocomposite is more crystalline and more thermally stable. A surface Plasmon absorption band may be obtained starting with the optical absorption during 380nm, which demonstrates that silver nanoparticles are exhibited in the Polyaniline grid. The optical band hole of nanocomposite abates for the expanding substance of silver nanoparticles. As a recent update, photoluminescence has been watched on Polyaniline–Silver nanocomposite over that on immaculate Polyaniline. The electrical conductivity of Polyaniline–Silver nanocomposite expands for the expansion of silver nanoparticle content over that of immaculate Polyaniline. This will be a basic best approach by which optical and electrical properties of Polyaniline can be improved toward doping with suitableness nanoparticles (Zeijlmans, 1999). The displayed work has chosen Polyaniline "around separate directing polymers" due to its simple amalgamation procedure, great organic and thermal stability, low-cost price and high conductivity and because it exhibits useful electrical, optical, attractive and chemical properties.

2.3 Effects of Nanoparticles on SPR Sensor

The effects of the SNR and sensitivity of SPR the sensors accepted model, the nanoparticle compound model and the suggested model have unmistakably exhibited the sorts of bimetallic setup to show their restrictions. The characteristics of the magic utilized parameters are numerical aperture=0. 24, center diameter D=600μm, sensing area period L=10mm, and downright bimetallic thickness=50nm. The numerical simulations were performed, and the sensitivity as well as the SNR to the two probes have ascertained that the circular silver and gold nanoparticles in the compound layer are about the same size RAg= RAu=10nm. Parameter's determinations are as demonstrated in Table 1 and Table 2.

Table 1. Sellmeier Coefficients for Optical Fiber Silica Core

Parameter	A_1	A_2	A_3	B_1	B_2	B_3
Numeric Value	0.7	0.9	0.4	0.07	9.9	0.12

Table 2. Metal Parameters for SPR Sensor

Parameter	**Silver**	**Gold**
High frequency dielectric constant, ε_∞	2.48	7.0
Plasma frequency, WP	1.35×10^{16} rad/s	1.40×10^{16} rad/s
Damping frequency, Wd (bulk)	7.62×10^{13} rad/s	3.78×10^{13} rad/s
Fermi velocity, Vf	1.40×10^{6} m/s	1.40×10^{6} m/s

The sensitivity of the proposed model which utilizes the novel nanocomposite material and nanoparticle compound SPRs models builds with expanding the silver fraction. However, the sensitivity of the traditional SPR model abates for expanding the silver portion. It is recognized that the sensitivity of the recommended model which utilizes the novel nanocomposite material will be higher than both the traditional and nanoparticle compound SPR models. Figure (1) illustrates the variety of SNR for the silver portion x on the conventional nanoparticle compound and the proposed SPR models. The SNR of the conventional and nanoparticle SPRs models expands for expanding the silver portion x. In any case, the SNR of the recommended model which employs the novel nanocomposite material SPR sensor declines with expanding the silver portion x. It is recognized that toward suggested model, SNR will be higher than the later nanoparticle compound SPR sensor. However, SNR values will be to a great extent most elevated to the conventional in case of the examination of the nanoparticle compound instance. The SNR of the alloy-based sensor can be further enhanced by expanding the molecule size. Thus, as the molecule extent increases, the thunder bend gets narrower because it's dependent upon shifting, and hence, SNR increments. To an accepted extent, the SNR is high in examination by a compound body of evidence for any portion of silver. To utilize the novel material, SNR can be higher than the nanoparticle alloy.

Figure 1. Effect of Silver fraction on SNR for conventional nanoparticle alloy and nanocomposite material-based fiber optic SPR models

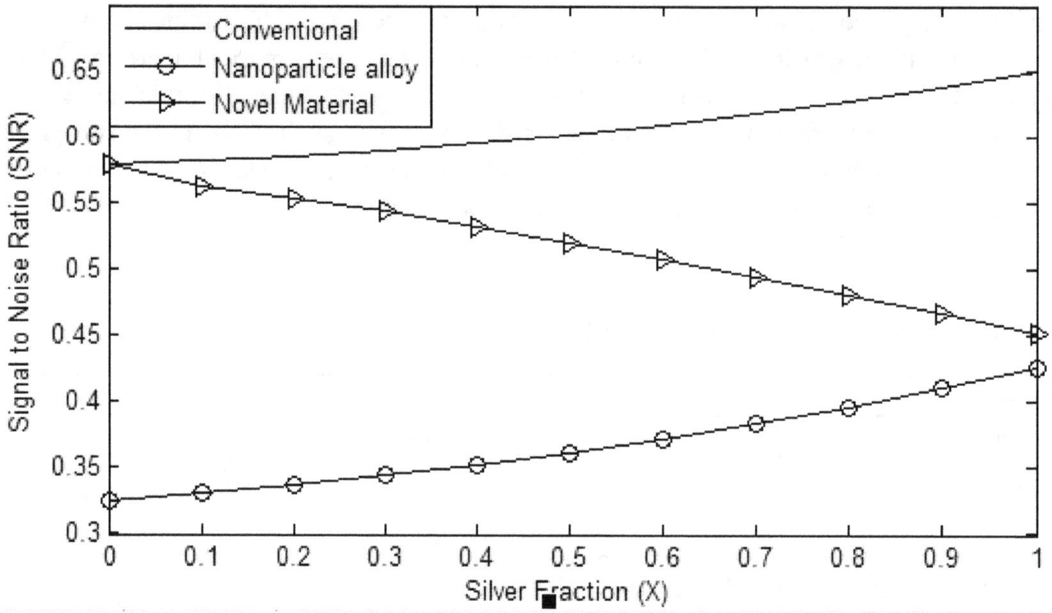

The sensitivity of suggested model which employs the novel nanocomposite material and nanoparticle compound SPR sensors expands with expanding the sensing layer refractive index, but the sensitivity of customary SPR sensor abates for expanding the sensing layer refractive ns. It is observed that the sensitivity of the proposed model which employs the novel nanocomposite material will be higher than both the conventional and nanoparticle compound models. The SNR of the conventional and nanoparticle compound SPR sensors expands for expanding the sensing layer refractive ns, but the SNR of the recommended model which employs the novel nanocomposite material abates. It is also inferred that the SNR of the suggested model which employs the novel nanocomposite materials is higher than the late nanoparticle compound model. In any case, the sensitivity of the recommended model which employs the novel nanocomposite material and the nanoparticle compound SPR sensors expands with expanding the wavelength. Hence, the sensitivity of the conventional model abates for expanding the wavelength. Accordingly, the SNR of the recommended model which employs the novel nanocomposite material abates with expanding the wavelength. It may be discovered that the SNR for the novel nanocomposite material can be higher than the previous nanoparticle compound sensors.

The sensitivity of the suggested model which employs the novel nanocomposite material and the nanoparticle compound models declines for expanding the refractive file for core and the sensitivity of customary SPR sensor increments for expanding the refractive file of center n1. The SNR for the refractive file of the center n1 in the conventional nanoparticle compound and the proposed model which utilizes the novel nanocomposite material SPR sensors are demonstrated in Fig. (2). The SNR of the conventional and nanoparticle compound models declines with expanding the refractive file of center n1. However, the SNR of the proposed model which utilizes the novel nanocomposite material increments for expanding the refractive file of center n1. It is recognized that in the novel nanocomposite material, SNR is higher than the previous SPR sensor which employs the nanoparticle compound. In the nanoparticle compound SPR sensor, SNR abates with expanding sensitivity. Toward recommended model of SPR sensor, SNR declines with expanding sensitivity. However, it is observed that in the recommended model, the SNR is higher than previous nanoparticle compound sensors. Hence, the SNR at the customary model has the most astounding values starting with both the nanoparticle compound and our proposed models, but the sensitivity of the recommended model has the most noteworthy values from conventional and nanoparticle alloy models.

Figure 2. Effect of refractive index of core on SNR for conventional nanoparticle alloy, and nanocomposite material

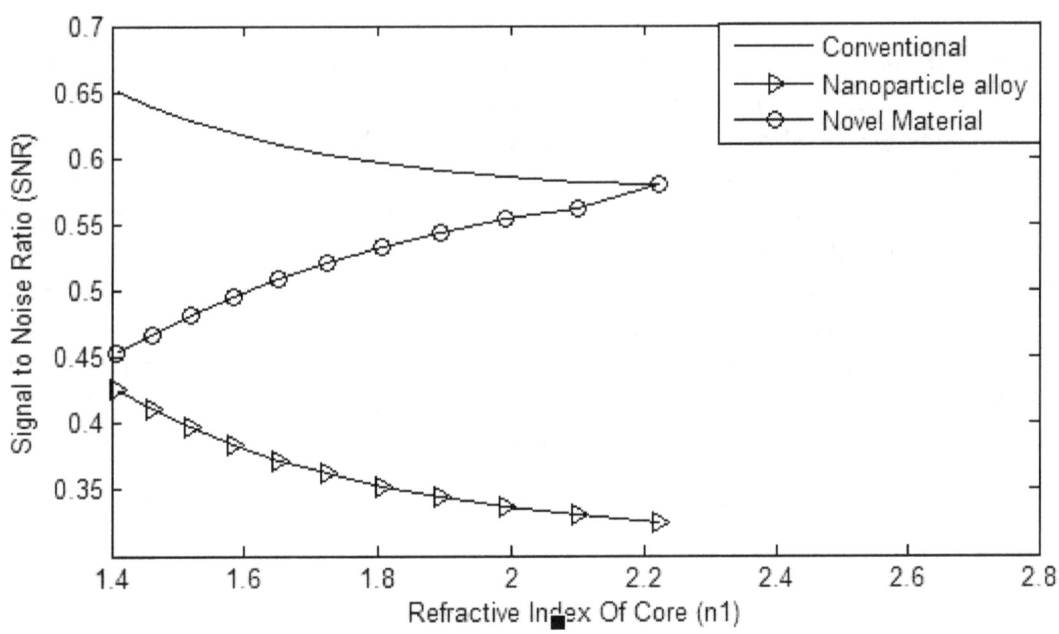

3 ORGANIC LIGHT EMITTING DIODE OLED

In general, the OLED model gives a functional device for encouraging device configuration. Numerical models, in particular, can hold future benefits. Y. Kawabe et al. (Kawabe et al., 1998) suggested a model for single layer devices over which the present density–voltage (J-V) characteristics were acquired numerically. This model indicates the present dependence on the barrier height in the metal-organic interface. Park et al. suggested that the instrument of the improved electron injection originated from bringing down the barrier height in the middle of the cathode and emanation layer (Carter et al., 1997; Friend, 2002; Kim et al., 2004; Parker, 1994; Zeng, 2000). In this chapter, a proposed novel model has mimicked the device characteristics of OLEDs holding insulating buffer layers. This model may be in view of Poisson's mathematical statement and Fowleer–Nordhein FN tunneling system. A tunneling figure is acquainted to give the impact of the insulating buffer layer under attention. The present densities are recreated for separate thicknesses of the buffer layers. Enhancing the injection of electrons from the cathode for an ITO/MEH-PPV/Al light-emanating diode has been conveyed toward inserting a ZnO buffer layer between the Al cathode and the polymeric emissive layer (Baraton, 1998; Bernius & Inbasekaran, 2000; Gouda et al., 2010; Thabet, Mobarak, & Abozeid, 2010; Thabet, Mobarak, & Samir, 2010; Zeijlmans, 1999). Moreover, the optical properties of MEH-PPV films have been enhanced by blending TiO2 for the MEH-PPV polymer layer. Then, the impacts of the suggested nano-materials on the (J_e–V) characteristics of the OLED structure have been investigated.

3.1 Theory and Parameters of OLED

The device operation of an organic LED can be dictated toward three processes: charge injection, charge transport and recombination. An unbalanced charge injection results in an excess of one carrier type, and thus in a current flow which practically does not contribute to the light emission. The vicinity of a contact barrier which hinders the addition injection of one kind of charge carrier will show a low efficiency. Hence, the decrease of the contact barriers prompts a larger current and a higher light-output in an equivalent voltage. Thus, in order to maximize the OLED efficiency, the work function of the cathode and anode should both be close to the conduction and valence bands of the PPV, respectively. An unbalanced charge transport also affects the device efficiency of OLED. The recombination transforms of electrons and gaps in the animated polymer layer simplify determines the light yield and in turn the efficiency of the OLED. The impact of a buffer layer on the carrier injection and the interfaces in the middle of emanation layer and the two electrodes are noteworthy. Consequently, the double-carrier injection can be viewed as in this model, and the injection is driven toward FN tunneling component which indicates the association between the middle of the present and potential barrier height toward two interfaces. Likewise, the direction of the anode position can be x=0, and the cathode one can be x=d. The current density can be expressed by the following equation (Friend, 2002):

$$J_0 = FN(E) = k_1 \frac{E^2}{\phi} \exp\left(-k_2 \frac{\phi^{1.5}}{E}\right) \tag{6}$$

Where:

$$k_1 = \frac{2.2e^2}{8h} \text{and}\, k_2 = \frac{8}{3h}\left(2em\right)^{0.5}$$

e is the elementary charge, h is Planck's constant, **m** is the effective mass of an electron or a hole (for simplicity, m*=m0), ϕ is the potential barrier height, and E is the electric field at the interface. When the carriers are injected, the carrier density in emission layer assume to be non-uniform. This means that the carrier density is the largest at the interface between the electrode and the emission layer and therefore declines exponentially inside. The total current density is the sum of current densities of electrons and holes:

$$J_{total}(E_0) = J_h + J_e \tag{7}$$

Where $\mathbf{J_h}$ and Je are the function of electric field $\mathbf{E_0}$. When other parameters are known as the voltage, V is also a function of E_0. Insertion of the different dielectric constant nanolayers between the emitting layer and the cathode can change the potential distribution, as shown in Fig. 3.

Figure 3. The potential drop distribution of OLEDs devices

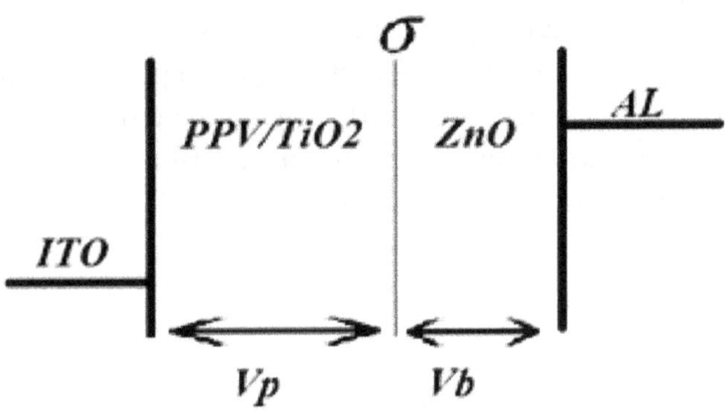

Figure *3* illustrates the potential distribution of OLEDs devices, where layers p and b represent the emitting MEH-PPV with TiO2 composite and the thin insulating layer ZnO respectively. If the total potential V is applied between the two electrodes, the potential drop in each layer at the steady state can be expressed as follows:

$$V = V_p + V_b \tag{8}$$

Where:

$$V_p = \frac{\left(\frac{\varepsilon_b}{d_b}\right)V - \sigma}{\left(\frac{\varepsilon_p}{d_p}\right) + \left(\frac{\varepsilon_b}{d_b}\right)} \text{ and } V_b = \frac{\left(\frac{\varepsilon_p}{d_p}\right)V - \sigma}{\left(\frac{\varepsilon_p}{d_p}\right) + \left(\frac{\varepsilon_b}{d_b}\right)}$$

Where subscripts p and b refer to polymer and buffer, respectively, d is the thickness of each layer and σ is the interface charge density caused by charge confinement at the interface between the emission layer and the insulating layer. The probability of electron tunneling will increase due to reduced energy $\left(\phi_{\mathbf{e}}^{'} = \phi_{\mathbf{e}} - \mathbf{V_b}\right)$ of OLED device. On the other hand, the insulating layer may cause an additional barrier for electron injection. The tunneling factor T means the probability of electron injection, just as the meaning of transmission is based on this ideal. By introducing an additional concept of this model based on quantum theory, the transmission factor of square potential barrier can be described as follows (Kim et al., 2004; Zeng, 2000):

$$T \propto \phi_b exp\left(-\frac{2a}{l}\sqrt{2m_0\phi_b}\right) \tag{9}$$

Where, ϕb is the barrier height of the buffer layer and is the width of the barrier.

For simplicity, consider that the buffer layer as one extent square potential barrier and the thickness and barrier stature of the buffer layer have significant impacts on the tunneling figure. It might be recognized that BaO, LiF, and ZnO are utilized as buffer layers for comparison with the ascertained reliance of the tunneling element on the thickness of the buffer layer. There may be a relationship between the middle of the implications of the tunneling figure t and the transmission component of the square potential barrier. By increasing FNe toward t, the changed present thickness of electrons injected starting with cathode is obtained.

3.2 Nano-Tech OLED

The conventional structural outline of the OLED device under contemplation are indicated as in Fig. (4 (a)) in this structure. The indium tin oxide ITO layer capacities as a transparent cathode permit the light created inside the diode to clear out of the device. The highest point cathode will be helpfully shaped by thermal evaporation of a metal. OLED operation is achieved when the diode is biased sufficiently to achieve injection of positive and negative charge carriers from opposite electrodes. Catch for oppositely accused transporters inside the region of the polymer layer might afterward bring about photon emanation (Baraton, 1998; Bernius & Inbasekaran, 2000; Zeijlmans, 1999). Thus, the recommended model structure can be indicated clinched alongside Fig. (4 (b)) to incorporate upgrading the performance of PPV-based devices toward composites made with polymers and nano-oxide particles, such as silicon oxide SiO_2 or titanium oxide TiO_2.

Figure 4. OLED Structure

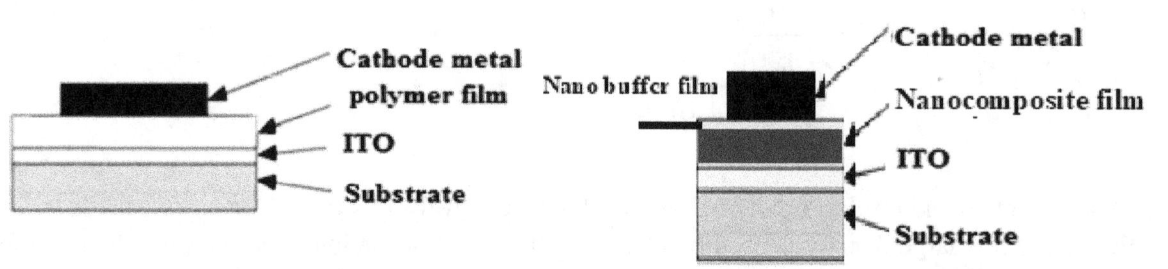

(a) Structure of a single layer OLED

(b) Structure of suggested model of single layer OLED

Utilizing nanocomposites has been accepted on expansion of the electrical conduction of the polymer in addition and moving forward its stability which is of prime vitality as previously mentioned; OLED. For poly(2-methoxy-5-(20-ethylhexyloxy)-1,4phenylenevinylene), MEH-PPV mixed for SiO_2 was discovered to have a progressed conductivity as contrasted with the uncovered polymer. It is recognized that blends utilizing TiO_2 nanoparticles don't break the PPV conjugation and have the potential of gas sensing applications. Furthermore, the suggested structure has come about from the certainty that the nano-buffer for bringing down the dielectric constant further expands the present thickness because of diminishing the potential drop in the emissive polymer layer. The current thickness is a work of the drive voltage in the ahead bias to a single-layer (ITO/PEDOT: PSS/MEH-PPV/TiO2/ZnO/Al) device. Thus, the ahead bias is characterized by the illustration of a certain voltage on the hole-injecting ITO contact. Similarly, as a dynamic layer, 100nm emit light at best possible turn-on voltage OLED with PEDOT: PSS. Turn-on voltage will be characterized here as the voltage during which the Electro radiance EL begins surpassing the foundation signal.

3.3 Effects of Nanoparticles in OLED

In order to elucidate the impact of the embedded encasing buffer layer on cathode, two device structures have been recreated for comparison, (ITO/PEDOT: PSS/MEH-PPV/Al) customary device, and (ITO/PEDOT: PSS/MEH-PPV/TiO_2/ZnO/Al) proposed device. The outcomes examined in these sections impact nano support dielectric constant, nanocomposite clinched alongside the potential drop of OLEDs, the tunneling element of the thickness of the buffer layer and inserting nano materials in OLEDs. In addition, the potential drop over the buffer layer has an impact on expanding the present thickness of electrons for diverse nano-buffer materials. As indicated previously, the nano-buffer ZnO with more level dielectric constant may further expand the potential drop over the buffer layer and diminish the potential drop crosswise over the emissive polymeric layer. ZnO with more level dielectric constant may be superior to SiO_2 and LiF, which is the transform of the most noteworthy esteem of buffer potential drop and present thickness of electrons toward consistent go thickness starting with 0 to 0.5 nm. Using of MEH-PPV blended with SiO_2 or TiO_2 nanocomposite with different dielectric constant of 3.36 and 13.39 respectively. It is found that the diminishing to the potential drop over polymer layer can be seen toward diminishing the vitality barrier stature for electrons. MEH-PPV mixed TiO_2 nanocomposite provides for a polymer potential drop more level than the reference test that organized just ITO/MEH-PPV/Al and MEH-PPV mixed SiO_2. The potential drop over the emanation layer has been identified with the layer thickness and the dielectric constant. By inserting a slim buffer layer, the slop of band bowing in the emanation layer will be brought down and the probability for electron tunneling will be expanded because of the lessened vitality barrier of OLED devices. The thickness and barrier stature of the buffer layer have imperative impacts on the tunneling figure. In any case, BaO, LiF and ZnO are utilized as buffer layers to comparison; their fill in capacities will be 2. 9eV, 3. 5eV and 4. 45eV separately. It has been explained that the tunneling element t declines exponentially as the thickness of the buffer layer expands. The iotas of the support material diffuse under the near-surface region, and the interfacial region has an estimated scale for regarding a few angstroms. Therefore, the thickness of the buffer layer must be brought down over the wanted worth. The tunneling component of ZnO can be bigger than that for BaO and LiF. Additionally, it is recognized that the ZnO buffer layer can be thicker than the others toward the ideal strength of electron injecting current thickness. OLED present thickness builds in all cases with nano-buffer insertion. It is envisaged that the electron injection can be improved toward the nanolayer insertion ZnO superior to BaO which brings higher dielectric constant for recognizing that the reality of gap injection cathode ITO is the same to every last one of the devices. The current thickness increments of nanolayer embedded devices are basically ascribed to higher electron injection toward the MEH-PPV/nanolayer interface.

4 THIN-FILM TRANSISTORS TFTS

In this section, the impact of C_{bottom} on the surface potential is observed with the recognition of the prompted charges on the sourball and channel districts, as well as the other potentials created because of these charges in the channel region. The low methodology temperature of polymer composites and nano-oxide particles hold potential benefits for low-cost applications.

4.1 Theory and Parameters of TFTs

A schematic cross-sectional perspective of an MOSFET with a high-k entryway dielectric is demonstrated in Fig. (5 (a)), with the fringing field lines starting with the bottom of the entryway cathode of the channel and sourball districts. For simplicity, accept that hardware field lines are as indicated in Fig. (5 (b)), comparative of the methodology utilized within prior investigations to the opposite segments of parasitic capacitance (Kumar & Chaudhry, 2004; Reddy & Kumar, 2005).

Figure 5. Cross-sectional view of an SOI-MOSFET and Fringing field (Kumar & Chaudhry, 2004; Reddy & Kumar, 2005)

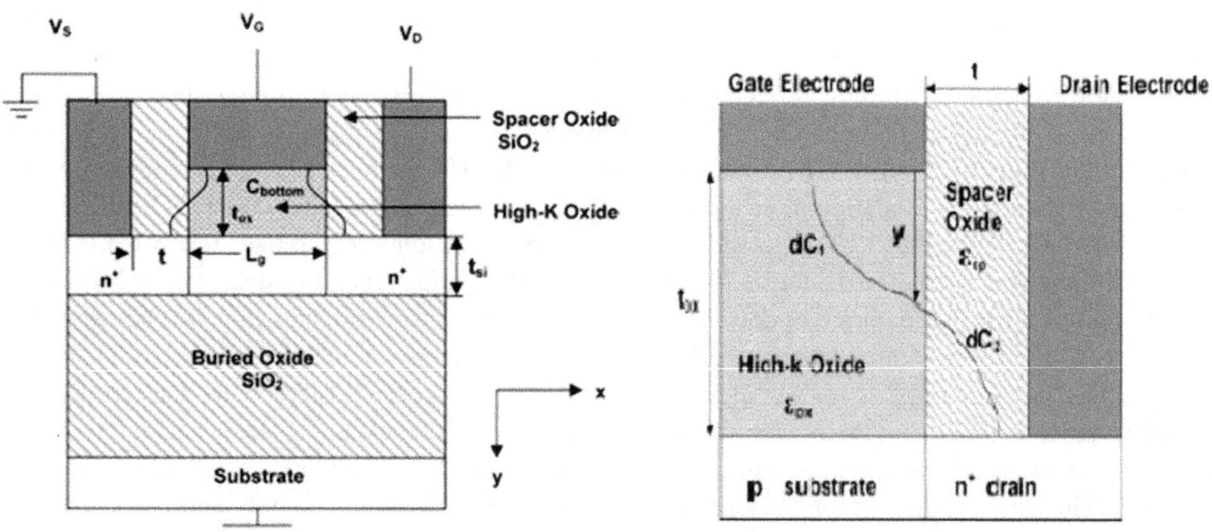

(a) Cross-sectional view of an SOI-MOSFET *(b) Fringing field of an SOI-MOSFET*

In an MOSFET, the field lines originating from the bottom of the charged plate only will contribute to the potential in the channel; therefore, only half of the coulomb potential due to these charges is considered. Then, the potential at a distance x from the uniformly charged plate is given in Fig. (6).

$$V(x)=\frac{1}{2}\int_{0}^{t}\int_{-\frac{w}{2}}^{\frac{w}{2}}\frac{\sigma dx_1 dx_2}{4k_{si}\sqrt{(x+x_1)^2+(x_2)^2}} \tag{10}$$

Where: ksi is the dielectric constant of silicon. A model for the surface potential along the channel for a fully depleted SOI-MOSFET assumes a simple parabolic potential profile in the vertical direction (Chen & Liao, 2008; Rébiscoul et al., 2010):

$$V(x)=\frac{\tilde{A}}{8k_{si}}\left((x+t)\ln\left(\frac{\sqrt{(x+t)^2+\left(\frac{w^2}{4}\right)}+\left(\frac{w}{2}\right)}{\sqrt{(x+t)^2+\left(\frac{w^2}{4}\right)}-\left(\frac{w}{2}\right)}\right)-(x)\ln\left(\frac{\sqrt{(x+t)^2+\left(\frac{w^2}{4}\right)}}{\sqrt{(x+t)^2-\left(\frac{w^2}{4}\right)}}\right)+(w)\ln\left(\frac{\sqrt{x^2+\left(\frac{w^2}{4}\right)}+t+\sqrt{x^2+t^2+2t\sqrt{x^2+\left(\frac{w^2}{4}\right)}}}{\sqrt{x^2+\left(\frac{w^2}{4}\right)}+x}\right)\right) \tag{11}$$

Figure 6 Potential due to a uniformly charged plate ()

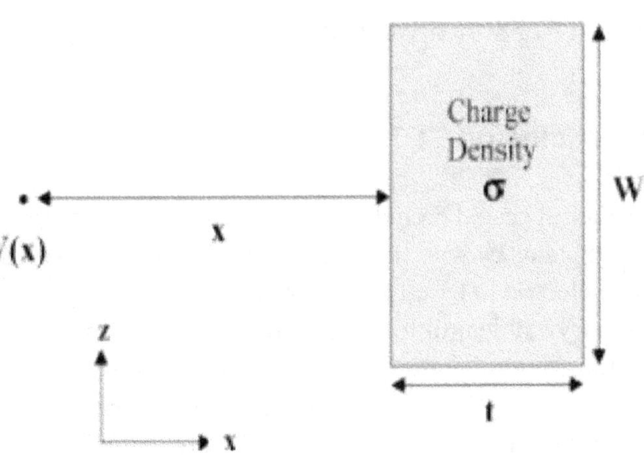

3.1 Nano-Tech TFTs

The ever-increasing demands of high performance obliges powerful innovation scaling and inventive plans. Hence, estimation of circuit performance is essential in the plan stage to guarantee best possible dependability. Fringing capacitances is a standout amongst the significant supporters on performance corruption and indicator integument issues in sub-100nm circuits. Subsequently, an exact Nano-Tech model about fringing capacitances is vital to foresee the performance preceding creation (Thabet et al., 2011). The traditional entryway dielectric structure outline of the TFT devices under consideration is indicated in Fig. (7 (a)) In this structure, contracting the conventional MOSFET Past innovation hub obliges innovations to smother parasitic fringing capacitances and short channel impacts. Nano-Tech model structure has been investigated as indicated in Fig. (7(b)) It includes entryway dielectric with high k strength for upgrading the performance TFT devices by utilizing novel nanocomposites with high dielectric constant worth k_{ox}=25 for lanthanum oxide La_2O_3 mixed with polyvinyl phenol PVP, as opposed to silicon dioxide. Therefore, the interceding material between the entryway and S/D contacts is precisely viable in unwinding the prerequisites on the virtual sourball velocity.

Figure 7. Structure TFTs gate dielectric

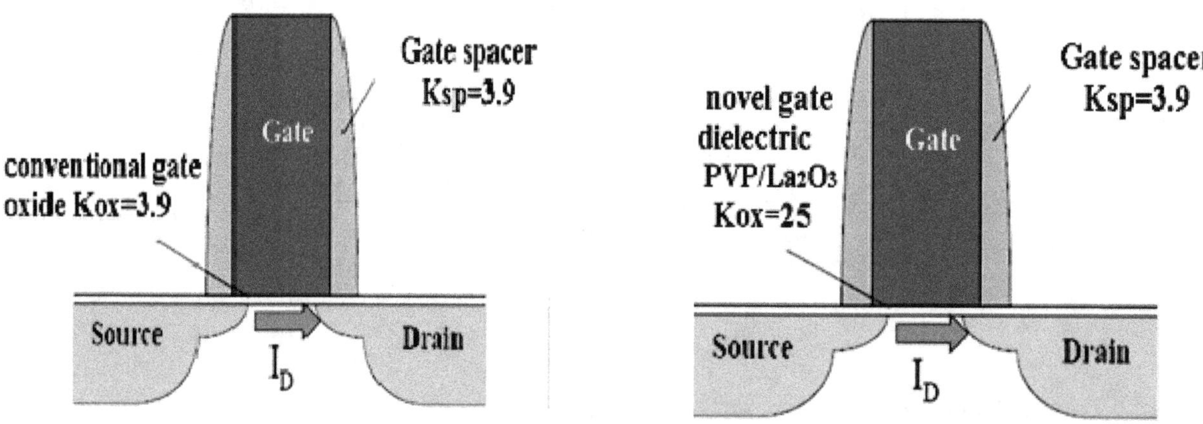

(a) Structure conventional TFTs gate dielectric. (b) Structure of suggested TFTs gate dielectric

3.2 Effects of Nanoparticles on TFTs

Computed comes about in the suggested model are demonstrated as stated previously. It is clear that the fringing capacitance C_{bottom} expands by inserting the k_{ox} novel nanocomposite material much higher than the conventional entryway dielectric. It has also been indicated the variety of inside fringing capacitance as a work for normalized entryway length to novel and customary entryway dielectrics. Additionally, it has been indicated that the fringing capacitance C_{bottom} declines for expanding the entryway period starting with 10nm to 80nm. However, PVP/La_2O_3 provides higher fringing capacitance than SiO_2 for a provided for channel segregation racial bias VD=0.05V. It is recognized that the novel nanocomposite entryway dielectric provides for higher surface potential than the accepted entryway dielectric constant with expanding the entryway spacer dielectric constant. Regarding the surface potential which has been raised to an entryway insulin hosting a higher dielectric constant PVP/La_2O_3, the impact of the interior border capacitance has been indicated in Fig. (8). The hypothetical model effects depict that the surface potential expands with high k entryway dielectric material which builds inward border capacitance ahead against separation along channel entryway. Fig. (9) clarifies that the novel nanocomposite material PVP/La_2O_3 has less edge voltage than the illustration compared for accepted SiO_2 of the same width. Similarly,

Figure 8. Surface potential variation along the channel for $k_{ox} = 25$

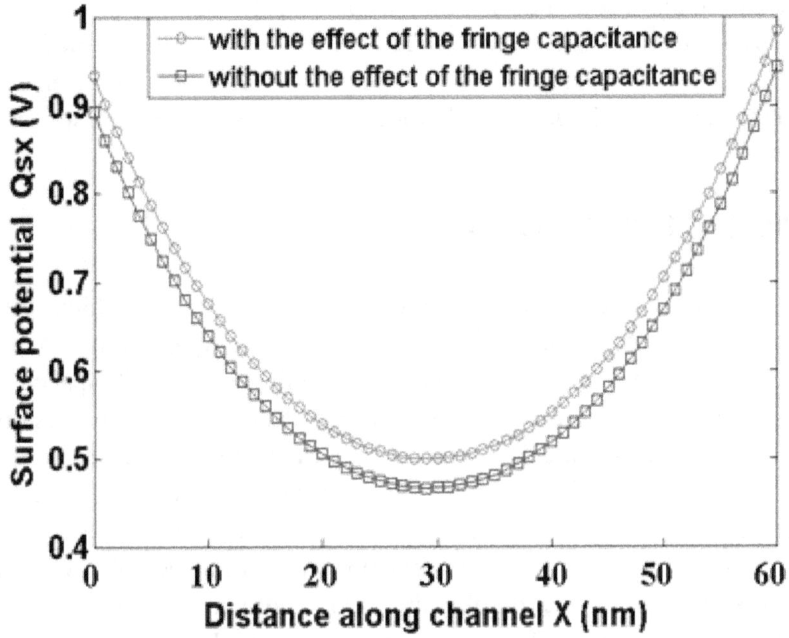

Figure 9. Influence of k_{ox} *value of gate dielectric film on the threshold voltage of TFT*

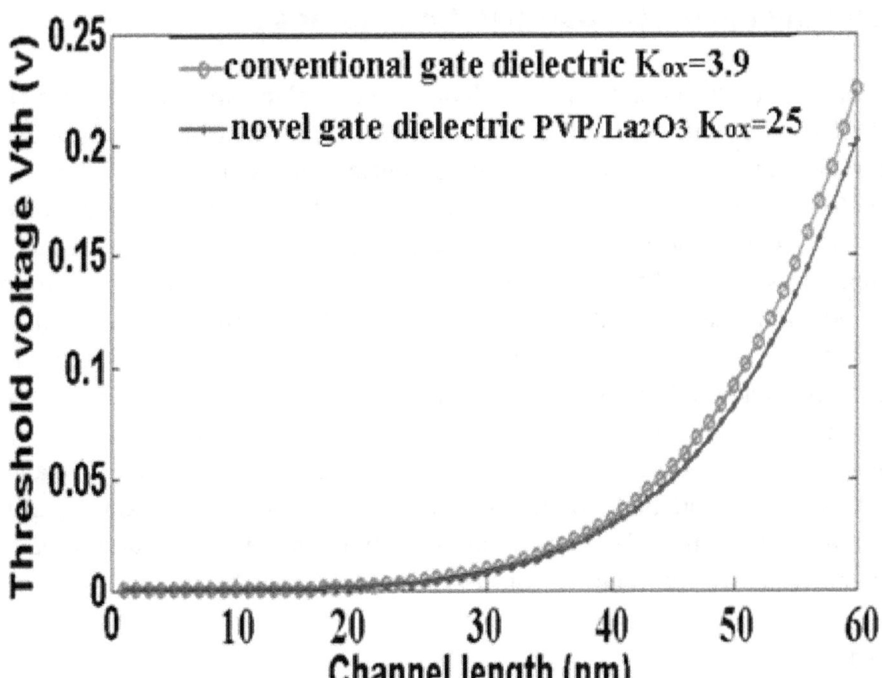

as an aftereffect of high permittivity of nanocomposite materials, the fringing entryway capacitances C_{bottom} has a significantly great role as TFT devices, as previously stated. Thus, the high permittivity of nanocomposite material builds the surface potential of the prompted charges in the channel and well-spring regions because of the fringing field lines starting with the bottom of the entryway cathode. The drop in the edge voltage can be as high as 25mV for Lg=60nm as ε_{ox} expands starting with 3. 9 to 25.

5 COMPACT MICROSTRIP PATCH ANTENNA

Compact Microstrip patch antenna includes a microstrip opening antenna which is exceptionally little and lightweight and builds the bandwidth of the antenna. Microstrip opening antennas (slot antennas eager toward a strip line) are extensively utilized within military (aircraft, spacecraft, satellites, and missiles) and industry (mobile radio and remote correspondence systems) provisions. These sorts of opening antennas have various guaranteeing offers. For instance, they bring light weight, little size, and low expense. In addition, they are effectively coordinated for planar and nonplanar surfaces and have a significant degree of opportunity in their configuration. However, they have huge radiation clinched alongside a portion of undesired directions, and this radiation will be actually undesirable to some applications (Baraton, 1998; Bernius & Inbasekaran, 2000; Zeijlmans, 1999). In the displayed work, a legitimate correlation has been conveyed for the proposed models with proposed nanocomposite materials and conventional rectangular microstrip antenna. Furthermore, this part has demonstrated the change in stature for substrate and width of patch by utilizing novel suggested nanocomposite materials which create RT Duriod 5880 with nanoparticle fumed Silica, aluminum oxide and Graphite.

5.1 Theory and Design of Microstrip Patch Antenna

A microstrip antenna consists of conducting patch on a ground plane divided by dielectric substrate for thickness h and relative permittivity ε_r. The conducting patch can take any shape, but rectangular and circular configurations are the most commonly used configurations. The majority commonplace substrates bring a dielectric constant for the range of 2.2 to12. Thick substrates with high dielectric constant are alluring similarly as the gatherings give exceptional efficiency, larger bandwidth and loosely bound fields to radiation into space, and the high dielectric constant substrates are for the most part favored for the greatest radiation (Kicklbick, 2008; Kwok & Xu, 2002). Slim substrates with high dielectric constant (2. 2-12) are generally connected to microwave meandering due to their minimized undesired radiation and coupling, and the littler sizes of their components (Kicklbick, 2008). The period and the width of a patch can be found toward a mixture of simulation Investigation and experimental methods. Patch width has a minor effect on the resonant frequency and radiation pattern of the antenna. A larger patch width increases bandwidth and radiation efficiency. For correct excitation, a particular case can decide a patch width w better than the patch period L without energizing the undesired modes. Hence, the width can be ascertained from (Gupta et al., 2010; Kwok & Xu, 2002):

$$W = \frac{c}{2f_r}\left(\frac{\varepsilon_r + 1}{2}\right)^{-0.5} \qquad (12)$$

Additionally, the patch length determines the full frequency, and is an incredulous parameter in configuration due to the intrinsic limited data transfer capacity of the patch. The period is found by figuring the half-wavelength strength et cetera subtracting a little period with the available fringing fields impact. Thus, the period can be provided in (Gupta et al., 2010; Kwok & Xu, 2002) as takes after:

$$l = \frac{c}{2 f_r \sqrt{\varepsilon_{eff}}} - 2\Delta l \tag{13}$$

$$\Delta l = 0.412\, h \frac{\left(\varepsilon_{eff} + 0.3\right)\left(\frac{W}{h} + 0.264\right)}{\left(\varepsilon_{eff} - 0.258\right)\left(\frac{W}{h} + 0.813\right)} \tag{13.i}$$

$$\varepsilon_{eff} = \frac{\varepsilon_r + 1}{2} - \frac{\varepsilon_r - 1}{2}\left(1 + \frac{12t}{W}\right)^{-0.5} \tag{13.ii}$$

A paramount outline step will decide a suitableness dielectric substrate of fitting thickness h and reduction digression. The substrate dielectric constant ε_r plays a role comparable to the substrate thickness. A high esteem of ε_r for the substrate will expand the fringing field toward the patch periphery and observe the radiated force. Information for microstrip antenna with ground plane opening has been expressed in addendum I and the size of the opening.

5.2 Nano-Tech Microstrip Patch Antenna

In this research, the taking after nanoparticles has been chosen for upgrading the performance of microstrip patch antenna. Fumed silica may be a standout amongst the practically essential fillers utilized within insulating materials, coordinated circuit circuits, electric components, conductors, and a lot of different applications. Primary favorable circumstances of fumed silica are costive and bring an incredible impact on the properties, such as viscosity, firmness and strength. Suggested nanocomposite fumed silica for RT Duriod 5880 has been high dielectric constant to compact size, basic creation and build radiation. Aluminum oxide can be the crew for inorganic mixes for the chemical formula Al_2O_3. It is an amphoteric oxide and is commonly referred to as alumina, corundum, as well as many other names, reflecting its widespread occurrence in nature and industry. It's practically noteworthy for its use in the the processing of aluminum metal, in spite of its utilization as abrasive because of its hardness and the illustration of a recalcitrant material due to its high softening point. Graphite is an electrical conductor (semimetal). It is, consequently, advantageous in such provisions as arc lamp electrodes, and it is the large portion stable manifestation of carbon under standard states. Therefore, it is utilized within thermos science as a standard state for characterizing the hotness of creation of carbon exacerbates. Graphite can be recognized in the most noteworthy review of coal recently as anthracite and it is also known as metaanthracite, although it may not be regularly utilized as fuel due to the direct result of its being tricky to be lighted (Carter et al., 1997; Thabet & Hassan, 2011; Zeng, 2000).

5.3 Effects of Nanoparticles in Microstrip Patch Antenna

Performance optimizations are targeted towards utilizing the industry of CST microwave studio and FDTD technobabble. Reenactment comes about with CST microwave studio are checked for FDTD method. From the reproduction of a patch antenna for FDTD, it has been regarded that the 3d FDTD strategy is a great technobabble for predicating electric field proliferation. The FDTD technobabble can be used to produce total recurrence reactions for no progress in demonstrating. Likewise, it gives a close complete result of Maxwell's equations in a 3D model. Figure (10) shows the profit reduction S_ (1,1) for microstrip antenna by utilizing new nanocomposite material (RT Duriod 5880 for 20% wt fumed Silica), during the same full recurrence 1800 mHz of microstrip antenna with opening and during the same full recurrence 915 mHz of microstrip antenna without opening. It is recognized that utilizing recommended nanocomposite materials have enhanced the radiation efficiency, bandwidth and surface wave force of microstrip antenna. Utilizing nanocomposite materials (RT Duriod 5880 with aluminum oxide and Graphite) have improved the radiation, data transfer capacity and surface wave force of microstrip antenna for opening. Nanocomposite materials (RT Duriod 5880 with fumed silica and Graphite) have improved the radiation, bandwidth and surface wave force of microstrip antenna without opening. The sizes of the proposed models are easier over customary model sizes during these frequencies. Figure (11) illustrates the give back misfortune S_(1,1) of microstrip antenna by utilizing new nanocomposite

Figure 10. $S_{1,1}$ of compact antenna with nanocomposite materials

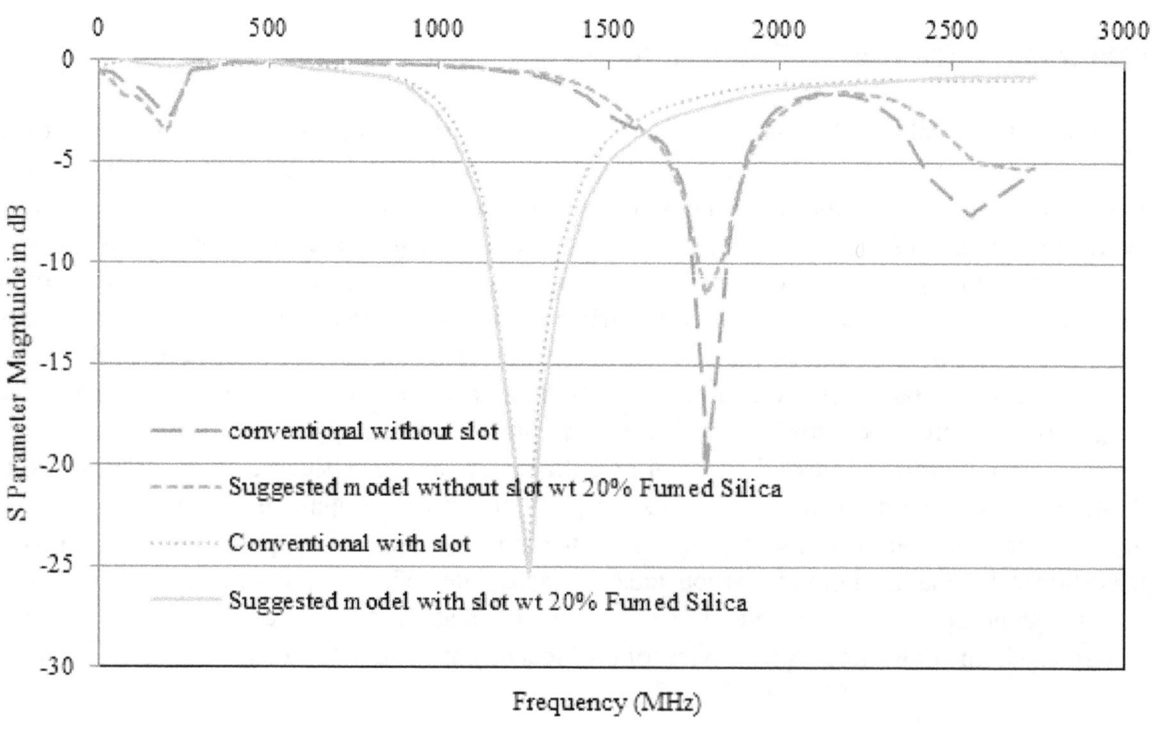

Figure 11. $S_{1,1}$ for compact antenna with nanocomposite materials

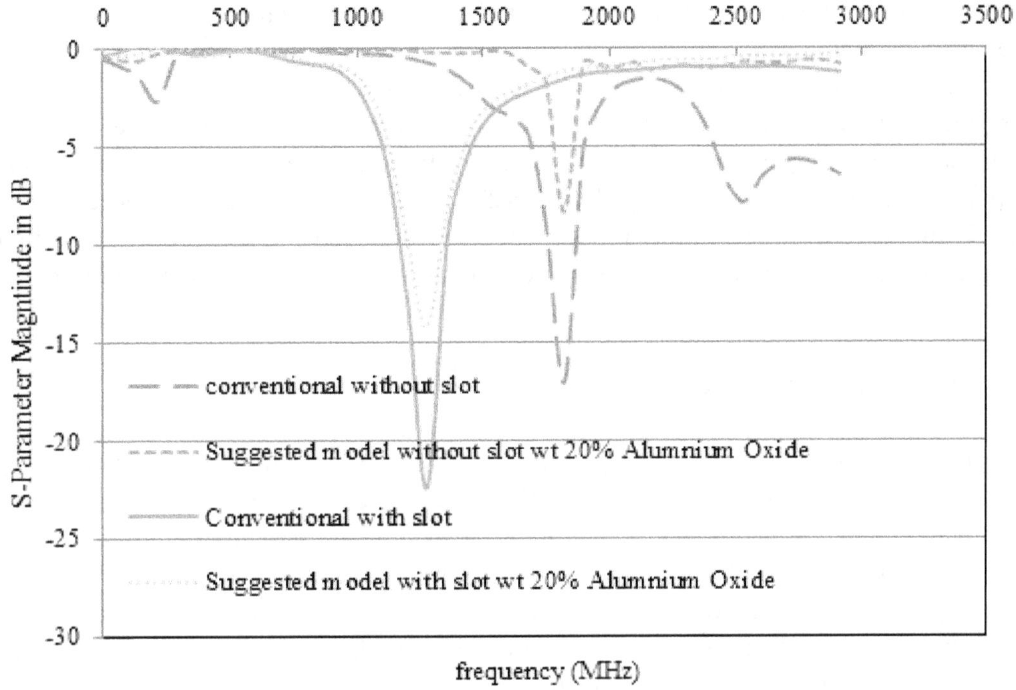

Figure 12. $S_{1,1}$ of compact antenna with nanocomposite materials

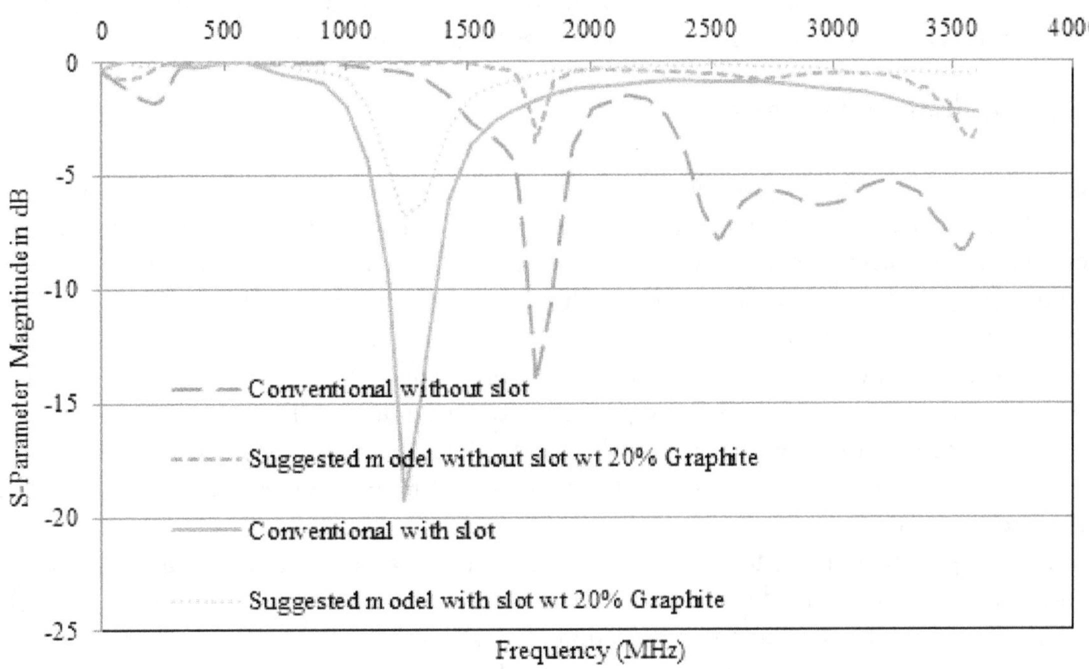

material (RT Duriod 5880 with 20% wt. aluminum Oxide), toward the same full recurrence 1800 mHz of accepted microstrip antenna for opening and toward the same full traditional recurrence 915 mHz of microstrip antenna without slot.

On the other hand, Fig. (12) illustrates the return misfortune S_(1,1) of microstrip antenna with new nanocomposite material (RT Duriod 5880 with 20% wt. Graphite), at the same full recurrence 1800 mHz for microstrip antenna for opening and toward the same full recurrence 915 mHz of microstrip antenna without opening. The radiation efficiency, data transfer capacity and surface wave force of microstrip antenna are being increased, while other measurements for microstrip antenna have been diminished. Compact microstrip antenna has investigated high performance, light weight, simple creation and low cost.

6 SEMICONDUCTORS APPLICATIONS

The rising field of nanotechnology offers possibilities for examining the key chemical and physical standards through the nanometer size scales. At the same time, it also offers new research parkways new stage advances, such as nano-devices, polymer-transistor, nanoscale light emitting diodes headed and diode array, resistive memories and nano-sensors. A particular case that presents challenges in the nanotechnology zone is the creation and intercontinental of nano-devices around semiconducting surfaces, such as silicon wafer. Leading polymers are one potential population of materials for this requisition. Directing polymers to that point of broad investment to provisions starts with electronic devices on mechanical actuators. Present photolithography, micro contact printing, format synthesis and examining electrochemical microlithography systems manage directing polymer microstructures. These strategies have noteworthy impediments for patterning structures of <100 nm extents. The proficiency of direct-write as example polymeric materials with intriguing electronic and electro-optical properties during the nanoscale grants many chances, since an extensive mixture of monomers/polymers are accessible. Here, over organic materials are utilized as animated semiconductor materials.

6.1 Semiconductors Devices

A large portion of revolutionary provisions of nanotechnology has been in the semiconductor regions. Likewise, the semiconductor roadmaps gaze out towards 2015 and beneath 20 nm features. Hence, the need for distinctive structures may be turning into evident. When we move to Ultrasuede et cetera X-beam lithography, there is no place to try picture ultimo little offers. Envision doping a carbon or silicon nanotube, covering it for contrastingly doped materials and amassing it in an exhibit. Envision also making quantum specks that can store a solitary electron charge, envision trapping iotas inside a nanotube and utilizing the electron turn on to make a quantum processing device. There is an extensive number of potential routes for new computing, capacity and optical devices. The devices settling on right away are very maladroit compared with the created semiconductor innovation organization. Anyhow, they will without doubt move forward.

A quit offering on that one instance of a semiconductor innovation that is generating great interest will be the nuclear bunch affidavit innovation pioneered toward Nano bunch devices NCD from Christchurch, New Zealand. The innovation organization effects start with the joining of two great secured technologies, nuclear group affidavit and the sort of lithography utilized at present in semiconductor fabrication. The engineering organization revolves around a capability to manufacture nanoscale wire, such as structures toward the gathering of directing nanoparticles. The attractions of this approach are:

1. Electrically directing nanowires can be structured utilizing just simple and straight-ahead techniques: Group affidavit and generally low determination lithography.
2. The coming about nanowires is organically associated with electrical contacts.
3. Electrical current can be passed along the nanowires starting with the minute of their framing.
4. No control of the groups may be obliged to structure the nanowire in case the wire is "self-assembled", utilizing a standout amongst two systems depicted beneath.
5. The width of the nanowire can be regulated by the span of the bunch that is picked by complex publicizing machine simulations and is used to outline device geometries that guarantee that a single wire-like way will be structured between the contacts close to the permeation edge.

To a great possibility, an electrically directing nanowire which is organically associated with electrical contacts may be hence shaped with no compelling reason to control particles separately or utilize complex creation systems. The width of the wire can be controlled by the extent of the stored particles. Items are under improvement with few organizations. Bundling ic and MEMS devices bundling for propelled devices is setting off will be problematic, since temperature will be the adversary of ultimo fine features which might be undoubtedly wrecked by thermal dispersion or differential extension. In the shorter term, moved forward fillers to mol mixes (which are officially ~90% filler) and under fills might prompt finer thermal and electrical conductivity. (Gray, 1997). The utilization of high thermal conductivity carbon nanotubes and jewel-like films for thermal conductivity through copper twice will give acceptable advantageous results as will CTE matched fillers with conductive and dielectric properties.

Nano-electronics and Nanophysics mean in length haul scrutinize on the elaboration and characterization of fundamental materials for tomorrow's hardware. It may be vital to seek investigations on the taking after subjects (non-exhaustive list) of single electron devices memories, transistors lithography toward the Nanometric scale: X-rays, electrons, SEM – close field spectroscopes connected to the characterization and the creation of nano-systems affidavit of self-organized layers looking into semiconductors: Grafted sub-atomic layers, elaboration of wires, quantum spots and super lattices of 3d crystallites. Inevitable requisition of the optoelectronic silicon coupling of semiconductors for organic atoms of a provided for alternately purpose with living atoms interfaces between semiconductors and dielectrics, superconductors or attractive materials. Requisition with new devices (e. G. Turn transistors) – mechanical switch or memory of nuclear or atomic span utilize the momentous properties of fullerenes, nanotubes and different manifestations of carbon improvement for new proper (fault tolerant) circuit architectures quantum registering. Regarding non electronic micro- and nano-devices, this is a quickly stretching field with micro-electro-mechanical systems MEMS or significantly micro-opto-electromechanical systems MOEMS (pressure sensors, accelerometers, micro motors), micro fluidic devices (e. g. Micro-pumps, micro-valves), compound or biochemical sensors, in addition to optical micro devices. These realizations have benefited from the innovations starting with microelectronics. Here again, parts identified with slim layers of semiconductors are fundamental (Grabert & Devoret, 1992). What's to come lies in the configuration and acknowledgment of new sorts of devices, of something like that known as keen systems, in the quest for particular provisions. Previously, it was inaccessible, but one can anticipate in the future advancement towards a decrease in clinched alongside dimensions in the end down of the nanometric scale. The acquaintanceship of nanoelectronics systems may prompt micro- and nanorobotics.

6.2 Organic and Polymer Transistors

Organic and polymer transistor-based circuits are being investigated for an amount of low-cost, large-area applications; especially those which are perfect for adaptable plastic circuits. The organic materials that have been utilized in the illustration animated semiconductor materials incorporate both sublimed and result transformed semiconductors, such as pentacene, oligothiophenes, hexadecafluorocopper and polythiophene. This decision about materials opens up few possibilities to create incorporated information for preparing innovations in view of organic transistors to different extensive areas and minimal effort applications. Organic field-effect transistors OFETs have been recommended for provisions, such as those that show switches (Al-Sarawi et al., 1998), presentation drivers, radio-frequency ID number FID tags, and sensors. Organic/polymer transistors are also incorporated for optical devices, such as light-emitting diodes LEDs, electrophoretic cells and fluid crystals to know only a couple of samples. Over late years, the multifaceted nature of circuits produced for organic transistors has increased, and the 1st vast scale (864 transistor) reciprocal circuits are being created. Analysts collect information on one active matrix shows and electronic paper for hundreds of transistors or more. The speed of ring oscillators is eight away on abundance of 100 kHz, and regarding the clock pace of clocked successive circuits for example, registers may be in kHz. Thus, there is a need to create advances for moderately quick circuits (~100 kHz clock rate) to use clinched alongside RFID tags and show drivers. It is restricted to fulfill this with an integral innovation organization brushing pentane p channel transistors (mobility regularly 0. 3-2 〖cm^2/Vs) with a n-channel innovation organization using materials for electron portability of abundance 0. 1 cm^2/Vs. A couple of such materials are accounted for over late years, with a lot of people more presently and constantly synthesized. Calculations show that with such materials, ring oscillators with frequencies more stupendous than alternately equivalent to 1 mHz and successive circuits for clock rates of 100 kHz are possible.

Over this light, we highlight developments of organic reciprocal circuits. There are few preferences for outlining circuits with p-channel gap leading and n-channel electron leading transistors. On account of Si, integral metal-oxide-semiconductor CMOS circuits scatter energy basically at the same time the constituent transistors are exchanging from one state to the other. There is next to no static energy dissipation, which implies that CMOS circuits expend significantly less control over whichever n-MOS alternately p-MOS circuits. In organic circuits, the same holds correct for such measurements, and in addition, simulations have indicated that integral circuits require considerably less control. However, on account of organic CMOS, control dispersal will be overwhelmed by spillage ebbs and flows. Different favorable circumstances for integral circuits incorporate the simplicity of out design, more stupendous speed, a significant part of preferred resistance with clamor effects, and more stupendous tolerance of variability and shifts in transistor operating aspects. The Head disservice of organic CMOS is that the electron-transporting semiconductor materials, which are necessary for n-channel transistors, have a tendency to be more organically delicate than p-channel transistor materials. Couple A has seen significant developments in recent years, moving forward both the accuse transporter versatility (which impacts the speed) and the soundness of n-channel FET materials.

6.3 Single-Electron Transistors and Circuits in Silicon

Single-electron devices give acceptable attempts on controlling the electrons in nanoscale electronic systems decisively while holding similarity for extensive scale coordinated circuit LSI circuit innovation organization. Single-electron transistors operating toward room temperature have nevertheless been created previously from silicon-on-insulator, polycrystalline silicon, and nanocrystalline silicon material. Memory phones in which states are saved are characterized by solitary electrons and few-electron memory Mobile arrays for incorporated feeling amplifiers to be shown. Few-electron rationale entryways utilizing novel 'binary-decision diagram' rationale circuits are likewise created. These devices investigate the ultimacy breaking points of electronic systems and raise the probability of LSI circuits for nanoscale and single-electron elements. Late progresses in incorporated information preparing innovation organization have prompted a diminishment in the span of electronic devices under the nanometer scale. Metal-oxide-semiconductor transistors MOSFETs with entryway lengths of few many nanometers are right away been fabricated, raising the plausibility of expansive builds in the number of transistors with respect to a chip. However, assuming that the base characteristic extent is decreased underneath 10 nm quantum mechanical for example, such as the tunneling device performance. The scaling-down of devices additionally prompts a diminishment in the number of electrons accessible for advanced exchanging operations. Ultimately, a couple of main electrons may be accessible to exchanging and measurable variances in the normal number of electrons to forestall the definition of clear advanced states. The coulomb barricade or the single-electron charging, which considers the exact control of small numbers of electrons, gives an elective working guideline to nanometer-scale devices. In addition, the diminishment in the number of electrons in an exchanging move incredibly diminishes the control dissipation, raising the plausibility of significantly higher levels of circuit combination. Coulomb barricade impacts were seen initially in the low-temperature conduction of thin, granular metallic films (Cantatore, 2003). The single-electron transistor SET uses this effect to control the charging of a small 'island' electrode by one electron at a time. The basic SET, an island electrode is isolated from the source and drain electrodes by tunneling potential barriers with capacitance c_1 and c_2.

A gate electrode couples to the island electrostatically via the gate capacitance c_G. If the tunnel resistances are greater than the quantum resistance, the electronic states on the island are relatively localized, according to the following Eq.s, where the total capacitance of the island is:

$$C_1 + C_2 + C_G = C \sim 10^{-15}F \tag{14}$$

or less, then the charging energy:

$$E_C = {e^2}/{2C} \tag{15}$$

The addition of a single electron on the island can be larger than the thermal energy at cryogenic temperatures, and electronic conduction occurs only if the applied drain-source voltage can overcome E_C. As the applied voltage increases further in increments of $2V_c$, the average number of electrons on the island increases one by one. If the tunnel resistances are very different, the characteristics are stepped. The gate electrode can also control the island charge. As V_{GS} is varied, E_c is overcome periodically and I_{DS} oscillates with a period e/CG, leading to single-electron conductance oscillations. Here, every swaying corresponds to charging toward a solitary electron. However, the single-island set can be touchy to actual fragmentary progressions in the background or 'offset' charge close to the island. An answer can utilize SETs with various tunnel junctions (MTJ) and islands, in case the coulomb hole is the whole of cash of the coulomb holes of every last one of islands. If the incitation in the counterbalance charge defeats the coulomb barricade in a special case of the islands, the aggregate coulomb hole voltage decreases toward the best portion. This makes the MTJ set considerably less touchy with counterbalance accuse and for more amazing useful hugeness. SETs utilize metal islands and electrodes and it may be perfect on a basic level with CMOS transforming. However, though the situated can be created in silicon, similarity with CMOS techniques, such as high-temperature oxidation and strengthening is enhanced significantly. It is likewise simpler to fuse on-chip MOSFET interface circuits on exchanging information, starting with the single-electron and only the chip.

6.4 Nanowire Semiconductors

The ability to rationally integrate NWs with controlled electronic properties and to collect NWs under standard arrays promptly enable uses to investigate these NW structures in an assortment of practical device arrays. A capable and adaptable strategy to the numerical result of the semi-classical boltzmann transport mathematical statement to a barrel shaped geometry has been considered and tried. A vital goal for starting this computational fill is to comprehend and examine finite-size impacts and the conductance for extensive (>100 nm diameter) semiconductor nanowires. This report card has concentrated on a portion of the consideration of the magneto resistive effects, the existing semi traditional hypotheses include prohibitive assumptions and it may be insufficient to illustrate the exceptionally late test discoveries. The numerical investigations are conveyed out in a Maxwellian administration for an isotropic and explanatory electron scattering connection. Calculations were aggravated in the relaxation-time approximation, with spiral variety of the unwinding time to incorporate the impact of the surface. The diminishment to conductivity has been registered because of improved surface scattering for wires about different radii and unwinding times. Surface impacts are the majorly maintained when electrons investigate the wire completely as an aftereffect of long unwinding times or high engineering. An attractive field could keep electrons from straying of the surface and scattering; this can be done when the electrons bring in length diffusing times that prompt long periods of binding attractive energy in the middle of dissipating occasions. Currently, the numerical plan is no doubt adjusted to ponder the particular instance of silicon nanowires. Semiconductor nanowires will attract increasing interest because of their potential to serve the illustration essential fabricating obstructs clinched alongside a base up creation plan for nanoscale devices and circuits. In order to achieve this goal, it is important to control the nanowire composition, size, conductivity and other physical properties, amass the wires under normal device geometries and empower the present stream between nanoscale and microscale structures. Hence, it is crucial to comprehend the effect of confined size and surface impacts on the electrical properties of these structures. In this program, a versatile, format-based approach of the amalgamation of metal/semiconductor/metal nanowires is created. This strategy uses nanoporous membranes; the illustration substrates to nanowire utilize semiconductor vapor-liquid-solid growth and metal cathode position. After removing the wires from the membrane, a field assisted assembly method is used to attract and align the nanowires on to pre-patterned contact pads on a substrate. This strategy uses a non-uniform electric field to polarize the nanowires and actuate the arrangement through dielectrophoretic. The structures and systems formed in this system are used to do nitty gritty test investigations of electrical transport over semiconductor nanowire and metal semiconductor nanocontacts. Hypothetical models for band structure and carrier dissipating are likewise formed for gathering information of the key transport components in the wires (Collier, 2000).

6.5 Nanowire Synthesis

The nanowire amalgamation method formed in this program, combines two different methodologies which are at present utilized for nanowire fabrication: Template-directed union using nanoporous materials and metal catalyzed vapor-liquid-solid growth. In the principal approach, nanoporous materials are utilized as templates for nanowire amalgamation by means of pore filling systems, such as the cathode position or compound vapor affidavit. The nanowire diameter is decided by the pore size, which can go starting with more or less than10 nanometers to the more stupendous of over 200 nanometers. The regulate affidavit for silicon under a nanopore utilizing compound vapor deposition, however, yields polycrystalline material. Alternatively, vapor-liquid-solid growth has been used to process single precious stone semiconductor nanowires such as, silicon and gallium arsenide. In this approach, gold can be used to type a fluid eutectic compound for silicon during 363°C, which, upon super saturation, nucleates the development of a silicon wire. The nanowire diameter, in this case, is confirmed by the extent of the beginning gold molecule. Upon joining together, the two techniques are unable to develop silicon nanowires up to 30 microns long for regulated diameters extending from 80 nanometers to 200 nanometers (Venkataraman et al., 2006).

6.6 Nanocontact Formation

Fabricating multilayered metal/semiconductor/metal nanowires by cathode posits the metals, such as cobalt, rhodium and platinum under the film pores former to electrodepositing gold and performs vapor-liquid-solid growth of silicon. By utilizing the right measure of gold in the pore, the gold sets out along the tip of the developing silicon nanowire. Alternate metal hails under contact with silicon and reacts by structuring a metal silicide, bringing about a multilayer, such as cobalt silicide/silicon/gold. The alumina membranes are at that point carved away with arrival of the nanowires to structural and electrical characterization. The flexibility of this creation method offers the potential to select different metal/semiconductor combinations, different wire diameters, and distinctive doping densities to encourage the consideration of the extent of impacts clinched alongside metal/semiconductor contacts. The technobabble additionally permits us to join together the semiconductor nanowires of diverse conductivity and kind to process exceptional structures with expanded purpose and diminished measure.

6.7 Nanowire Assembly and Electrical Characterization

The electrical properties of the nanowires synthesized in this system are described by utilizing a variable temperature of 4-point resistivity and gate-dependent source-to-drain present estimations. After the nanowires are developed and evacuated from the template, they are adjusted in the middle of pairs of predefined electrodes on an insulating substrate to encourage electrical characterization. This gathering technique, which depends on powers prompted on the nanowires because of their polarization in a non-uniform electric field, has been used to adjust to high yield silicon nanowires for diameter extending from 80 – 250 nm and period 4 – 15 mm. Taking after alignment, electron shaft lithography is used to design two extra metal contacts, which serve as the voltage probe throughout the 4-point estimation. We will be measuring now the 2- and 4-point resistivity of undoped and boron-doped silicon nanowires as a capacity for temperature to investigate the impact of nanowire diameter and dope on their electrical transport properties. Additionally, ready specimens hold numerous metal entryways for measuring the reliance of the source-to-drain current on the connected entryway voltage with focus on the doping kind (n- or p-) of the undoped and eagerness doped nanowires. These back-gated tests are additionally used to assess the field impact properties of the silicon nanowire devices for use in electronic circuit applications, which incorporate high-density rationale and heterogeneously incorporated sensors. We are usually utilizing comparative systems to ponder the large-area ohmic contact framing of the nanowire tips also to disconnect electrically silicide-silicon nanocontact junctions.

6.8 Crossed Nanowire Devices

Immediate gathering of exceptionally coordinated circuit utilitarian electronic circuits in view of NWs obliges the advancement of new device ideas for versatile device setup and high yield gathering of these devices for controllable utilitarian properties. The crossed NW grid speaks to a Perfect setup, since the discriminating device measurement is typically characterized toward the cross-point and could be promptly scaled down on nanometer level. The crossed NW setup itself will be characteristically a versatile construction modeling and, hence, empowers the probability of monstrous system combination. Moreover, the crossed NW grid will be a versant structure and can be arranged under an assortment of incredulous device elements, such as diodes and transistors. Electrical estimations of such crossed junctions show clear present correction crosswise over the intersection and straight current conduct technique previously mentioned and the single one NWs, demonstrating the creation of p-n diode during the crossing side of the point. Significantly, the three basic FET device measurements are regularly characterized in the nanometer scale in the collected crossed NW-FETs (cNW-FETs): (1) A nanoscale channel width dictated toward the diameter of the dynamic NW (~2–20 nm contingent upon picked NW); (2) a nanoscale channel period characterized toward the crossed entryway NW diameter (~10 nm); and (3) a nanoscale entryway dielectric thickness decided toward the NW surface oxide (~1 nm). These unique nanoscale device measurements prompt significantly moved forward device characteristics, such as high gain, high speed and low control dispersal. Moreover, the neighborhood NW entryway empowers freely addressable FET arrays; and accordingly empowers exceedingly coordinated nano-circuits.

7 INTEGRATED NANOSCALE ELECTRONICS AND OPTOELECTRONICS

Semiconductor nanowires NW represent a perfect gas system to investigate low dimensional material science and is relied upon to assume a paramount part as both interconnects and utilitarian device components to nanoscale electronic and optoelectronic devices. Here, a survey of an arrangement about the way developments characterizes another standard of base up amassing incorporated nanosystems utilizing the semiconductor NW fabricating pieces. Hence, a general approach is principally presented to the union of a wide extend of semiconductor NWs with decisively regulated chemical composition, physical dimension, and electronic and optical properties utilizing a metal cluster-catalyzed vapor–liquid–solid growth instrument. Subsequently, depiction of normal methodologies for the hierarchic gathering of NW fabricating obstructs under practical devices and mind-boggling architectures regarding electric field or micro-fluidic stream. Following an examination of an assortment of new nanoscale electronic device ideas including crossed NW p-n diode and crossed NW field impact transistors FETs, proliferation gathering for these versatile crossed NW device components empowers an inventory of coordinated circuit structures, including rationale entryways and computational circuits. Lastly, depiction of an extensive variety of photonic and optoelectronic devices, including nanoscale light-emitting diodes nanoLEDs, multicolor headed arrays, coordinated circuit nanoLEDs–and nanoFET arrays is presented. The potential requisition of these nanoscale light sources to compound and living analyses will be examined. Contingent upon the conductance modulation, the doping kind and assess the transporter portability in individual NWs utilizing standard transistor as follows:

$$\frac{di}{dvg} = \mu\left(C/L^2\right)V_{sd} \ \& \ C = 2\pi\varepsilon\varepsilon_0 L / ln\left(\frac{2h}{r}\right) \tag{16}$$

Where μ is the carrier mobility, C is the capacitance, ε is dielectric constant, h is the thickness of the SiO_2 dielectric, L is the length, and r is radius of the NW. In this way, we have characterized a broad range of NW materials including p-type Si, n-type GaN, CdS, and InP NWs.

In all cases, the NW materials show excellent carrier mobility comparable to bulk materials, which demonstrates the high strength of these materials. Additionally, the conveyed crazy optical investigations should explore the photograph radiance properties of these NW materials (Lew et al., 2002). Photograph radiance estimations were made on singular NWs utilizing a home-based micro radiance instrument flying. Investigations looking into an expansive go of compound semiconductor NWs including GaN, CdS, CdSeS, and CdSe indicated clean radiance of spectra maxima of ~370, ~510, ~600, and ~700 nm, separately (Lew & Redwing, 2001), which are reliable for close to band edge emanation of these materials, and hence exhibit fantastic optical properties of these NW materials. Additionally, photograph radiance investigations with respect to InP NWs of variable sizes indicated deliberate blue-shift of emanation crest from ~900, with ~810 nm as the diameter of the InP NWs lessened starting with 50 to 10 nm. This can be attributed to quantum restriction impact. Therefore, there can be time permits to control the emanation wavelength of these NW materials toward whichever fluctuating the chemical creation or physical dimension, in this way putting forth the best adaptability for decisive tune.

7.1 Nanoscale Photonics and Optoelectronics

The accessibility of an extensive variety of NW materials promptly permits us to pick materials for separate properties with tailor device capacities to an aa way that is interesting to the base up gathering methodology and to nanoscale electronics. The wide range of optically animated III-V and II-VI bunch intensify semiconductor NW materials engaging as building squares for miniaturized photonic and optoelectronic devices. With this end, it has amassed an assortment of photonic devices, including nanoscale LEDs and diode array, single NW optical waveguide, cavity, and laser.

7.2 Nanoscale LED

The past p-n diode portrayal can be acquired by crossing a p- and n-type nw. To regulate the band hole semiconductors, such as InP, the p-n diode likewise manifests the foundation to the basic optoelectronics devices, including LEDs and laser diodes (LDs). With survey of our nanoscale devices and whether they can carry on, we concentrate on the electroluminescence (EL) starting with crossed nw p-n junctions. Significantly, el can be promptly observed starting with these nanoscale junctions for forward bias. An 3d plot of the el force level taken from an ordinary nw p-n diode toward ahead bias indicates the emitted light hails starting with a perspective similar to source, and moreover, correlation for el and PL pictures recorded on the same test indicate that the position of the el most extremely corresponds with the crossing side of the point in the PL picture. This information along these lines shows that the emitted light undoubtedly goes starting with the crossed NW p-n intersection. El spectra (peaked 820 nm) recorded from the cNW-LEDs show blue-shifts relative to the greater part band hole of InP (925 nm). The blue-shifts are expected to some degree for the quantum restriction of contribute, in spite of the fact that other elements might additionally help. Moreover, PL investigations have shown that the PL top can be efficiently blue-shifted as the NW diameter may be decreased, and accordingly, these outcomes give an acceptable method for regulating the color of the LEDs to a well-defined best approach. Indeed, el comes about recorded starting with p-n junctions collected starting with lower and higher diameter NWs indicating bigger and more diminutive blue-shifts. In addition, the emanation shade starting with NanoLEDs can be further shifted by utilizing synthetically different semiconductor NWs with different band holes. The base up gathering methodologies permit adaptable mix of artificially dissimilar nanoscale building squares that might generally be structurally or artificially contrary to a consecutive growth methodology ordinary for planar creation. This ability ought to empower gathering of nanostructures with capacities not promptly obtained toward different systems and offer new opportunities. For example, if misused, the capacity to structure nanoLEDs with nonemissive Si NW hole-injectors will amass multicolor arrays comprising n-type GaN, CdS, and CdSe NWs crossing a solitary p-type SiNW. The ability to seamlessly collect and incorporate separate materials together and freely tune the emanation starting with every nanoLED offers considerable potential transforming particular wavelength sources and shows a critical venture at coordinated nanoscale photonic circuits. Although lithography sets the reconciliation scale of these multicolor arrays, there is a chance for time permits to make a significant part denser nanoLED arrays by means of (1) controlled development of adjusted NW super grid structures (Chen et al., 2000), (2) specific gathering of separate semiconductor materials (Chen et al., 2003). The potential for coupling the base up gathering of nanophotonic devices together with main down created silicon structures has also been investigated, since this coupling Might gatherings give another approach to present proficient photonic abilities under incorporated silicon hardware. With this purpose, a mixture top-down/bottom-up approach can be utilized: (1) utilizing lithography to example p-type silicon wires on the surface of a silicon-on-insulator (SOI) substrate, et cetera (2) amassing n-type emissive NWs for the highest priority of silicon structures to structure arrays comprising about p-n junctions at cross-points. Conceptually, this mixture structure is basically the same as the crossed NW structures portrayed (Datta et al., 1997).

8 RESISTIVE MEMORIES

A standout amongst the practically fascinating possibilities of sub-atomic hardware is to take advantage of the delicate way for organic particles. Upon a provided for excitation, particles can experience conformational progressions. Assuming that two distinctive conformations are connected with two diverse conductivity levels of the molecule, this impact can be used to settle atomic switches and memories. Such an impact may be normal over π-conjugated oligomers utilized as sub-atomic wires, whether a standout amongst the monomer is turned away starting with a planar conformity of the atom or not (Chen et al., 1999). Bending particular case monomer breaks the conjugation along the backbone, hence diminishing the accuse exchange efficiency along the atom. This has been experimentally observed for a little atomic wire where the national unit might have been substituted with redox moieties. Different atoms with a portion transform of the redox moieties have exhibited on/off proportion of something like 1. 5 during RT. The exchanging conduct of these mixes embedded in an alkanethiol SAM may have been additionally watched by STM. With separate innate conduct technique of the atoms starting with the molecule/metal interface, the same sorts of atoms have been measured looking into different test beds (Loppacher et al., 2003). However, it will not be solidly settled that this exchanging conduct technique is exclusive because of the particles. Recently, the Lindsey's gathering indicated that the time permits component will, in turn, become irregular and brief, softening up the compound connection between the middle of the atom and the gold surface. Furthermore, this purpose may still be liable to level-headed discussion. Catenane and rotaxane are a population of atoms synthesized to show a bi-stable conduct. In brief, these atoms are aggravated in two parts; one permitted to move around alternately along the opposite particular case (e.g., a ring around a rod, two interlocked rings). These particles receive two different conformations relying upon their redox states, evolving the redox state triggers the uprooting of the portable and only the structure will minimize the aggregate vitality. This sort of atom can be used to manufacture sub-atomic memories. A Market Mix Modeling (MMM) intersection utilizing a LB monolayer of these particles blended for phospholipid corrosive indicated an acceptable electrical bi-stable conduct technique in room temperature. The exchanging conduct technique will be probable because of the framing and breaking of metallic micro-filaments presented in spite of the monolayer through the top banana metal evaporation. The vicinity of such filaments will not be precise. Moreover, alert has to be given when definitively ascribing the memory impact. The illustration is actually because of the vicinity of the particles. At the same time of hosting poor performance instead at the moment, these exhibits permit us to imagine the advancing period of mixture electronics, the place atomic cross-bar memories such as these ones will be tended to by multiplexer/demultiplexer, and along these lines, a particular case is created for the standard semiconductor CMOS innovations. The point of such atomic cross bar memories is: (i) a low cost; (ii) a high combination density; (iii) a defect-tolerant architecture; (iv) a simple post-processing onto a CMOS circuitry; and (v) a low force utilization. For instance, it has been measured that vitality of 50 zJ (or 0. 3 eV) is the addition to turn the dibutyl-phenyl side gathering of a single porphyrin atom (Aravamudhan & Bhansali, 2008).

9 MEMORY RAM

Although memory phenomena have been mulled over in heft organic materials, the working standards can't actually be scaled down of the sub-atomic level clinched alongside heft properties as answerable for sign transduction. Chen et al. (Chen et al., 2000) and reed et al. (Chen et al., 2003) exhibited nanoscale electronically programmable and erasable memory devices using atomic SAM and a memory cell pertinent to an arbitrary access memory ram. Give or take 1000 particles are sandwiched between au contacts shaped by a chemisorbed thiol/Au contact with respect to the lowest part and a dissipated au contact over the highest point. The whole dynamic area has a diameter of 30–50 nm. The vital characteristic of the creation procedure will be the utilization of a nanoscale device area, which provides a climb to a little amount of self-amassed atoms and likewise dispenses with pinhole and other abandon components that hamper monolayer electronic transport estimations. At the same time, the vicinity of the amide bunch progresses the introductory conductivity states of the molecule, well on the way brought on by the separate dipole dissemination it introduces of the atom. They suggested that the diverse conductivity states are brought on toward separate trap levels involved by electrons in the atom. Separate trap levels involved toward electrons on an atom can mean whichever separate electronic states for separate accuse circulation (such as different dipole distribution) or different conformity states for an atom (such as rotational conformity and conceivable rotational conformation), or both. These trap levels can cause a separate electronic distribution, prompting different conductivity states of the atom. The primary certain pulse configures the state of the mobile toward composing a bit, and the second and third sure pulses peruse the cell. The third pulse (and the resulting perused pulses not demonstrated here for simplicity) shows that the cell is hearty and proceeds to hold the state (up as far as possible of the bit maintenance time, this showing highlights the dramatically long touch maintenance time). The negative pulse erases the bit, resetting the Mobile. The second set of four pulses repeats this pattern, and many hours of constant operation have been observed of no corruption clinched alongside performance.

10 CHEMICAL SENSOR AND BIOSENSOR

Many individual sorts of sensors utilizing leading polymers have been portrayed for both compound and living utilization. The oxidation level of leading polymers can be effortlessly impacted by their intrinsic reversible doping-dedoping (oxidation-reduction) mechanisms, creating the varieties clinched alongside conductivity, color, mass, and volume. Thus, the leading polymers are skilled at exhibiting touchy reactions to particular compounds or living species. Furthermore, the sensitivity of leading polymers can be custom-made toward attaching practical gatherings of the polymer chains or incorporating proper counter ions during the polymerization. Owing to these characteristics, substantial exertion has been committed to the creation of sensor devices in view of leading polymers. In recent years, different directing polymer nanomaterials have been utilized to identify an extensive variety of chemical and living species. It is foremost that the nanostructures including nanorods, nanofibers, and nanotubes do not just give a high surface-to-volume proportion as well as permit fast dissemination of analyses under and out of the material. The majority of habitually utilized directing polymers for creating new sensor innovations are polypyrrole nanowires PPy and Polyaniline PANI. After all, PPy nanomaterials are connected on different sensors, such as poisonous gas noses, non-toxic gas noses, fragrance sensors, stickiness sensors and microbial noses. For example, carbon nanotube CNT PPy gas sensors were created through a simple chemical polymerization accompanied toward spin-casting onto pre-patterned electrodes PPy which may have been uniformly covered on the single divider CNTs to build the particular surface range. CNT-PPy sensors indicated an n-type conduct attributed to the anion doping in PPy throughout the compound polymerization. The nanowire sensor shows unrivaled progressions of safety compared with micron or submicron wire. This Look into can be further developed from the creation of PPy nanowire biosensor to label-free bio-organic bias sensing. The one-stage consolidation of utilitarian living particles under PPy nanowire during its electro polymerization can be a discriminating point of a novel creation strategy through the CNT and silicon nanowire biosensors that oblige post-synthetic adjustment and positioning. Recently, PANI nanofiber has been broadly connected to different sensor exhibitions. The nanofiber can be kept on the bury digitated electrodes in the structure of fiber networks, and their safety progressions have been monitored clinched alongside as ongoing. PANI nanofiber sensors exhibit the moved forward reactions of different analyses. On the other hand, the compound sensors made of a solitary PANI nanowire were created toward a method of scanned-tip electro turning affidavit. A fascinating procedure is formed so as to permit a simple optical identification of the iodide. This technique is in the view of a conformational change of conjugated spine of cationic poly (3-alkoxy- 4-methylthiophene) upon tying for an iodide anion. It can be likewise accounted for that the electro active, water-soluble, and cationic PT can be utilized for identifying DNA to diagnose different ailments.

11 SUPER CAPACITOR

Super capacitors materials as leading polymers have preferences, such as high capacitive vitality thickness and low material cost. On the other hand, they also have hindrances, such as bring down life cycle and moderate particle transport. For this reason, the nanocomposite electrodes in view of CNT covered with leading polymers bring are broadly utilized for super capacitors (Pandolfo & Hollenkamp, 2006). The high surface zone and conductivity of CNT improved the redox property of leading polymers. However, the structure and capacitance of the nanocomposites are organically powerless of the manufactured course. Electrochemical Growth techniques are constrained by the measure of polymer is stored on the CNT without blocking the electrolyte channels, and chemical oxidation polymerization strategies can endure from amassing of the polymer saved on the CNTs. The particular capacitance of every impostor and geometric zone of the nanocomposite can be as high as192 f g^{-1} and 1.0 f Cm^{-2}, separately. These values are in any event two times higher than the individuals of accepted PPy novel into a film. At the same time, bucky paper can be predominantly created of micropores, pyrrole dealt with specimens are primarily made of meso- and macropores. Correlating the capacitance for surface area, pore size, and pore extent distribution, the macropore has committed a noteworthy commitment of the capacitance performance of these materials. The electrochemical quartz Precious stone microbalance can be used to explain the affidavit and inter-capitular vein conduct technique of PPy-CNT and PPy-chloride slim layer (Snook & Chen, 2004). Interestingly, the capacitive charging current can be watched on the impostor gram during certain exchanging potential, and this aftereffect affirms the redox pseudo capacitive conduct of PPy. Carbon nanofiber has critical points of interest as far as low fabricate cost and mass number compared with CNT. Recently, PANI-coated carbon nanofiber can be promptly created utilizing one-stage VDP. The particular capacitance of PANI-coated carbon nanofiber is defenseless of the thickness of the PANI layer.

12 FORECAST AND RECOMMENDATIONS

Including polyaniline/silver nanocomposite has upgraded the performance of SPR sensor. The proposed model sensitivity is higher than the accepted and late nanoparticle compound sensors. The SNR of the recommended model is higher than the late nanoparticle compounds and brings down in the traditional sensor. The reaction for nanoparticle-based sensor is quicker than the conventional sensor. The speedier reaction of nanoparticle sensors permits the investigations of macromolecular target analyses that are allowed with customary sensors, but the novel nanocomposite material SPR sensor requires speedier reaction than customary and nanoparticle SPR. The novel model in this chapter has exhibited the Electro-Luminescence el efficiency for OLEDs which has been fundamentally enhanced by inserting a slim insulating layer for more level dielectric constants. Utilizing the PPV mixed with nanoparticles of TiO_2 has progressed conductivity and optical properties of OLED, as contrasted with the uncovered polymer. The vicinity of the low dielectric constant nanolayer of ZnO at the Al–MEH-PPV interface makes a significant bringing down of the electron injection barrier height, which has been attributed to the moved forward adjusting of charge injection. The ZnO nanolayer acquires an exceedingly progressed luminance, as well as brought down the turn-on edge voltage compared with the single layer OLED device. The produced model in this part has shown utilizing novel nanocomposite PVP/La_2O_3 as high k dielectric material to move forward the performance of TFT devices by expanding the surface potential and diminishing the edge voltage. The recommended model with high permittivity's kox and stupendous physical thicknesses of the element kox/$kSiO_2$ keeps immediate entryway tunneling and diminishing entryway spillage current. Novel nanocomposite materials made of RT Duriod 5880 for nanoparticles aluminum oxide and graphite have improved the plan of microstrip antenna for opening. Novel nanocomposite materials are made of RT Duriod 5880 with nanoparticles fumed silica and Graphite bring improving the configuration of microstrip antenna without opening. Height and width of the recommended microstrip antenna models are smaller than those of the accepted rectangular microstrip antenna for and without opening. The radiation efficiency, data transfer capacity and surface wave force of microstrip antenna have been improved by utilizing the recommended new nanocomposite materials for simple fabrication, high performance and low expense. The sorts and concentrations of metal oxide nanoparticles influence the morphological tenet and optical properties of the arranged nanocomposites and diminish the band hole vitality. Expanding the proportion of metal oxides (ZnO and SiO_2) proportion increments the absorption of the readied polyvinyl chloride nanocomposites, while expanding the metal oxide (Al_2O_3) proportion increments the transmission of the readied nanocomposites. In any case, the sorts and concentrations of inorganic nanoparticles are viable variables for regulating optical light characterization for polyvinyl chloride and can upgrade the optical light properties of polyvinyl chloride toward nanotechnology strategies as an investment phase.

REFERENCES

Al-Sarawi, S. F., Abbott, D., & Franzon, P. D. (1998). A Review of 3-D Packaging Technology. IEEE Trans. CPMT – Part B, 21(1). doi:10.1109/96.659500

Aravamudhan, S., & Bhansali, S. (2008). Development of micro-fluidic nitrate-selective sensor based on doped-polypyrrole nanowires. *Sensors and Actuators. B, Chemical, 132*(2), 623–630. doi:10.1016/j.snb.2008.01.046

Baraton, M. I. (1998). Transparent Conductive Oxide Materials: Financial Stakes and Technological Challenges. *International Journal of Nanotechnology, 356*, 9.

Bernius, M.T., & Inbasekaran, M. (2000). A Hybrid Organic–Inorganic Electrode for Enhanced Charge Injection or Collection in Organic Optoelectronic Devices. *Advanced Material Journal, 12*, 17-37.

Cantatore, E. (2003). Circuit yield of organic integrated electronics. *IEEE International Solid-State Conference ISSCC Dig. Tech. Papers*, 382. 10.1109/ISSCC.2003.1234346

Carter, S. A., Scott, J. C., & Brock, P. J. (1997). Enhanced luminance in polymer composite light emitting devices. *Applied Physics Letters Journal, 71*(9), 1145–1147. doi:10.1063/1.119848

Chaudhry, A., & Kumar, M. J. (2004). Controlling Short-Channel Effects in Deep Submicron SOI MOSFETs for Improved Reliability: A Review. *IEEE Transactions on Device and Materials Reliability, 4*(1), 99–109. doi:10.1109/TDMR.2004.824359

Chen, F.-C., & Liao, C.-H. (2008). Improved Air Stability of N-Channel Organic Thin-Film Transistors with Surface Modification on Gate Dielectrics. *Applied Physics Letters*, *93*(10), 1063–1066. doi:10.1063/1.2980421

Chen, J., Reed, M. A., Rawlett, A. M., & Tour, J. M. (1999). *Large on-off ratios and negative differential resistance in a molecular electronic device* (Vol. 286). Science.

Chen, J., Wang, W., Reed, M. A., Price, D. W., & Tour, J. M. (2000). Room-temperature negative differential resistance in nanoscale molecular junctions. *Applied Physics Letters*, *77*(8), 1224–1226. doi:10.1063/1.1289650

Chen, Y., Jung, G.-Y., Ohlberg, D. A. A., Li, X., Stewart, D. R., Jeppesen, J. O., Nielsen, K. A., Stoddart, J. F., & Williams, R. S. (2003). Nanoscale molecular witch crossbar circuits Nanotechnology. Academic Press.

Collier, C. P. (2000). *Catenae-based solid state electronically reconfigurable switch* (Vol. 289). Science.

Datta, S., Tian, W., Hong, S., Reifenberger, R., Henderson, J. I., & Kubiak, C. P. (1997). Current–voltage characteristics of self-assembled monolayers by scanning tunneling microscopy. Physical Review Letters, 79(13), 2530–2533. doi:10.1103/PhysRevLett.79.2530

Friend. (2002). Pi-electronic and Electrical Transport Properties of Conjugated Polymer Nanocomposites PPV with Homogeneously Dispersed Silica Nanoparticles. *The Journal of Chemical Physics*, *116*, 6782. doi:10.1063/1.1416682

Gouda, O., Mobarak, Y. A., & Samir, M. (2010). A Simulation Model for Calculating the Dielectric properties of Nanocomposite Materials and Comprehensive Interphase Approach. *14th International Middle East Power Systems Conference (MEPCON-2010)*, 151-156.

Grabert, H., & Devoret, M. H. (1992). *Single Charge Tunneling Coulomb Blockade Phenomena in Nanostructures*. Advanced Science Institutes. doi:10.1007/978-1-4757-2166-9

Gray, F. M. (1997). *Polymer Electrolytes. The Royal Society of Chemistry.*

Gupta, K., Jana, P. C., & Meikap, A. K. (2010). Optical and electrical transport properties of Polyaniline silver nanocomposite. *Synthetic Metals*, *160*(13-14), 1566–1573. doi:10.1016/j.synthmet.2010.05.026

Jing, S., Xing, S., Yu, L., & Zhao, C. (2007). Synthesis and characterization of Ag/polypyrrole nanocomposites based on silver nanoparticles colloid. *Materials Letters*, *61*(23-24), 4528–4530. doi:10.1016/j.matlet.2007.02.045

Jong Seol Yuk, Æ. (2006). Sensitivity enhancement of spectral surface plasmon resonance biosensors for the analysis of protein arrays. *European Biophysics Journal*, *35*(6), 469–476. doi:10.100700249-006-0054-x

Kawabe, Y., Morrell, M. M., Jabbour, G. E., Shaheen, S. E., Kippelen, B., & Peyghambarian, N. (1998). A Numerical Study of Operational Characteristics of Organic Light-Emitting Diodes. *Journal of Applied Physics*, *84*(9), 5306–5314. doi:10.1063/1.368779

Kicklbick, V. (2008). *Improved Transport and Photo Stability of Polymer Thin Films by Boron Doped Freestanding Silicon* (Vol. 28). Nanocrystals Progress in Polymer Science.

Kim, Y. E., Park, H., & Kim, J. J. (2004). Effect of Polymer-Insulating Nanolayers on Electron Injection in Polymer Light-Emitting Diodes. *American Institute of Physics*, *84*(10), 1783.

Kumar, M. J., & Chaudhry, A. (2004). Two-Dimensional Analytical Modelling of Fully Depleted DMG SOI MOSFET and Evidence for Diminished SCEs. *IEEE Transactions on Electron Devices*, *51*(4), 569–574. doi:10.1109/TED.2004.823803

Kwok, H. L., & Xu, J. B. (2002). A Model for Exciton Formation in Organic Electroluminescent Devices. Solid-State Electronics, 46(5), 645-650. doi:10.1016/S0038-1101(01)00336-7

Lew, K. K., & Redwing, J. M. (2001). Growth characteristics of silicon nanowires synthesized by vapor-liquid-solid growth in nanoporous alumina templates. *Journal of Crystal Growth*, *13*, 1489.

Lew, K. K., Reuther, C., Carim, A. H., Redwing, J. M., & Martin, B. R. (2002). Template-directed vapor-liquid-solid growth of silicon nanowires. *Journal of Vacuum Science & Technology. B, Microelectronics and Nanometer Structures: Processing, Measurement, and Phenomena: An Official Journal of the American Vacuum Society*, *20*(1), 38. doi:10.1116/1.1430240

Loppacher, C., Guggiesberg, M., Pfeiffer, O., Meyer, E., Bammerlin, M., Lüthi, R., Schlitter, R. G., Tang, J. K. H., & Joachim, C. (2003). Direct determination of the energy required to operate a single molecule switch. *Physical Review Letters*, *90*(6), 066107. doi:10.1103/PhysRevLett.90.066107 PMID:12633309

Mohapatra. (2003). Modeling of Parasitic Capacitances in Deep Submicrometer Conventional and High-K Dielectric MOS Transistors. *IEEE Trans. Electron Devices, 50*(4), 959–966.

Pandolfo, A. G., & Hollenkamp, A. F. (2006). Carbon properties and their role in supercapacitors. *Journal of Power Sources*, *157*(1), 11–27. doi:10.1016/j.jpowsour.2006.02.065

Parker, I. D. (1994). Carrier Tunneling and Device Characteristics in Polymer Light Emitting Diodes. *Journal of Applied Physics*, *75*(3), 1656–1666. doi:10.1063/1.356350

Ramachandran, G. K. (2003). *A bond-fluctuation mechanism for stochastic switching in wired molecules* (Vol. 300). Science.

Rébiscoul, D., Favier, S., Barnes, J.-P., Maes, J. W., & Martin, F. (2010). Characterization of the Annealing Impact on La2O3/HfO2 and HfO2/La2O3 Stacks for MOS Applications. *Microelectronic Engineering*, *87*(3), 278–281. doi:10.1016/j.mee.2009.06.013

Reddy, G. V., & Kumar, M. J. (2005). A New Dual-Material Double-Gate (DMDG) Nanoscale SOI MOSFET-Two-Dimensional Analytical Modeling and Simulation. *IEEE Transactions on Nanotechnology*, *4*(2), 260–268. doi:10.1109/TNANO.2004.837845

Snook, G. A., & Chen, G.Z. (2004). Achieving high electrode specific capacitance with materials of low mass specific capacitance. *Journal of Electro Chemical Analysis, 568*, 135.

Thabet, A., Atef, M., & Mobarak, Y. A. (2011). *Enhanced TFTs Model Using Novel Nano-Composite High-K Gate Dielectric*. In 14th Power Electronics and Applications European Conference "EPE-2011", Birmingham, UK.

Thabet, A., & Hassan, A. (2011, November). Design of Compact Microstrip Patch Antenna with and Without Ground Plane Slot using New Nano-Composite Materials. *Journal of Engineering Sciences, Assiut University*, *39*(6), 1375–1385.

Thabet, A., & Mobarak, Y. A. (2010). Dielectric Characteristics of New Nanocomposite Industrial Materials. *International Conference on High Voltage Engineering and Application (ICHVE-2010)*, 568-571.

Thabet, Mobarak, & Abozeid. (2010). Exponential Power Law Model for Predicting Dielectric Constant of New Nanocomposite Industrial Materials. *International Conference on Materials Imperatives in the New Millennium*.

Thabet, Mobarak, & Samir. (2010). Novel Industrial Materials Enhanced Dielectric Characteristics. *International Conference on Materials Imperatives in the New Millennium, (MINM-2010)*.

Venkataraman, L., Klare, J. E., Nuckolls, C., Hybertsen, M. S., & Steigerwald, M. L. (2006). Dependence of single molecule junction conductance on molecular conformation. *Nature*, *442*(7105), 904–907. doi:10.1038/nature05037 PMID:16929295

Zeijlmans, P. A. (1999). Evaporation Gap in LiF for Electronic Excitations by Slow Antiprotons. *Physical Review B: Condensed Matter and Materials Physics*, *59*, 1095.

Zeng, J. Y. (2000). *Quantum Mechanics* (Vol. 1). Science Publishing Company.

Conclusion

At the end of this book, we conclude that nanotechnology is one of the most important technologies nowadays and in the future. It has become at the forefront of the most important sciences because of its importance in improving products, treating diseases and serving mankind in all areas of life, in addition to giving hope for scientific revolutions in the future in physics, chemistry, biology, engineering, etc. Consequently, we must attempt to take advantage of the distinct properties of nanomaterials to bring about innovations and inventions that benefit humanity in the areas of peace, as well as accelerate and facilitate life and combat malignant diseases for which science has not reached a radical treatment, in addition to other services. Since nanoscale is the focus of interest in science nowadays, we hope that manufactures will witness a great interest in such field and that our country will become one of the most sought-after countries in researching this technology and its newness so that we can catch up with such scientific progress, unleash scientific energies in people's minds to prove their worth and competence.

Ahmed Thabet Mohamed
Aswan University, Egypt

About the Author

Ahmed Thabet Mohamed was born in Aswan, Egypt in 1974. He received the BSc from Faculty of Energy Engineering (FEE) Electrical Engineering degree in 1997, and MSc (FEE) Electrical Engineering degree in 2002 both from Faculty of Energy Engineering, Aswan, Egypt. PhD degree had been received in Electrical Engineering in 2006 from El-Minia University, Minia, Egypt. He joined with Electrical Power Engineering Group of Faculty of Energy Engineering in Aswan University as a Demonstrator at July 1999, until; he held Professor position at October 2017 up to date. His research interests focus in the areas of analysis and developing electrical engineering models and applications, investigating novel nanotechnology materials via addition nano-scale particles and additives for usage in industrial branch, electromagnetic materials, electroluminescence and the relationship with electrical and thermal ageing of industrial polymers. A lot of mobility's has investigated for supporting his research experience in UK, Finland, Italy, USA and KSA …etc. On 2009, he had been a Principle Investigator of a funded project from Science and Technology Development Fund "STDF" for developing industrial materials of ac and dc applications by nano-technology techniques through it was published many scientific researches interested in developing industrial applications using the latest technologies in nanotechnology. He has been established 1st Nano-Technology Research Centre in the Upper Egypt, hence, he filed several patents in the field of various engineering applications. He has high quality publications which have been published and under published in cited international journals and conferences. On 2018, he has been held a full professor position in Engineering of College at Qassim University in Kingdom of Saudi Arabia up to date. Web-site Details: https://www.researchgate.net/profile/Ahmed_Thabet_Mohamed/research https://scholar.google.com/citations?user=40yhB0kAAAAJ&hl=th

Index

M

N

O

P

Q

S

T

U

W

Z

www.ingramcontent.com/pod-product-compliance
Lightning Source LLC
LaVergne TN
LVHW062144300926
849161LV00010B/2617